Advances in Intelligent Systems and Computing

Volume 227

T0138162

Series Editor

J. Kacprzyk, Warsaw, Poland

For further volumes:
http://www.springer.com/series/11156

Michael Emmerich · André Deutz · Oliver Schütze
Thomas Bäck · Emilia Tantar
Alexandru-Adrian Tantar · Pierre del Moral
Pierrick Legrand · Pascal Bouvry
Carlos Coello Coello
Editors

EVOLVE - A Bridge between Probability, Set Oriented Numerics, and Evolutionary Computation IV

International Conference held at Leiden
University, July 10–13, 2013

Springer

Editors

Michael Emmerich
Leiden Institute of Advanced Computer Science
Leiden University,
Leiden, The Netherlands

André Deutz
Leiden Institute of Advanced Computer Science
Leiden University, Leiden, The Netherlands

Oliver Schütze
CINVESTAV-IPN
Depto. De Computación
Mexico City, Mexico

Thomas Bäck
Leiden Institute of Advanced Computer Science
Leiden University, Leiden, The Netherlands

Emilia Tantar
Luxembourg Centre for Systems Biomedicine
University of Luxembourg
Belval, Luxembourg

Alexandru-Adrian Tantar
Computer Science and Communications
    Research Unit
University of Luxembourg
Luxembourg

Pierre del Moral
Bordeaux Mathematical Institute
Université Bordeaux I
Talence cedex, France

Pierrick Legrand
Université Bordeaux Segalen,
UFR Sciences et Modélisation
Bordeaux, France

Pascal Bouvry
Faculty of Sciences, Technology and
    Communication
Computer Science and Communication Group
University of Luxembourg
Luxembourg

Carlos Coello Coello
CINVESTAV-IPN
Depto. de Computatción
Mexico City, Mexico

ISSN 2194-5357   ISSN 2194-5365   (electronic)
ISBN 978-3-319-01127-1   ISBN 978-3-319-01128-8   (eBook)
DOI 10.1007/978-3-319-01128-8
Springer Cham Heidelberg New York Dordrecht London

Library of Congress Control Number: 2012944264

Printed on acid-free paper

Springer is part of Springer Science+Business Media (www.springer.com)

Preface

The overarching goal of the EVOLVE international conference series is to build a bridge between probability, statistics, set oriented numerics and evolutionary computing, as to identify new common and challenging research aspects and solve questions at the cross-sections of these fields. There is a growing interest for large-scale computational methods with robustness and efficiency guarantees. This includes the challenge to develop sound and reliable methods, a unified terminology, as well as theoretical foundations.

In this year's edition of EVOLVE held at LIACS, Leiden University, The Netherlands (evolve2013.liacs.nl) major themes are machine learning, probabilistic systems, complex networks, genetic programming, robust and diversity-oriented and multiobjective optimization using evolutionary methods as well as set-oriented numerics and cell-mapping. The range of topics is also well reflected in the spectrum of special sessions, organized by internationally renowned experts: *Set Oriented Numerics* (Michael Dellnitz and Kathrin Padberg), *Evolutionary Multiobjective Optimization* (Oliver Schütze), *Genetic Programming* (Pierrick Legrand, Leonardo Trujillo and Edgar Galvan), *Probabilistic Models and Algorithms: Theory and Applications* (Arturo Hernández-Aguirre), *Diversity-oriented Optimization* (Vitor Basto Fernandes, Ofer Shir, Iryna Yevseyeva, Michael Emmerich, and André Deutz), *Complex Networks and Evolutionary Computation* (Jing Liu), *Robust Optimization* (Massimiliano Vasile), and *Computational Game Theory* (Rodica I. Lung and D. Dumitrescu).

The twenty peer reviewed contributions discuss overarching themes and present answers to questions, such as: How to assess performance? How to design reliable and efficient computational algorithms for learning, genetic programming, and multiobjective/multimodal/robust optimization? What are the sources of complexity and problem difficulty? How to build a frameworks for set-oriented

search? Therefore these proceedings present cutting edge research in these related fields and will serve as a stepping stone towards a more integrated view of
advanced computational models and methods.

Leiden, Mexico City, Luxembourg, and Bordeaux   Michael Emmerich
April 2013   André Deutz
                                                                                  Oliver Schütze
                                                                                  Thomas Bäck
                                                                                  Emilia Tantar
                                                              Alexandru-Adrian Tantar
                                                                                  Pierre del Moral
                                                                                  Pierrick Legrand
                                                                                  Pascal Bouvry
                                                              Carlos Coello Coello

Organization

EVOLVE 2013 is organized by the *Leiden Institute for Advanced Computer Science (LIACS), Faculty of Science*, Leiden University, the Netherlands in cooperation with INRIA, France, CINVESTAV, Mexico, University of Bordeaux, INRIA Bordeaux Sud-Ouest, and the University of Luxembourg, Computer Science and Communication Research Unit.

Executive Committee

General Chair

Michael Emmerich	Leiden University, nl

Local Chairs

André Deutz	Leiden University, nl
Thomas Bäck	Leiden University, nl

Series Chairs

Oliver Schütze	CINVESTAV-IPN, Mexico City, mx
Emilia Tantar	University of Luxembourg, lu
Alexandru Tantar	University of Luxembourg, lu
Pierrick Legrand	University of Bordeaux 2, fr
Pierre del Moral	INRIA Bordeaux Sud-Ouest, fr
Pascal Bouvry	University of Luxembourg, lu

Local Support

Marloes van der Nat	Leiden University, nl
Irene Nooren	Leiden University, nl
Zhiwei Yang	Leiden University, nl
Edgar Reehuis	Leiden University, nl

Publicity Chair

Rui Li   Leiden University, nl

Art Design

Adriana Martínez   KREAPROM, Mexico City, mx

Advisory Board

Enrique Alba   University of Málaga, es
François Caron   INRIA Bordeaux Sud-Ouest, fr
Frédéric Ciérou   INRIA Rennes Bretagne Atlantique, fr
Carlos A. Coello Coello   CINVESTAV-IPN, mx
Michael Dellnitz   University of Paderborn, de
Frédéric Guinand   University of Le Havre, fr
Arnaud Guyader   Université Rennes 2, INRIA Rennes Bretagne
                                   Atlantique, fr
Arturo Hernández-Aguirre   CIMAT, Guanajuato, mx
Günter Rudolph   TU Dortmund, de
Marc Schoenauer   INRIA Saclay – Île-de-France, Université Paris
                                   Sud, fr
Franciszek Seredynski   Polish Academy of Sciences, Warsaw, pl
El-Ghazali Talbi   Polytech'Lille University of Lille 1, fr
Marco Tomassini   University of Lausanne, ch
Massimiliano Vasile   University of Strathclyde, uk

Special Session Chairs

Oliver Schütze   CINVESTAV, Mexico City, mx
Michael Dellnitz   University of Paderborn, de
Kathrin Padberg   TU Dresden, de
Pierrick Legrand   University of Bordeaux 2, fr
Leonardo Trujillo   ITT, mx
Edgar Galvan   Trinity College, ie
Arturo Hernández-Aguirre   CIMAT, Guanajuato, mx
Vitor Basto Fernandes   CIIC, pt
Ofer Shir   IBM Research, il
Iryna Yevseyeva   Newcastle University, uk
Michael Emmerich   Leiden University, nl
André Deutz   Leiden University, nl
Jing Liu   Xidian University, cn
Rodica Ioana Lung   Babes-Bolyai University, ro
Dumitru Dumitrescu   Babes-Bolyai University, ro
Massimiliano Vasile   University Strathclyde, uk

Program Committee

Conference Chair

Michael Emmerich   Leiden University, nl

Program Chairs

Michael Emmerich   Leiden University, nl
André Deutz   Leiden University, nl
Oliver Schütze   CINVESTAV, mx
Thomas Bäck   Leiden University, nl
Emilia Tantar   University of Luxembourg, lu
Alexandru Tantar   University of Luxembourg, lu
Pierrick Legrand   University of Bordeaux 2, fr
Pierre del Moral   INRIA Bordeaux Sud-Ouest, fr
Pascal Bouvry   University of Luxembourg, lu

All full papers were thoroughly peer reviewed. We thank the referees for their voluntary effort.

Referees

Josiah Adeyemo
Gideon Avigad
Vitor Manuel Basto
  Fernandes
Urvesh Bhowan
Nohe R. Cazaez-Castro
Francisco Chicano
Brendan Cody-Kenny
Nareli Cruz
Liliana Cucu-Grosjean
Luis Gerardo De La Fraga
Michael Dellnitz
André Deutz
Jianguo Ding
Christian Domínguez
  Medina
Dumitru Dumitrescu
Enrique Dunn
Erella Eisenstadt
Michael Emmerich
Francisco Fernandez

Edgar Galvan
Arturo Hernández
  Aguirre
Ahmed Kattan
Timoleon Kipouros
Joanna Kolodziej
Angel Kuri-Morales
Adriana Lara
Pierrick Legrand
Rui Li
Jing Liu
Francisco Luna
Rodica Ioana Lung
Gabriel Luque
Nashat Mansour
Jörn Mehnen
James McDermott
Nicolas Monmarché
Sanaz Mostaghim
Boris Naujoks
Sergio Nesmachnow

Gustavo Olague
Kathrin Padberg-Gehle
Eunice E. Ponce-de-Leon
Marcus Randall
Edgar Reehuis
Eduardo Rodriguez-Tello
Günter Rudolph
Thomas Schaberreiter
Christoph Schommer
Ofer Shir
Ignacio Segovia
  Dominguez
Oliver Schütze
Sara Silva
Leonardo Trujillo
Cliodhna Tuite
Massimiliano Vasile
Sergio Ivvan Valdez
Fatos Xhafa
Iryna Yevseyeva

Sponsoring Institutions

LIACS, Faculty of Science, Leiden University, nl
CINVESTAV-IPN, mx
NWO, nl
University of Luxembourg, lu
University of Bordeaux, fr
INRIA Bordeaux Sud-Ouest, fr

Conference Logo

Contents

Genetic Programming

Robust Optimization

Constructing a Solution Attractor for the Probabilistic Traveling Salesman Problem through Simulation

Weiqi Li

University of Michigan – Flint 303 E. Kearsley Street,
Flint, MI Postal Code 48502, U.S.A.
weli@umflint.edu

Abstract. The probabilistic traveling salesman problem (PTSP) is a variation of the classic traveling salesman problem and one of the most significant stochastic network and routing problems. This paper proposes a simulation-based multi-start search algorithm to construct the solution attractor for the PTSP and find the best *a priori* tour through the solution attractor. A solution attractor drives the local search trajectories to converge into a small region in the solution space, which contains the most promising solutions to the problem. Our algorithm uses a simple multi-start local search process to find a set of locally optimal *a priori* tours, stores these tours in a so-called hit-frequency matrix E, and then finds a globally optimal *a priori* tour in the matrix E. In this paper, the search algorithm is implemented in a master-worker parallel architecture.

Keywords: Stochastic optimization, global optimization, simulation, probabilistic traveling salesman problem, local search, parallel algorithm.

1 Introduction

The classic traveling salesman problem (TSP) is defined as: Given a set of n cities and an $n \times n$ cost matrix C in which $c(i, j)$ denotes the traveling cost between cities i and j ($i, j = 1, 2, \ldots, n$; $i \neq j$). A tour π is a closed route that visits every city exactly once and returns at the end to the starting city. The goal is to find a tour π^* with minimal traveling cost.

In the real world, many optimization problems are inherently dynamic and stochastic. Such problems exist in many areas such as optimal control, logistic management, scheduling, dynamic simulation, telecommunications networks, genetics research, neuroscience, and ubiquitous computing. As real-time data in information systems become increasingly available with affordable cost, people have to deal with more and more such complex application problems. For the TSP under dynamic and stochastic environment, the number of cities n can increase or decrease and the cost $c(i, j)$ between two cities i and j can change with time. In this paper we consider only the case in which the number of cities n changes with time t. Therefore, the TSP can be defined as:

M. Emmerich et al. (eds.), *EVOLVE - A Bridge between Probability, Set Oriented Numerics, and Evolutionary Computation IV*, Advances in Intelligent Systems and Computing 227,
DOI: 10.1007/978-3-319-01128-8_1, © Springer International Publishing Switzerland 2013

$$\min \quad f(\pi) = \sum_{i=1}^{n_t-1} c(i,\ i+1) + c(n_t,\ 1) \qquad (1)$$

$$\text{subject to} \quad n_t \in N$$

where n_t is the number of cities at time t and N is the set of all potential cities existing in the problem. If we want to design an algorithm and its purpose is to continuously track and adapt the changing n through time and to find the currently best solution quickly, that is, to re-optimize solution for every change of n, the TSP is defined as a dynamic TSP. If we treat the number of cities n as a random variable and wish to find an *a priori* tour through all N cities, which is of minimum cost in the expected value sense, the TSP becomes a probabilistic TSP (PTSP). In a PTSP, on any given realization of the problem, the n cities present will be visited in the same order as they appear in the *a priori* tour, i.e., we simply skip those cities not requiring a visit. The goal of an algorithm for PTSP is to find a feasible *a priori* tour with minimal expected cost [1, 2].

The PTSP was introduced by Jaillet [2, 3], who examined some of its combinatorial properties and derived a number of asymptotic results. Further theoretical properties, asymptotic analysis and heuristic schemes have been investigated by [4-7]. Surveys of approximation schemes, asymptotic analysis, and complexity theorems for a class of a priori combinatorial optimization problems can be found in [1, 8].

Formally, a PTSP is defined on a complete graph $G = (V, A, C, P)$, where $V = \{v_i: i = 1, 2, \ldots, N\}$ is a set of nodes; $A = \{a(i, j): i, j \in V, i \neq j\}$ is the set of edges that completely connects the nodes; $C = \{c(i, j): i, j \in A\}$ is a set of costs associated with the edges; $P = \{p_i: i \in V\}$ is a set of probabilities that for each node v_i specifies its probability p_i of requiring a visit. In this paper, we assume that the costs are symmetric, that is, traveling from a node v_i to v_j has the same cost as traveling from node v_j to v_i. We also assign v_1 as the depot node with the presence probability of 1.0. Each non-depot node v_i is associated with a presence probability p_i that represents the possibility that node v_i will be present in a given realization. Based on the values of presence probability (p_i) of non-depot nodes, two types of PTSP can be classified: the homogeneous and heterogeneous PTSP. In the homogeneous PTSP, the presence probabilities of non-depot nodes are all equal ($p_i = p$ for every non-depot node v_i); in the heterogeneous PTSP, these probabilities are not necessarily the same.

Designing effective and efficient algorithms for solving PTSP is a really challenging task, since in PTSP, the computational complexity associated with the combinatorial explosion of potential solutions is exacerbated by the stochastic element in the data. The predominant approach to finding good solutions for PTSP instances has been the adaptation of TSP heuristics [5-8]. In general, researchers use two techniques in their search algorithms: analytical computation and empirical estimation [9]. The analytical computation approach computes the cost $f(\pi)$ of an *a priori* tour π, using a closed-form expression. Empirical estimation simply estimates the cost through placeMonte Carlo simulation. Birattari et al. [9] discussed some limitations on analytical computation technique and suggested that the empirical estimation approach can overcome the difficulties posed by analytical computation.

In recent years, many local search and metaheuristic algorithms such as ant colony optimization, evolutionary computation, simulated annealing and scatter search, using analytical computation or empirical estimation approach, have been proposed to solve the PTSP [4, 7, 10-21]. Bianchi et al. [22] provided a comprehensive overview about recent developments in the metaheuristic algorithms field. In this paper, we propose a new simulation-based algorithm to solve PTSP. This paper is organized as follows. In section 2 we discuss the placeMonte Carlo sampling approximation. In section 3, we briefly describe the construction of solution attractor for TSP. In section 4, we describe the proposed simulation-based algorithm and discuss some experimental results. Section 5 concludes this paper.

2   Monte Carlo Sampling Approximation

The sampling approximation method is an approach for solving stochastic optimization problem by using simulation. The given stochastic optimization problem is transformed into a so-called sample average optimization problem, which is obtained by considering several realizations of the random variable and by approximating the cost of a solution with a sample average function [23]. In the context of the PTSP, this method has been shown to be very effective [9, 24].

In sampling approximation search, a stochastic optimization problem is represented by a computer simulation model. Simulation models are models of real or hypothetical systems, reflecting all important characteristics of the system under studied. Perhaps one of the best known methods for sampling a probability distribution is the placeMonte Carlo sampling technique, which is based on the use of a pseudo random number generator to approximate a uniform distribution. Currently, the placeMonte Carlo sampling approximation method is the most popular approach for solving stochastic optimization problems. In this technique the expected objective function of the stochastic problem is approximated by a sample average estimate derived from a random sample. We assume that the sample used at any given iteration is independent and identically distributed, and that this sample is independent of previous samples. The resulting sample average approximating problem is then solved by deterministic optimization techniques. The process can be repeated with different samples to obtain candidate solutions along with statistical estimates of their optimality gaps [18]. Sampling approximation via simulation is statistically valid in the context of simulation as the underlying assumptions of normality and independence of observations can be easily achieved through appropriate sample averages of independent realizations, and through adequate assignment of the pseudo-random number generator seeds, respectively.

In the case of PTSP, the elements of the general definition of the stochastic problem take the following format: a feasible tour π is an *a priori* tour visiting once and only once all N cities, and the random variable n is extracted from an N-variate Bernoulli distribution and prescribes which cities need being visited. This leads in total of 2^N possible scenarios for N cities. A nave way to calculate the expected cost of a solution would be to sum over all possible scenarios the cost of the *a posteriori* solution in this scenario multiplied with the probability for this scenario. Obviously the summation over 2^N terms is computationally intractable for reasonable values of N. Another more efficient way is using placeMonte Carlo sampling: instead of summing over all possible scenarios, we could sample M ($M < 2^N$) scenarios of using the known probabilities and take the average over the costs of the *a posteriori* tours for the sampled scenarios. Therefore, the cost $f(\pi)$ of a PTSP tour π can be empirically estimated on the basis of a sample $f(\pi, n_1), f(\pi, n_2), \ldots, f(\pi, n_M)$ of costs of *a posteriori* tours obtained from M independent realizations $n_1, n_2, \ldots, n_M$ of the random variable n:

$$\hat{f}_M(\pi) = \frac{1}{M} \sum_{i=1}^{M} f(\pi, n_i) \qquad (2)$$

$\hat{f}_M(\pi)$ denotes the average of the objective values of the M realizations on the *a priori* tour π, which gives us an approximation for the estimated cost for tour π. Clearly, $\hat{f}_M(\pi)$ is an unbiased estimator of $f(\pi)$. A search algorithm for PTSP is looking for the optimal tour π^* which has the smallest estimated objective value $\hat{f}_M(\pi^*)$, that is,

$$\pi^* \in \arg\min\{\hat{f}_M(\pi_1), \hat{f}_M(\pi_2), \ldots\} \qquad (3)$$

The optimal value $\hat{f}_M(\pi^*)$ and the optimal tour π^* to the PTSP provide estimates of their true counterparts.

Because we obtain only estimates using this way, it may not be possible to decide with certainty whether tour π_i is better than tour π_j, which may frustrates the search algorithm that tries to move in an improving direction. In principle, we can eliminate this complication by making so many replications at each iterative point that the performance estimate has essentially no variance. In practice, this could mean that we will explore very few iteration due to the time required to simulate each one. Therefore, in a practical sampling approximation algorithm, the test if a solution is better than another one can only be done by statistical sampling, that is, obtaining a correct comparison result only with a certain probability. The goal now is to get a good average case solution and the expected value of the objective is to be optimized. The way simulation approximation is used in an optimization algorithm largely depends on the way solutions are compared and the best solutions among a set of other solutions is selected.

The number of realizations M should be large enough for providing a reliable estimate of the cost of solutions but at the same time it should not be too large otherwise too much time is wasted. The appropriate number of realizations M depends on the stochastic character of the problem at hand. The larger the probability that a city is to be visited, the less stochastic an instance is. In this case, the algorithms that obtain the best results are those that consider a reduced number of realizations and therefore explore more solutions in the unit of time. On the other hand, when the probability that a city is to be visited is small, the instance at hand is highly stochastic. In this case, it pays off to reduce the total number of solutions explored and to consider a larger number of realizations for obtaining more accurate estimates [23, 25, 26].

There are many sampling strategies available. For the PTSP, common sampling strategies include (1) the same set of M realizations is used for all steps of the iteration in the algorithm; (2) a set of M realizations is sampled anew each time an improved solution is found; and (3) a set of M realizations is sampled anew for each comparison of solutions. The first strategy is a well-known variance-reduction technique called the method of common random numbers (CRN). CRN takes advantage of the same set of random numbers across all alternatives for a given replication. CRN is typically designed to induce positive correlation among the outputs of each respective alternative for a given replication, thereby reducing the variance of the difference between the mean alternative point estimators. One of the practical motivations for using CRN in a search algorithm is to speed up the sample average computations. However, one major problem with CRN is that the iterates of the algorithm may be "trapped" in a single "bad" sample path. Second and third strategies are called variable-sample method. Resampling allow the iterates of the algorithm to get away from those "bad" sample paths. Another advantage of a variable-sample scheme is that the sample sizes can increase along the algorithm, so that sampling effort is not wasted at the initial iteration of the algorithm [27].

Several researchers have been proposed estimation-based algorithms to deal with the PTSP, using local search or metaheuristics [21, 24-28]. This paper introduces a new optimization approach by using solution-attractor construction in the context of placeMonte Carlo simulation. Our algorithm includes optimization and simulation in a parallel iterative process in order to gain the advantages of optimization (exact solution), simulation (stochasticity) and speed (parallel processing).

3   Solution Attractor Construction

Our approach uses a parallel multi-start search procedure to construct the solution attractor for the PTSP. Due to the NP-hardness and intractability of the combinatorial optimization problems, heuristic search techniques have become a popular means to find reasonable good solutions to these problems. Many search heuristics have been based on or derived from a general technique known as local search [29]. A *search trajectory* is the path followed by the search process in the solution space as it evolves with time. Local search techniques are locally convergent. The final solution usually is a locally optimal solution. Local optimality

6   W. Li

depends on the initial solution and the neighborhood function that is used in the search process. In order to overcome local optimality, heuristic search usually require some type of diversification to avoid a large region of the solution space remaining completely unexplored. One simple way to achieve this diversification is to start the search process from several different initial points. Multi-start heuristics produce several local optima, and the best overall is the algorithm's output. Multi-start search helps to explore different areas in the solution space and therefore it generates a wide sample of the local optima.

For some optimization problem such as TSP, these multi-start search trajectories will converge to a small region in the solution space. From dynamic system perspective, this small region is called a *solution attractor* [30]. The solution attractor of a heuristic search algorithm on an optimization problem is defined as a subset of the solution space that contains the whole solution space of the end points (local optima) of all local search trajectories. Fig. 1 illustrates the concepts of local search trajectories and solution attractor. Since the globally optimal point is a special case of locally optimal points, it is expected to be embodied in the solution attractor. Li [30-32] developed a multi-start search approach to construct the solution attractor for a TSP and applied this approach to tackle multi-objective TSP and dynamic TSP. This paper applies this attractor-construction technique to solve PTSP.

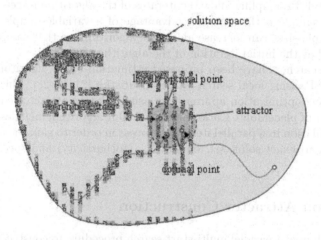

Fig. 1. The concepts of local search trajectories and solution attractor

For a TSP instance, the solution space contains all tours that a salesman may traverse. If we start several distinct search trajectories from different initial points, after letting the search process run for a long time, these trajectories would settle into the same attractive region (the solution attractor) if the problem has only one optimal solution. Fig. 2(a) presents a sequential procedure for constructing the solution attractor of local search in a TSP instance and

Fig. 2(b) shows a parallel procedure implemented in a master-worker architecture. The construction procedure is very straightforward: generating K locally optimal tours, storing them into a matrix (called *hit-frequency matrix E*), removing some unfavorable edges in E if necessary, and then finding all tours contained in E.

The hit-frequency matrix E plays a critical role in the construction of solution attractor. When each search trajectory reaches its locally optimal point, it leaves its "final footprint" in E. That is, E is used to record the number of hits on the edge $a(i, j)$ of the graph by the set of K locally optimal tours. Therefore, E can provide the architecture that allows individual tours to be linked together along a common structure and generate the structure for the global solution. This structure can help us to find the globally optimal tour.

```
1  procedure TSP_Attractor_Workers(C)
2  begin
3     for worker processor 1, ..., K
4        π_i = Initial_Tour() ;
5        π_j = Local_Search(π_i) ;
6        sends π_j to master processor ;
7  end
```

```
1  procedure TSP_Attractor(C)
2  begin
3     repeat
4        π_i = Initial_Tour() ;
5        π_j = Local_Search(π_i) ;
6        Update(E(π_j)) ;
7     until number_of_local_optima = Q ;
8     E = Remove_Noise(E) ;
9     Exhausted_Search(E) ;
10 end
```

```
1  procedure TSP_Attractor_Master(E)
2  begin
3     initialize E ;
4     repeat
5        receive a tour from a worker processor
6        update E ;
7     until number_of_local_optima = Q
8     Exhausted_Search(E) ;
9  end
```

(a)   (b)

Fig. 2. The procedure for constructing solution attractor of local search in TSP

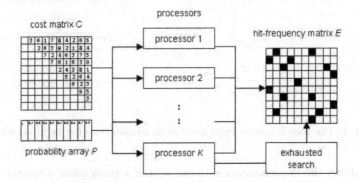

Fig. 3. Schematic structure of the search system for the PTSP

4   The Proposed Simulation-Based Search System

4.1   The Parallel Search System

We implemented our multi-start search system into a parallel search system. Parallel processing can be useful to efficiently solve difficult optimization problems, not only by seeding up the execution times, but also improving the quality of the final solutions. Fig. 3 sketches the basic idea of our parallel search system for the PTSP. This search system contains K processors and bears intrinsic parallelism in its features. Based on a common cost matrix C and probability array P, this system starts K separate search trajectories in parallel. When a search trajectory reaches its locally optimal solution, the processor stores the solution in the common hit-frequency matrix E. The processor starts a new search if more computing time is available. Finally, at the end of the search, the matrix E is searched by an exhausted search process and the best solution in the attractor is outputted as the optimal solution.

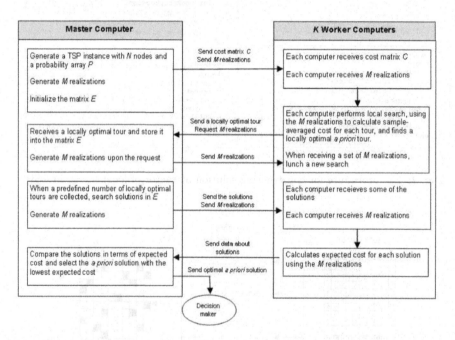

Fig. 4. The search system implemented in a master-worker architecture

Fig. 4 shows the implementation of our search system using a master-worker architecture. In our master-worker architecture, one computer serves as a master and other computers as workers. The master computer generates a TSP instance with N nodes and a probability array P, initializes the hit-frequency matrix E, and sends a copy of the cost matrix C to each of the worker computers.

The multi-search task is distributed to K worker computers. Then based on the probability values in the array P, the master computer generate a set of M realizations $(m_1, m_2, \ldots, m_M)$. Each of the M realizations is an array that contains binary values, where a value "1" at position i in the array indicates that node v_i requires being visited whereas a value "0" means that it does not require being visited. The master computer sends the set of M realizations to the worker computers. Depending on the setting, the master computer can send the same set of M realizations to all worker processors or a different set of M realizations to each of the worker computers. When receiving the set of M realizations, each worker computer independently performs its local search: it randomly generates an initial a priori tour, calculates the sample-averaged cost by using the M realizations, and then generates a new a priori tour, calculates its sample-averaged cost by using the same M realizations; if this new a priori tour has lower average cost, the current a priori tour is replaced by this better a priori tour; otherwise another new a priori tour is generated and compared with the current a priori tour. When a worker computer reaches a locally optimal a priori tour, it sends the tour to the master processor and requests a new set of M realizations from the master processor. When the master processor receives a locally optimal a priori tour from a worker processor, it stores the tour into the matrix E. It then generates a new set of M realizations and sends it to the requesting worker computer. In such was, the multi-start search is performed by the K worker computers in parallel. When the worker computer receives the new set of M realizations, it lunches new local search. When a predefined number of locally optimal a priori tours are collected from the worker computers and stored in the matrix E, these locally optimal tours form a solution attractor for the problem. The master computer sends a signal to worker computers to stop their searching. It then launches an exhausted search process in the matrix E. After finding all tours in E, the master computer sends these tours to the worker computers with a common set of M realizations; each worker computer receives different tours. The worker computers use the set of M realizations to calculate the sampling average and standard deviation for each of the tours, and send these values back to the master computer. The master computer compares these tours in terms of sampling average and finally outputs the best a priori tour.

4.2 The Experimental Problem

Due to the novelty of the field of PTSP there is no test problem instance available to be used for implementing a new search algorithm and assessing its suitability, effectiveness and efficiency. We design a test problem for our experiments.

The design of a test problem is always important in designing any new search algorithm. The context of problem difficulty naturally depends on the nature of problems that the underlying algorithm is trying to solve. In the context of solving PTSP, our test problem was designed based on several considerations. First, the size of problem should be large, since TSP instances as small as 200 cities are now considered to be well within the state of the global optimization art. The instance must be considerable large than this for us to be sure that the heuristic approach is really called for. Second, there is no any pre-known information related to the result of the experiment, since in a real-world problem one does not usually have any knowledge of the solutions. Third, when dealing with stochastic combinatorial problems, randomness in both process and data means that the underlying model in the algorithm is suitable for the modeling of natural problems. We should formulate a problem that goes nearer to real world conditions. Last, the problem instance should be general, understandable and easy to formulate so that the experiments are repeatable and verifiable.

We set up a TSP instance with $N = 1000$ cities. In the cost matrix C, $c(i, j)$ is assigned a random integer number in the range $[1, 1000]$, where $c(i, j) = c(j, i)$. A probability array P contains a set of probability values that assigns probability p_i to city i. We specify city 1 as the depot node with $p_1 = 1.0$. The probability of each non-depot city is generated from a uniform random number in a certain range. Therefore, our test problem is a heterogeneous PTSP.

4.3   The Experiments

Our experiments were conducted on a network of 6 PCs: Pentium 4 at 2.4 GHz and 512 MB of RAM running Linux, interconnected with a Fast Ethernet communication network using the LAM implementation of the MPI standard. The search system was developed using Sun Java JDK 1.3.1. The network was not dedicated, but was very steady. In our parallel search system, one computer serves as a master computer and other five computers as worker processors. Our network architecture and algorithms are asynchronous by nature, meaning that the processors do not have a common clock by which to synchronize their processing and calculation.

Our experiments relied heavily on randomization. All initial tours were randomly constructed. We used simple 2-opt local search in our search system. During the local search, the search process randomly selected a solution in the neighborhood of the current solution. A move that gave the first improvement was chosen. The local search process in each search trajectory terminated when no improvement could be achieved after 1000 iterations.

Because our search system uses simulation, it is not possible to decide with certainty whether a solution is better than another during the search. This can only be tested by statistical sampling, obtaining a correct comparison result only with a certain probability. In other words, the simulation estimates should be accompanied with some indication of precision. The first decision we had to make in our experiment was to choose an appropriate sample size M (number of realizations) in our simulation. The accuracy factors we considered include desired precision of results, confidence level and degree of variability. We used the following equation to determine our sample size [33, 34]:

$$M = \frac{V(1-V)}{\frac{A^2}{Z^2} + \frac{V(1-V)}{P}} \tag{4}$$

where M is the required sample size; P is the size of population; V is the estimated variance in population, which determines the degree to which the attributes being measured in the problem are distributed throughout the population; A is the desired precision of results that measures the difference between the value of the sample and the true value of the real population, called the sampling error; Z is the confidence level that measures the percentage of the samples would have the true population value within the range of chosen precision; and P is the size of population.

In our search system, we have two phases of simulation. In the first phase, the worker computers use simulation to calculate sample averages and compare tours. In this phase, we choose the sampling error $A = \pm 3\%$ and confidence level $Z = 1.96$ (95% confidence). Because our PTSP instance is heterogeneous, we use variability $V = 50\%$. Using Eq. (4) we calculate $M = 1067$. In the second phase, the workers use simulation to calculate sample averages for the tours found in the matrix E, and then the master computer uses these information to order the tours. In this phase, we choose the sample error $A = \pm 3\%$, confidence level $Z = 2.57$ (99% confidence) and $V = 50\%$. We calculate $M = 1835$. Therefore, in our experiment, we used $M_I = 1100$ realizations in the first phase and $M_{II} = 1850$ in the second phase.

We generated a TSP instance and a probability array P, in which the value of p_i was generated from a uniform random number in the range $[0.1, 0.9]$. When the master computer collected 30 locally optimal tours from the worker computers, the search system stopped searching. The master computer applied an exhausted-search procedure on E, and found 36 tours in E. The master computer sent these tours to the worker computers. Four worker computers got 7 tours and one worker computer got 8 tours. The worker computers calculate the sampling averages and standard deviations for these tours. Table 1 lists the five best tours found in E. We can see that the expected cost of the best tour is 4889 with standard deviation 201.

Table 1. The five best torus found in the matrix E

Tour	Sampling Average	Standard Deviation
1	4889	201
2	4902	224
3	5092	199
4	5413	213
5	5627	209

Then we ran the same TSP instance in the search system four more times. Table 2 lists the results of these five trials. The table shows the number of tours found in the matrix E, the sampling average of the best tour found in each of the trials and total computing time consumed by each of the computers. We compared these five best tours and found that they were the same tour; even it bore a different sampling average value in each of trials. This best tour is probably a globally optimal *a prior* tour for the PTSP instance.

Table 2. Results of five trials on the same TSP instance

Trial	Number of Tours in E	Sampling Average for Best Tour	Standard Deviation	Time (seconds)					
				Master	Worker1	Worker2	Worker3	Worker4	Worker5
1	36	4889	201	1007	1578	1549	1561	1468	1593
2	35	4872	209	1200	1563	1600	1589	1564	1465
3	36	4850	195	1157	1480	1533	1474	1593	1572
4	34	4873	199	1004	1550	1472	1543	1569	1530
5	35	4880	202	1058	1461	1486	1542	1473	1609

In another experiment, we used the same TSP instance but generated three different probability arrays P_1, P_2 and P_3. The values in P_1, P_2 and P_3 were generated from the range [0.1, 0.4], [0.3, 0.7] and [0.6, 0.9], respectively. For the probability array P_1, we ran the search system five times. The search system outputted three different best tours in these five trials. It indicates that, when a PTSP instance becomes more stochastic, our search system has more difficulty to find the globally optimal *a priori* tour. Then we ran search system five times for the probability array P_2, the search system outputted the same best tours with different sampling average values. Last we ran search system five times for the probability array P_3, the search system also outputted the same best tour. Obviously, when a PTSP instance is less stochastic or its average probability is 50/50, our search system may be able to find the globally optimal *a priori* tour. The results of this experiment are shown in Table 3.

Table 3. The results of experiment on different probability arrays

Trial	Number of Tours in E	Sampling Average for Best Tour	Standard Deviation	Time (seconds)					
				Master	Worker1	Worker2	Worker3	Worker4	Worker5
P_1									
1	38	1201	275	968	1261	1319	1208	1232	1304
2	37	1180	271	992	1290	1302	1258	1244	1326
3	39	1260	269	985	1239	1211	1215	1309	1269
4	42	1238	267	965	1237	1283	1278	1319	1211
5	39	1189	279	956	1309	1256	1304	1251	1245
P_2									
1	37	5580	166	993	1492	1345	1491	1454	1330
2	37	5606	175	980	1428	1381	1444	1307	1340
3	39	5554	164	1063	1465	1142	1313	1342	1493
4	36	5581	171	1044	1429	1384	1397	1475	1430
5	37	5561	161	995	1432	1369	1481	1330	1382
P_3									
1	34	7047	156	1201	1502	1558	1561	1563	1478
2	35	7049	154	1156	1583	1496	1558	1537	1525
3	33	7012	165	1127	1570	1487	1476	1471	1561
4	36	7009	161	1211	1495	1474	1589	1535	1589
5	37	7108	163	1124	1533	1505	1538	1538	1574

For the PTSP instance with the probability array P_1, we were wondering if we could improve the search quality by increasing the number of realizations in the search process. We set $M_I = 1100$ and $M_{II} = 1850$ and ran the problem instance 10 times. We found five different best tours in the ten trials and five trials outputted the same tour. Then we change $M_I = 3000$ and $M_{II} = 3000$, rand the problem instance 10 times again. We found three different best tours in the ten trials and seven trials outputted the same tour. Then we did the same procedure using $M_I = 6000$ and $M_{II} = 6000$. This time we found two different best tours and nine trials gave us the same tour. This experiment indicates that we can improve our search by increasing the number of realizations in the search process. Table 4 lists the experiment results.

Table 4. Search results in three different M settings

Setting	Number of Best Tours in 10 Trials	Number of Trials Having the Same Tour	Average Standard Deviation
$M_I = 1100$ $M_{II} = 1850$	5	5	268
$M_I = 3000$ $M_{II} = 3000$	3	7	247
$M_I = 6000$ $M_{II} = 6000$	2	9	226

5 Conclusion

This paper describes a new search algorithm for PTSP, using Monde Carlo simulation and being implemented in a parallel-processing architecture. The search process and simulation for realizations are performed by several parallel processors. If only one processor conducts local search and simulation in a finite time, the search trajectories are trapped in some local valleys and the search system is not really ergodic. Our search system takes a large ensemble of placeMonte Carlo salesmen from different processors to construct the solution attractor; it produces global nature of output. Our search system bears intrinsic parallelism, flexibility and diversity.

Since the primary goal of this paper is to introduce a new search algorithm, demonstrate the general applicability to PTSP, and analyze its search behavior, we didn't spend time on comparison with other algorithms. Work which follows this paper will comprehend the performance analysis of the proposed algorithm and comparison of the algorithm with respect to other PTSP algorithms.

References

1. Bertsimas, D.J., Jaillet, P., Odoni, A.R.: A Priori Optimization. Operations Research 38, 1019–1033 (1990)
2. Jaillet, P.: Probabilistic Traveling Salesman Problems. Ph.D Thesis, Massachusetts Institute of Technology, MA, USA (1985)
3. Jaillet, P.: A Priori Solution of a Traveling Salesman Problem in Which a Random Subset of the Customers Are Visited. Operations Research 36, 929–936 (1988)
4. Bertsimas, D.J.: Probabilistic Combinatorial Optimization Problems. Ph.D Dissertation. Massachusetts Institute of Technology, MA, USA (1988)
5. Bertsimas, D.J., Howell, L.: Further Results on the Probabilistic Traveling Salesman Problem. Journal of Operational Research 65, 68–95 (1993)

6. Jézéquel, A.: Probabilistic Vehicle Routing Problems. Master Thesis, Massachusetts Institute of Technology, MA, USA (1985)
7. Rossi, F., Gavioli, F.: Aspects of Heuristic Methods in the Probabilistic Traveling Salesman Problem. In: Advanced School on Stochastics in Combinatorial Optimization, pp. 214–227. World Scientific, Hackensack (1987)
8. Jaillet, P.: Analysis of Probabilistic Combinatorial Optimization Problems in Euclidean Spaces. Mathematics of Operations Research 18, 51–70 (1993)
9. Birattari, M., Balaprakash, P., Stützle, T., Dorigo, M.: Estimation-Based Local Search for Stochastic Combinatorial Optimization Using Delta Evaluations: A Case Study on the Probabilistic Traveling Salesman Problem. INFORMS Journal on Computing 20, 644–658 (2008)
10. Bianchi, L.: Ant Colony Optimization and Local Search for the Probabilistic Traveling Salesman Problem: A Case Study in Stochastic Combinatorial Optimization. Ph.D Dissertation, Universite Libre de Bruxelles, Brussels, Belgium (2006)
11. Bianchi, L., Campbell, A.M.: Extension of the 2-p-opt and 1-shift Algorithms to the Heterogeneous Probabilistic Traveling Salesman Problem. European Journal of Operational Research 176, 131–144 (2007)
12. Bianchi, L., Gambardella, L.M., Dorigo, M.: An Ant Colony Optimization Approach to the Probabilistic Traveling Salesman Problem. In: Guervós, J.J.M., Adamidis, P.A., Beyer, H.-G., Fernández-Villacañas, J.-L., Schwefel, H.-P. (eds.) PPSN 2002. LNCS, vol. 2439, pp. 883–892. Springer, Heidelberg (2002)
13. Binachi, L., Knowles, J., Bowler, N.: Local Search for the Probabilistic Traveling Salesman Problem: Correction to the 2-p-opt and 1-Shift Algorithms. European Journal of Operational Research 16, 206–219 (2005)
14. Bowler, N.E., Fink, T.M., Ball, R.C.: Characterization of the Probabilistic Traveling Salesman Problem. Physical Review E 68, 1–7 (2003)
15. Branke, J., Guntsch, M.: Solving the Probabilistic TSP with Ant Colony Optimization. Journal of Mathematical Modeling and Algorithms 3, 403–425 (2004)
16. Campbell, A.M.: Aggregation for the Probabilistic Traveling Salesman Problem. Computers and Operations Research 33, 2703–2724 (2006)
17. Liu, Y.-H.: Solving the Probabilistic Traveling Salesman Problem Based on Genetic Algorithm with Queen Selection Scheme. In: Greco, F. (ed.) Traveling Salesman Problem, pp. 157–172. InTech (2008)
18. Liu, Y.-H., Jou, R.-C., Wang, C.-C., Chiu, C.-S.: An Evolutionary Algorithm with Diversified Crossover Operator for the Heterogeneous Probabilistic TSP. In: Torra, V., Narukawa, Y., Yoshida, Y. (eds.) MDAI 2007. LNCS (LNAI), vol. 4617, pp. 351–360. Springer, Heidelberg (2007)
19. Marinakis, Y., Migdalas, M., Pardalos, P.M.: Expanding Neighborhood Search GRASP for the Probabilistic Traveling Salesman Problem. Optimization Letters 2, 351–360 (2008)
20. Marinakis, Y., Marinakis, M.: A Hybrid Multi-Swarm Particle Swarm Optimization Algorithm for the Probabilistic Traveling Salesman Problem. Computers & Operations Research 37, 432–442 (2010)
21. Tang, H., Miller-Hooks, E.: Approximate Procedures of the Probabilistic Traveling Salesperson Problem. Transportation Research Record 1882, 27–36 (2004)
22. Bianchi, L., Dorigo, M., Gambardella, L.M., Gutjahr, W.J.: A Survey on Metaheuristics for Stochastic Combinatorial Optimization. Natural Computing 8, 239–287 (2009)
23. Kleywegt, A.J., Shapiro, A., Homen-de-Mello, T.: The Sample Average Approximation Method for Stochastic Discrete Optimization. SIAM Journal on Optimization 12, 479–502 (2001)

24. Verweij, B., Ahmed, S., Kleywegt, A.J., Nemhauser, G., Shapiro, A.: The Sample Average Approximation Method Applied to Stochastic Routing Problems: A Computational Study. Computational Optimization and Application 24, 289–333 (2003)
25. Balaprakash, P., Pirattari, P., Stützle, T., Dorigo, M.: Adaptive Sample Size and Importance Sampling in Estimation-Based Local Search for the Probabilistic Traveling Salesman Problem. European Journal of Operational Research 199, 98–110 (2009)
26. Balaprakash, P., Pirattari, P., Stützle, T., Dorigo, M.: Estimation-based Metaheuristics of the Probabilistic Traveling Salesman Problem. Computers & Operations Research 37, 1939–1951 (2010)
27. Homen-de-Mello, T.: Variable-Sample Methods for Stochastic Optimization. ACM Transactions on Modeling and Computer Simulation 13, 108–133 (2003)
28. Weyland, D., Bianchi, L., Gambardella, L.M.: New Approximation-Based Local Search Algorithms for the Probabilistic Traveling Salesman Problem. In: Moreno-Díaz, R., Pichler, F., Quesada-Arencibia, A. (eds.) EUROCAST 2009. LNCS, vol. 5717, pp. 681–688. Springer, Heidelberg (2009)
29. Aart, E., Lenstra, J.K.: Local Search in Combinatorial Optimization. Princeton University Press, Princeton (2003)
30. Li, W.: Seeking Global Edges for Traveling Salesman Problem in Multi-Start Search. Journal of Global Optimization 51, 515–540 (2011)
31. Li, W.: A Parallel Multi-Start Search Algorithm for Dynamic Traveling Salesman Problem. In: Pardalos, P.M., Rebennack, S. (eds.) SEA 2011. LNCS, vol. 6630, pp. 65–75. Springer, Heidelberg (2011)
32. Li, W., Feng, M.: A Parallel Procedure for Dynamic Multi-Objective TSP. In: Proceedings of 10th IEEE International Symposium on Parallel and Distributed Processing with Applications, pp. 1–8. IEEE Computer Society (2012)
33. Sudman, S.: Applied Sampling. Academic Press, New York (1976)
34. Walson, J.: How to Determine a Sample Size. Penn Cooperative Extension, University Park, PA (2001)

Unsupervised Classifier Based on Heuristic Optimization and Maximum Entropy Principle

Edwin Aldana-Bobadilla and Angel Kuri-Morales

Universidad Nacional Autónoma de México, Mexico City, Mexico
Instituto Técnologico Autónomo de México, México City, Mexico
ealdana@uxmcc2.iimas.unam.mx,
akuri@itam.mx

Abstract. One of the basic endeavors in Pattern Recognition and particularly in Data Mining is the process of determining which unlabeled objects in a set do share interesting properties. This implies a singular process of classification usually denoted as "clustering", where the objects are grouped into k subsets (clusters) in accordance with an appropriate measure of likelihood. Clustering can be considered the most important unsupervised learning problem. The more traditional clustering methods are based on the minimization of a similarity criteria based on a metric or distance. This fact imposes important constraints on the geometry of the clusters found. Since each element in a cluster lies within a radial distance relative to a given center, the shape of the covering or hull of a cluster is hyper-spherical (convex) which sometimes does not encompass adequately the elements that belong to it. For this reason we propose to solve the clustering problem through the optimization of Shannon's Entropy. The optimization of this criterion represents a hard combinatorial problem which disallows the use of traditional optimization techniques, and thus, the use of a very efficient optimization technique is necessary. We consider that Genetic Algorithms are a good alternative. We show that our method allows to obtain successfull results for problems where the clusters have complex spatial arrangements. Such method obtains clusters with non-convex hulls that adequately encompass its elements. We statistically show that our method displays the best performance that can be achieved under the assumption of normal distribution of the elements of the clusters. We also show that this is a good alternative when this assumption is not met.

Keywords: Clustering, Genetic Algorithms, Shannon's Entropy, Bayesian Classifier.

1 Introduction

Pattern recognition is a scientific discipline whose purpose is to describe and classify objects. The descriptive process involves a symbolic representation of

M. Emmerich et al. (eds.), *EVOLVE - A Bridge between Probability, Set Oriented Numerics,* 17
and Evolutionary Computation IV, Advances in Intelligent Systems and Computing 227,
DOI: 10.1007/978-3-319-01128-8_2, © Springer International Publishing Switzerland 2013

these objects called *patterns*. In this sense, the most common representation is through a numerical vector $\boldsymbol{x}$:

$$\boldsymbol{x} = [x_1, x_2, \ldots x_n] \in \Re^n \tag{1}$$

where the n components represent the value of the properties or attributes of an object. Given a pattern set X, there are two ways to attempt the classification: a) *Supervised Approach* and b) *Unsupervised Approach*.

In the supervised approach, $\forall \boldsymbol{x} \in X$ there is a class label $y \in \{1, 2, 3, \ldots, k\}$. Given a set of class labels Y corresponding to some observed patterns $\boldsymbol{x}$ ("training" patterns), we may postulate a hypothesis about the structure of X that is usually called the *model*. The model is a mathematical generalization that allows us to divide the space of X into k decision regions called *classes*. Given a model M, the class label y of an unobserved (unclassified) pattern $\boldsymbol{x}'$ is given by:

$$y = M(\boldsymbol{x}') \tag{2}$$

On the other hand, the unsupervised approach consists in finding a hypothesis about the structure of X based only on the similarity relationships among its elements. The unsupervised approach does not use prior class information. The similarity relationships allow to divide the space of X into k subsets called *clusters*. A cluster is a collection of elements of X which are "similar" between them and "dissimilar" to the elements belonging to other clusters. Usually the similarity is defined by a *metric* or distance function $d : X \times X \to \Re$.

In this work we discuss a clustering method which does not depend explicitly on minimizing a distance metric and thus, the shape of the clusters is not constrained by hyper-spherical hulls. Clustering is a search process on the space of X that allows us to find the k clusters that satisfy an optimization criteria. Mathematically, any criterion involves an objective function f which must be optimized. Depending on the type of f, there are several methods to find it. Since our clustering method involves an objective function f where its feasible space is, in general, non-convex and very large, a good optimization algorithm is compulsory. With this in mind, we made a comprehensive study [13] which dealt with the relative performance of a set of structurally different GAs and a non-evolutionary algorithm over a wide set of problems. These results allowed us to select the statistically "best" algorithm: the EGA [20]. By using EGA we may be sure that our method will displays high effectiveness for complex arrangements of X.

The paper is organized as follows: In Section 2, we briefly show the results that led us to select the EGA. Then we present the different sets of patterns X that will serve as a the core for our experiments. We use a Bayesian Classifier[2,4,8] as a method of reference because there is theoretical proof that its is optimal given data stemming from normal distributions. In this section we discuss the issues which support our choice. In Section 3 we discuss the main characteristics of our method and the experiments which show that it is the best alternative. In Section 4 we present our general conclusions.

2 Preliminaries

As pointed out above, a "good" optimization algorithm must be selected. We rest on the conclusions of our previous analysis regarding the performance of a set of GAs[13] . Having selected the best GA, we prove the effectiveness of our clustering method by classifying different pattern sets. To this effect, we generated pattern sets where, for each pattern, the class of the objects is known. Hence, the class found by our clustering method may be compared to the true ones. To make the problems non-trivial we selected a non-linearly separable problems. We discuss the process followed to generate these sets. Finally, we resort to a *Bayesian Classifier* [4] in order to show that the results obtained by our method are similar to those obtained with it.

2.1 Choosing the Best Optimization Algorithm

This section is a very brief summary of the most important results found in [13]. A set A of 4 structurally different GAs and a non-evolutionary algorithm (NEA) was selected in order to solve, in principle, an unlimited supply of systematically generated functions in $\Re \times \Re$(called unbiased functions). An extended set of such functions in $\Re \times \Re^2$ and $\Re \times \Re^3$ was generated and solved. Similar behavior of all the GAs in$A$ (within statistical limits) was found. This fact allowed us to hypothesize that the expected behavior of A for functions in $\Re \times \Re^n$ will be similar. As supplement, we tackled a suite of problems (approximately 50) which includes hard unconstrained problems (which traditionally have been used for benchmarking purposes) [19,3] and constrained problems [11]. Lastly, atypical GA-hard functions were analyzed [18,16].

Set of Algorithms. The set A included the following GAs: a)An elitist canonical GA (in what follows referred to as TGA [eliTist GA]) [21], b) A Cross generational elitist selection, Heterogeneous recombination, and Cataclysmic mutation algorithm (CHC algorithm) [5], c) An Eclectic Genetic Algorithm (EGA) [20], d) A Statistical GA (SGA) [23,12] and e) A non-evolutionary algorithm called RMH [17].

Table 1 shows the relative global performance of all algorithms for the functions mentioned. The best algorithm in the table is EGA.

Table 1. Global Performance

A_i	Unbiased	Suite	Atypical	Global Performance	Relative
EGA	9.64	8.00	4.48	7.37	100.00%
RMH	6.24	0.012	2.04	2.76	37.49%
TGA	1.35	1.16	4.77	2.43	32.91%
SGA	1.33	0.036	3.33	1.57	21.23%
CHC	2.12	0.08	2.10	1.43	19.44%

2.2   The Pattern Set

Given a set of patterns to be classified, the goal of any classification technique is to determine the decision boundary between classes. When these classes are unequivocally separated from each other the problem is separable; otherwise, the problem is non-separable. If the problem is linarly separable, the decision consists of a hyperplane. In Figure 1 we illustrate ths situation.

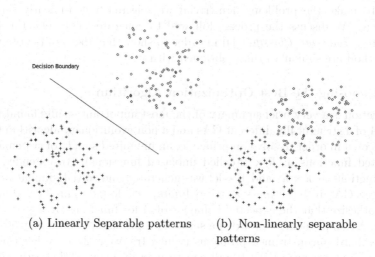

(a) Linearly Separable patterns   (b) Non-linearly separable patterns

Fig. 1. Decision boundary

When there is overlap between classes some classification techniques (e.g. Linear classifiers, Single-Layer Perceptrons [8]) may display unwanted behavior because decision boundaries may be highly irregular . To avoid this problem many techniques has been tried (e.g. Support Vector Machine [9], Multilayer Perceptrons [22]). However, there is no guarantee that any of this methods will perform adequately. Nevertheless, there is a case which allows us to test the appropriateness of our method. Since it has been proven that if the classes are normally distributed, a Bayesian Classifier yields the best possible result (in Section 2.3 we discuss this fact) and the error ratio will be minimized. Thus, the Bayesian Classifier becomes a good method with which to compare any alternative clustering algorithm.

Hence, we generated Gaussian pattern sets considering singular arrangements in which determining the decision boundaries imply non-zero error ratios. Without loss generality we focus on patterns defined in $\Re^2$. We wish to show that the results obtained with our method are close to those obtained with a Bayesian Classifier; in Section 3, the reader will find the generalization of our method for $\Re^n$.

Gaussian Patterns in $\Re^2$. Let X_j be a pattern set defined in $\Re^2$ and $C_i \subset X_j$ a pattern class. A pattern $\boldsymbol{x} = [x_1, x_2] \in C_i$ is drawn from a Gaussian distribution if its joint probability density function (pdf) is given by:

$$f(x_1, x_2) = \frac{1}{2\pi\sigma_{x_1}\sigma_{x_2}\sqrt{1-\rho^2}}e^{\left(-\frac{1}{2(1-\rho^2)}\left[\left(\frac{x_1-\mu_{x_1}}{\sigma_{x_1}}\right)^2 - 2\rho\frac{(x_1-\mu_{x_1})(x_2-\mu_{x_2})}{\sigma_{x_1}\sigma_{x_2}} + \left(\frac{x_2-\mu_{x_2}}{\sigma_{x_2}}\right)^2\right]\right)}$$

$$(3)$$

where $-1 < \rho < 1$, $-\infty < \mu_{x_1} < \infty$, $-\infty < \mu_{x_2} < \infty$, $\sigma_{x_1} > 0$, $\sigma_{x_2} > 0$. The value ρ is called the correlation coefficient.

To generate a Gaussian pattern $\boldsymbol{x} = [x_1, x_2]$, we use the *acceptance-rejection* method [1,10] which allows us to generate random observations (x_1, x_2) that are drawn from $f(x_1, x_2)$. In this method, a uniformly distributed random point (x_1, x_2, y) is generated and accepted iff $y < f(x_1, x_2)$. In Figure 2.1 we show different pattern sets obtained by applying this method with distinct statistical arguments in (3)

Fig. 2. Different Gaussian pattern sets with $\mu_{x_1} = \mu_{x_2} = 0.5, \sigma_{x_1} = \sigma_{x_2} = 0.09$. Each set was generated with different correlation coefficient: a. $\rho = 0.0$, b. $\rho = -0.8$, c. $\rho = 0.8$.

The degrees of freedom in (3) allow us to generate Gaussian pattern sets with varied spatial arrangements. In principle, we analyze pattern sets with the following configurations:

- Sets with disjoint pattern classes.
- Sets with pattern classes that share some elements (partial overlap between classes)
- Sets with pattern classes whose members may share most elements (total overlap).

We proposed these configurations in order to increase gradually the complexity of the clustering problem and analyze systematically the performance of our method. In the following subsections, we make a detailed discussion regarding the generation of sets with these configurations.

Gaussian Pattern Set with Disjoint Classes

Definition: Let X_1 be a pattern set with classes $C_i \subset X_1 \forall i = 1, 2...k$ which are drawn from a Gaussian distribution. X_1 is a set with disjoint classes if $\forall C_i, C_j \subset X_1, C_i \cap C_j = \phi$.

Based on the above definition, we generate two different sets where $x \in [0, 1]^2$ (in every set there are three classes and $|X| = 1000$). In Figure 3 we illustrate the spatial arrangement of these sets.

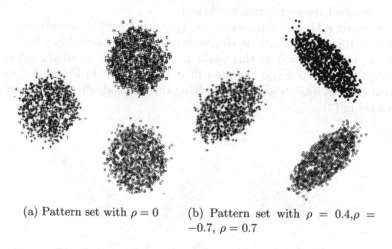

(a) Pattern set with $\rho = 0$ (b) Pattern set with $\rho = 0.4, \rho = -0.7, \rho = 0.7$

Fig. 3. Gaussian pattern sets with disjoint classes

Gaussian Pattern Set with Partial Overlap

Definition: Let X_2 be a pattern set with classes $C_i \subset X_2 \forall i = 1, 2...k$ which are drawn from a Gaussian distribution. X_2 is a set with partial overlap if $\exists C_i, C_j \subset X_2$ such that $C_i \cap C_j \neq \phi$ and $C_i \nsubseteq C_j$.

Based on the above definition, we generate two different sets where $x \in [0, 1]^2$ (in every set there are three classes and $|X| = 1000$). In Figure 4 we illustrate the spatial arrangement of these sets.

Gaussian Pattern Set with Total Overlap

Definition: Let X_3 be a pattern set with classes $C_i \subset X_3 \forall i = 1, 2...k$ which are drawn from a Gaussian distribution. X_3 is a set with total overlap if $\exists C_i, C_j \subset X_3$ such that $C_i \subseteq C_j$.

Based on the above definition, we generate two different sets where $x \in [0, 1]^2$ (in every set there are three classes and $|X| = 1000$). In Figure 5 we illustrate the spatial arrangement of these sets.

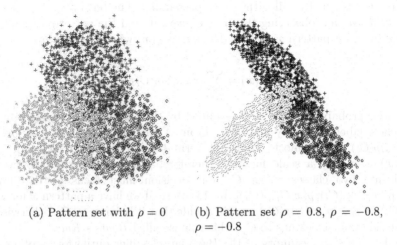

(a) Pattern set with $\rho = 0$ (b) Pattern set $\rho = 0.8$, $\rho = -0.8$,
$\rho = -0.8$

Fig. 4. Pattern sets with overlap classes

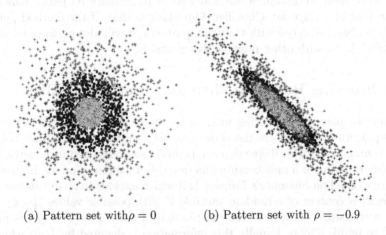

(a) Pattern set with $\rho = 0$ (b) Pattern set with $\rho = -0.9$

Fig. 5. Pattern sets with total overlap classes

2.3 Bayesian Classifier

If the objects in the classes to be clustered are drawn from normally distributed data, the best alternative to determine the decision boundary is using a Bayesian Classifier. The reader can find a extended discussion in [4,8].

Given a sample of labeled patterns $x \in X$, we can hypothesize a partitioning of the space of X into k classes C_i. The classification problem can be reduced to find the probability that given a pattern x, it belongs to C_i. From Bayes's theorem this probability is given by:

$$p(C_i|x) = \frac{p(x|C_i)p(C_i)}{p(x)} \tag{4}$$

where $p(C_i)$ is usually called the *prior* probability. The term $p(\boldsymbol{x}|C_i)$ represents the *likelihood* that observing the class C_i we can find the pattern $\boldsymbol{x}$. The probability to find a pattern $\boldsymbol{x}$ in X is denoted as $p(\boldsymbol{x})$ which is given by:

$$p(\boldsymbol{x}) = \sum_{i=0}^{k} p(\boldsymbol{x}|C_i)p(C_i) \tag{5}$$

The prior probability $p(C_i)$ is determined by the training pattern set (through the class labels of every pattern). From 4 we can note that the product $p(\boldsymbol{x}|C_i)p(C_i)$ is the most important term to determine $p(C_i|\boldsymbol{x})$ (the value of $p(\boldsymbol{x})$ is merely a scale factor). Therefore, given a pattern $\boldsymbol{x}$ to be classified into two classes C_i or C_j, our decision must focus on determining $max\,[p(\boldsymbol{x}|C_i)p(C_i), p(\boldsymbol{x}|C_j)p(C_j)]$. In this sense, if we have a pattern $\boldsymbol{x}$ for which $p(\boldsymbol{x}|C_i)p(C_i) \geq p(\boldsymbol{x}|C_j)p(C_j)$ we will decide that it belongs to C_i, otherwise, we will decide that it belongs to C_j. This rule is called *Bayes's Rule*.

Under gaussian assumption, the Bayesian classifier outperforms other classification techniques, such as those based on *linear predictor functions* [8]. We discuss our method assuming normality so as to measure its performance relative to that of a Bayesian Classifier. Our claim is that, if the method performs satisfactorily when faced with Gaussian patterns, it will also perform reasonably well when faced with other possible distributions.

3   Clustering Based on Shannon's Entropy (CBE)

We can visualize any clustering method as a search for k regions (in the space of the pattern set X), where the *dispersion* between the elements that belong to them is minimized. This dispersion can be optimized via a distance metric [15,7], a quality criterion or a membership function [14]. In this section we discuss an alternative based on *Shannon's Entropy* [24] which appeals to an evaulation of the *information content* of a random variable Y with possible values $\{y_1, y_2, ...y_n\}$. From a statistical viewpoint, the information of the event $(Y = y_i)$ is proportional to its likelihood. Usually this information is denoted by $I(y_i)$ which can be expressed as:

$$I(y_i) = log\left(\frac{1}{p(y_i)}\right) = -log\left(p(y_i)\right) \tag{6}$$

From information theory [24,6], the information content of Y is the expected value of I. This value is called Shannon's Entropy which is given by:

$$H(Y) = -\sum_{i=1}^{n} p(y_i)log\left(p(y_i)\right) \tag{7}$$

When $p(y_i)$ is uniformly distributed, the entropy value of Y is maximal. It means that all events in the probability space of Y have the same ocurrence probability and thus Y has the highest level of unpredictability. In other words, Y has the maximal information content.

In the context of the clustering problem, given an unlabeled pattern set X, we hypothesize that a cluster is a region of the space of X with a large information content. In this sense, the clustering problem is reduced to a search for k regions where the entropy is maximized. Tacitly, this implies finding k optimal probability distributions (those that maximize the information content for each regions). It is a hard combinatorial optimization problem that requires an efficient optimization method. In what follows we discuss the way to solve such problem through EGA. In principle, we show some evidences that allow us to think that our method based on entropy is succesful. We statistically show this method is the best.

3.1   Shannon's Entropy in Clustering Problem

Given an unlabeled pattern set X, we want to find a division of the space of the X into k regions denoted by C_i, where Shannon's Entropy is maximized. We consider that the entropy of C_i depends on the probability distribution of all possible patterns x that belong to it. In this sense, the entropy of C_i can be expressed as:

$$H(C_i) = \sum_{x \in C_i} p(x|C_i) log(p(x|C_i) \tag{8}$$

Since we want to find k regions C_i that maximize such entropy, the problem is reduced to an optimization problem of the form:

$$\text{Maximize: } \sum_{i=1}^{k} \sum_{x \in C_i} p(x|C_i) log(p(x|C_i)) \\ \text{subject to:} \\ p(x|C_i) > 0 \tag{9}$$

To find the C_i's that minimize (9) we resort to the EGA. We encoded an individual as a random sequence of symbols L from the alphabet $\sum = \{1, 2, 3...k\}$. Every element in L represents the class label of a pattern x in X such that the length of L is $|X|$. Tacitly, this encoding divides the space of X into k regions C_i as is illustrated in Figure 5.1.

Given this partititon, we can determine some descriptive parameters θ_i of the probability distribution of C_i (e.g. the mean or the standard deviation). Having determined θ_i and under the assumption that the probability distribution of C_i is known, the entropy of C_i can be determined. Complementarily, in Subection 3.3.2 we discuss a generalization making this assumption unnecessary.

3.2   Effectiveness of the Entropic Clustering for Gaussian Patterns in $\Re^2$

Based on the above, given an encoding solution L of a clustering problem (where X is a pattern set in $\Re^2$) we can determine k regions C_i with parameter $\theta_i = [\mu(C_i), \sigma(C_i)]$. Assuming that C_i is drawn from a Gaussian distribution, the value of $p(x|C_i)$ with $x = [x_1, x_2]$ is given by:

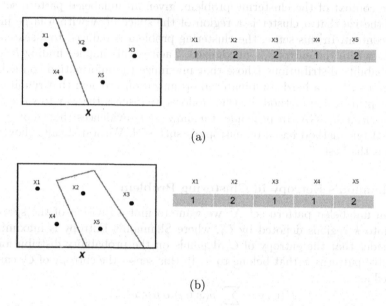

(a)

(b)

Fig. 6. Possible divisions of the space of a pattern set X in $\Re^2$ based on the encoding of an individual

$$\int_{x_1} \int_{x_2} f(x_1, x_2) \tag{10}$$

where $f(x_1, x_2)$ is the bivariate Gaussian density function (see Equation 3). Given such probability the entropy for each C_i can be determined, the fitness of every L is given by $\sum_{i=1}^{k} H(C_i)$. The optimal solution will be the individual with the best fitness value after G iterations. To measure the effectiveness of our method, we ran 100 times the EGA (with 150 individuals and $G = 300$), selecting randomly different Gaussian pattern sets (disjoint, partial overlap and total overlap) in $\Re^2$. In Table 2 we show the performance of CBE for these problems. The "performance value" is defined as the success ratio based on the class labels of the patterns known a priori. We also illustrate the performance displayed by the Bayesian Classifier for the same set of problems.

Table 2. Performance of CBE and BC for different Gaussian pattern sets in $\Re^2$

Algorithm	Disjoint	Partial Overlap	Total Overlap	Global	Relative
CBE	99.0	70.8	57.6	75.8	99.7%
BC	99.9	71.7	56.5	76.0	100%

We see that there is not an important difference between results of our method and the results of a Bayesian Classifier. This result allows us to ascertain that our method is as good as the BC. Recall that the BC displays the best possible performance under Gaussian assumption. Furthermore, it is very important to stress that our method is unsupervised and, hence, the pattern set X is unlabeled. CBE allows us to find the optimal value of θ_i for all C_i that maximize the objective function (see Equation 9). These results are promising but we want to show that our method performs successfully, in general, as will be shown in the sequel.

3.3   Comprehensive Effectiveness Analysis

In order to evaluate the general effectivennes of CBE, we generated systematically a set of 500 clustering problems in $\Re^n$ assuming normality. The number of clusters for each problem and the dimensionality were randomly selected (the number of clusters $k \sim U(2, 20)$ and the dimensionality $n \sim U(2, 10)$). Thus, we obtained an unbiased set of problems to solve through CBE and BC. Similarly, we also propose a method to generate systematically a set of clustering problems in $\Re^n$ without *any assumption regarding the pdf* of the patterns to be classified.

Effectiveness for Gaussian Patterns in $\Re^n$. We wrote a computer program that generates Gaussian patterns $x = [x_1, x_2, ...x_n,]$ through the *acceptance-rejection* method [1,10] given a value of n. Here, a uniformly distributed random point $(x_1, x_2, ...x_n, y)$ is generated and accepted iff $y < f(x_1, x_2, ..., x_n)$ where $f(x_1, x_2, ..., x_n)$ is the Gaussian density function with parameters μ and σ. Our program determines randomly the values of $\mu = [\mu_{x_1}, \mu_{x_2}..., \mu_{x_n}]$ and $\sigma = [\sigma_{x_1}, \sigma_{x_2}..., \sigma_{x_n}]$ such that $\mu_{x_i} \in [0, 1]$ and $\sigma_{x_i} \in [0, 1]$. In this way a cluster is a set of Gaussian patterns with the same values of μ and σ. and a clustering problem is a set of such clusters. The cardinality of a cluster is denoted by $|C_i|$ whose value was established as 200. It is important to note that the class label of every generated pattern was recorded in order to determine the performance or effectiveness of a classification process. We obtained a set of 500 different Gaussian clustering problems. To evaluate the performance of any method (CBE or BC) for such problems, we wrote a computer program that executes the following steps:

1. A set of $N = 36$ clustering problems are randomly selected.
2. A effectiveness value y_i is recorded for each problem.
3. For every N problems $\bar{y}_i$ is calculated.
4. Steps 1-3 are repeated until the values $\bar{y}_i$ are approximately normally distributed with parameter μ' and σ' (from *the central limit theorem*).

Dividing the span of means $\bar{y}_i$ in deciles, normality was considered to have been reached when:

1. $\chi^2 \leq 4$ and
2. The ratio of observations in the i-th decil $O_i \geq 0.5 \; \forall i$.

From Chebyshev's Inequality [25] the probability that the performance of an method denoted by τ lies in the interval $[\mu' - \lambda\sigma', \; \mu' + \lambda\sigma']$ is given by:

$$p(\mu' - \lambda\sigma' \leq \tau \leq \mu' + \lambda\sigma') \geq 1 - \frac{1}{\lambda^2} \tag{11}$$

where λ denotes the number of standard deviations. By setting $\lambda = 3.1623$ we ensure that the values of τ will lie in this interval with probability $p \approx 0.9$. Hence, the largest value of performance found (with $p \approx 0.95$ if we assume a symmetric distribution) by any method (CBE and BC) is $\mu + \lambda\sigma$. The results are shown in Table 3. These results allow us to prove statistically (with significance level of 0.05) that in $\Re^n$ our method is as good as the BC under normality assumption.

Table 3. Comparative average success ratio for Gaussian Problems

Algorithm	μ	σ	$\mu + \lambda\sigma$	Relative
CBE	75.53	3.22	85.71	99%
BC	76.01	3.35	86.60	100%

Effectiveness for Non-Gaussian Patterns $\Re^n$. To generate a Non-Gaussian patterns in $\Re^n$ we resort to polynomial functions of the form:

$$f(x_1, x_2, ..., x_n) = a_{m1}x_1^m + ... + a_{mn}x_n^m + ... + a_{11}x_1 + a_{1n}x_n \tag{12}$$

Given that such functions have larger degrees of freedom, we can generate many points uniformly distributed in $\Re^n$. As reported in the previous section, we wrote a computer program that allows us to obtain a set of 500 different problems. These problems were generated for random values of n and k with $|C_i| = 200$. As before, a set of tests with $N = 36$ was performed. As before, the number of samples of size N dependend on the distribution reaching normality. The results of this set of experiments is shown in Table 4. These results show that our method outperforms BC with a significant level of 0.05.

Table 4. Comparative average success ratio for non-Gaussian Problems

Algorithm	μ	σ	$\mu + k\sigma$	Relative
CBE	74.23	2.12	80.93	100%
BC	61.76	2.65	70.14	86%

4   Conclusions

The previous analysis, based on the solution of an exhaustive sample set, allows us to reach the following conclusions:

Entropic clustering (CBE) is reachable via an efficient optimization algorithm. In this case, based on previous work by the authors, one is able to take advantage of the proven efficiency of EGA. The particular optimization function (defined in (9)) yields the best average success ratio. We found that CBE is able to find highly irregular clusters in pattern sets with complex arrangements. When compared to BC's performance over Gaussian distributed data sets, CBE and BC have, practically, indistinguishable success ratios. Thus proving that CBE is comparable to the best theoretical option. Here we, again, stress that while BC corresponds to supervised learning whereas CBE does not. The advantage of this characteristic is evident. When compared to BC's performance over non-Gaussian sets CBE, as expected, displayed a much better success ratio. Based on comprehensive analysis (subsection 3.3), the conclusions above have been reached for statistical p values of $O(0.5)$. In other words, the probability of such results to persist on data sets outside our study is better than 0.95. Thus ensuring the reliability of CBE. Clearly above older alternatives.

References

1. Casella, G., Robert, C.P.: Monte carlo statistical methods (1999)
2. De Sa, J.M.: Pattern recognition: concepts, methods, and applications. Springer (2001)
3. Digalakis, J., Margaritis, K.: An experimental study of benchmarking functions for genetic algorithms (2002)
4. Duda, R., Hart, P., Stork, D.: Pattern classification, Section 10, P. 6. John Wiley, New York (2001)
5. Eshelman, L.: The chc adaptive search algorithm. how to have safe search when engaging in nontraditional genetic recombination (1991)
6. Gallager, R.G.: Information theory and reliable communication (1968)
7. Halkidi, M., Batistakis, Y., Vazirgiannis, M.: On clustering validation techniques. Journal of Intelligent Information Systems 17(2), 107–145 (2001)
8. Haykin, S.: Neural Networks: A Comprehensive Foundation, 2nd edn. (1998)
9. Hsu, C.W., Chang, C.C., Lin, C.J., et al.: A practical guide to support vector classification (2003)
10. Johnson, J.L.: Probability and statistics for computer science. Wiley Online Library (2003)
11. Kim, J.H., Myung, H.: Evolutionary programming techniques for constrained optimization problems (1997)
12. Kuri-Morales, A.: A statistical genetic algorithm (1999)
13. Kuri-Morales, A., Aldana-Bobadilla, E.: A comprehensive comparative study of structurally different genetic algorithms (sent for publication, 2013)
14. Kuri-Morales, A., Aldana-Bobadilla, E.: The Search for Irregularly Shaped Clusters in Data Mining. Kimito, Funatsu and Kiyoshi, Hasegawa (2011)

15. MacQueen, J., et al.: Some methods for classification and analysis of multivariate observations. In: Proceedings of the Fifth Berkeley Symposium on Mathematical Statistics and Probability, California, USA, vol. 1, p. 14 (1967)

16. Mitchell, M.: An Introduction to Genetic Algorithms. MIT Press (1996)

17. Mitchell, M., Holland, J., Forrest, S.: When Will a Genetic Algorithm Outperform Hill Climbing? In: Advances of Neural Information Processing Systems 6, pp. 51–58. Morgan Kaufmann (1994)

18. Molga, M., Smutnicki, C.: Test functions for optimization needs, pp. 41–42 (2005), http://www.zsd.ict.pwr.wroc.pl/files/docs/functions.pdf (retrieved March 11, 2012)

19. Pohlheim, H.: Geatbx: Genetic and evolutionary algorithm toolbox for use with matlab documentation (2012)

20. Rezaee, J., Hashemi, A., Nilsaz, N., Dezfouli, H.: Analysis of the strategies in heuristic techniques for solving constrained optimisation problems (2012)

21. Rudolph, G.: Convergence Analysis of Canonical Genetic Algorithms. IEEE Transactions on Neural Networks 5(1), 96–101 (1994)

22. Rumelhart, D.E., Hintont, G.E., Williams, R.J.: Learning representations by backpropagating errors. Nature 323(6088), 533–536 (1986)

23. Sánchez-Ferrero, G., Arribas, J.: A statistical-genetic algorithm to select the most significant features in mammograms (2007)

24. Shannon, C.E.: A mathematical theory of communication. ACM SIGMOBILE Mobile Computing and Communications Review 5(1), 3–55 (2001)

25. Steliga, K., Szynal, D.: On markov-type inequalities (2010)

Training Multilayer Perceptron by Conformal Geometric Evolutionary Algorithm

J.P. Serrano, Arturo Hernández, and Rafael Herrera

Center for Research in Mathematics
Computer Science Department
Guanajuato, Guanajuato, Mexico
{jpsr,artha,rherrera}@cimat.mx
http://www.cimat.mx

Abstract. In this paper, we propose a novel evolutionary algorithm to train the Multilayer Perceptron MLP which we have called Conformal Geometric Evolutionary Optimization Algorithm (CGEOA). The implementation is based on Conformal Geometric Algebra (CGA). CGA contains elements which represent points, point pairs, lines, circles, planes and spheres. CGA provides a concise way to perform transformations such as rigid euclidean transformations, reflections on spheres and scaling. The main operation of the algorithm is based on geometric inversions of points with respect to sphere. CGEOA is applied to train a Multilayer Perceptron (MLP) using several benchmark data sets. We present a comparison of our results with other proposals of the state of the art in Artificial Neural Networks (ANNs).

Keywords: Evolutionary Algorithms, Geometric Algebra, Conformal Geometric Algebra, Artificial Neural Networks.

1 Introduction

One of the main features of evolutionary algorithms is the reproduction operator. The way an algorithm searches the space mostly depends on this operator. The reproduction operator defines the algorithm, for instance, Particle Swarm Optimization (PSO) generates new individuals by the linear combination of vectors or particles (individuals) of the population. Another examples is Differential Evolution (DE) which creates new individuals by adding a mutation vector (computed as the difference of two random vectors) to a third vector. A geometric interpretation of reproduction operators has been recently studied by Moraglio [12]. In 2007, Moraglio [13] proposed a new operator called product geometric crossover which is introduced to generate new individuals by the convex linear combination of vectors. In 2009, Moraglio [14] generalized the DE algorithm to combinatorial search spaces extending its geometric interpretation to these spaces. Conformal Geometric Algebra (CGA), is introduced in this paper to map the population to the conformal space in which a number of computations is

M. Emmerich et al. (eds.), *EVOLVE - A Bridge between Probability, Set Oriented Numerics,* 31
and Evolutionary Computation IV, Advances in Intelligent Systems and Computing 227,
DOI: 10.1007/978-3-319-01128-8_3, © Springer International Publishing Switzerland 2013

performed. For example, the distance between two points and the distance point-plane. CGA is also used to determine whether a point is inside or outside a sphere. In our approach the population is mapped to the conformal space, and a specific operator for this search space are proposed. We define a new operator to perform the search in the conformal space, called the *inverse operator*. Albeit most of the evolutionary search and operations take place in the conformal space, the fitness function is evaluated in the Euclidean space.

CGA is an extension of the Geometric Algebra, also known as Clifford Algebra (William K. Clifford, [4]). Clifford algebra unifies the main properties of the Grassmann's exterior algebra and Hamilton's quaternions by using the geometric product. The real numbers, complex numbers and quaternions can be seen as sub-algebras of Clifford algebras. In 1960, David Hestenes [6,7] extended the Clifford algebra towards a new model called the conformal model. It has been shown to have useful practical applications in computer graphics, computer games, animation, machine learning, computer vision, and robotics [1,10,16,19,20,22]. Geometric objects such as lines, points, hyper-spheres are represented by sub-spaces of a vector space. Computing geometric features with geometric objects such as lines, points and hyperspheres is easily performed with the Conformal Geometric Algebra.

The goal of this paper is to apply the CGA to Evolutionary Algorithms for the design of new mutation and crossover operators. The paper is organized as follows: section 2 gives an introduction to GA, CGA and inversions; section 3 describes the implementation of the CGEOA; section 4 presents the implementation of the MLP using the CGEOA; section 5 and 6, present the experimental details, results and comparisons with other works respectively. Finally section 7 presents the conclusions.

2   Geometric Preliminaries

2.1   Geometric Algebra

The heart of geometric algebra is the geometric product, which is defined as the sum of the inner and outer products of vectors (more generally blades). More precisely, the geometric product of two vectors $a, b \in \mathbb{R}^n$ is defined by:

$$ab = a \bullet b + a \wedge b. \tag{1}$$

We use the symbol $\bullet$ to denote the inner/scalar product [5]. The outer product is the operation denoted by the symbol $\wedge$. The properties of the outer product are: antisymmetry, scaling, distributivity and associativity. The product $a \wedge b$ is called a bivector. The bivector $a \wedge b$ can be visualized as the parallelogram formed by a and b. The orientation of $b \wedge a$ will be the opposite to that of $a \wedge b$, due to the antisymmetry property. Therefore, the outer product is not commutative. The outer product is generalizable in higher dimensions, since it is associative. For instance, the outer product $(a \wedge b) \wedge c$ gives a trivector which can be visualized as a parallelepiped. In particular, the outer product of k linearly independent vectors is called a $k - blade$ and can be written as:

$$A_{<k>} = a_1 \wedge a_2 \wedge ...a_k =: \bigwedge_{i=1}^{k} a_i. \tag{2}$$

The grade $< k >$ of a blade is simply the number of vectors included in the outer product. A k-vector is a finite linear combination of k-blades. The set of all k-vectors is a vector space denoted by $\bigwedge^k \mathbb{R}^n$, of dimension $\binom{n}{k} = \frac{n!}{(n-k)!k!}$. If $\{e_1, ..., e_n\} \subset \mathbb{R}^n$ is an orthonormal basis for $\mathbb{R}^n$, then the set of k-blades

$$\{e_{i1} \wedge ... \wedge e_{ik} \mid i_{i1}, ..., i_k \in \{1, ..., n\} \text{ and } i_1 < i_2 < ... < i_k\} \tag{3}$$

forms a vector space basis for $\bigwedge^k \mathbb{R}^n$. The geometric algebra $Cl_n = Cl(\mathbb{R}^n)$ is the space of all linear combinations of blades of arbitrary grade, so that

$$Cl_n = \overset{0}{\bigwedge} \mathbb{R}^n \oplus \overset{1}{\bigwedge} \mathbb{R}^n \oplus \overset{2}{\bigwedge} \mathbb{R}^n \oplus ... \oplus \overset{n}{\bigwedge} \mathbb{R}^n, \tag{4}$$

where $\bigwedge^0 \mathbb{R}^n = \mathbb{R}$, $\bigwedge^1 \mathbb{R}^n = \mathbb{R}^n$. Thus, the dimension of the Geometric Algebra $Cl(\mathbb{R}^n)$ is

$$\sum_{k=0}^{n} \binom{n}{k} = 2^n. \tag{5}$$

A vector space basis of the Geometric Algebra Cl_n is given by the union of all the bases of the multivector spaces of every grade. The element $I = e_1 \wedge ... \wedge e_n$ is called the pseudoscalar. For example, the Geometric Algebra Cl of 3-dimensional space $\mathbb{R}^3$ has $2^3 = 8$ basis blades.

$$\underbrace{1}_{scalar}, \underbrace{e_1, e_2, e_3}_{vectors}, \underbrace{e_1 \wedge e_2, e_2 \wedge e_3, e_3 \wedge e_1}_{bivectores}, \underbrace{e_1 \wedge e_2 \wedge e_3}_{trivector} \equiv I. \tag{6}$$

2.2   Conformal Geometric Algebra

The conformal model provides a way to deal with Euclidean Geometry in a higher dimensional space. Let $e_0, e_1, ..., e_n, e_\infty$ be a basis of $\mathbb{R}^{n+2}$. We will consider the Euclidean points $x = x_1 e_1 + x_2 e_2 +, ..., x_n e_n \in \mathbb{R}^n$ to be mapped to points in $\mathbb{R}^{n+1,1}$ according into the following conformal transformation:

$$\mathbb{P}(x) = (x_1 e_1 + x_2 e_2 + ... + x_n e_n) + \frac{1}{2} x^2 e_\infty + e_0, \tag{7}$$

where $\mathbb{R}^{n+1,1}$ is a $(n + 2)$-dimensional real vector space that includes the Euclidean space $\mathbb{R}^n$ and has two more independent directions generated by two basic vectors e_0 and e_∞ [5]. Furthermore, it is endowed with a scalar product of vectors satisfying: $e_i \bullet e_j = 1$, for $i = j$, $e_i \bullet e_j = 0$, for $i \neq j$, $e_i \bullet e_\infty = 0$, for $i = 1, ..., n$, $e_i \bullet e_0 = 0$, for $i = 1, ..., n$, $e_\infty \bullet e_0 = -1$, and $e_\infty^2 = e_0^2 = 0$. Note that the Euclidean points $x \in \mathbb{R}^n$ are now mapped into points of the null cone in $\mathbb{R}^{n+1,1}$. The vectors e_0 and e_∞ represent the origin and a point at infinity respectively. The number x^2 is:

$$x^2 = \sum_{i=1}^{n} x_i^2. \tag{8}$$

A hyper-sphere in $\mathbb{R}^n$ is determined by its center $c \in \mathbb{R}^n$ and its radius $\rho \in \mathbb{R}$. Its conformal representation is the point:

$$\mathbb{S}(c,\rho) = \mathbb{P}(c) - \frac{1}{2}\rho^2 e_\infty. \tag{9}$$

The representation of the sphere s in *Outer Product Null Space* notation (OPNS) in $\mathbb{R}^n$, can be written with the help of $n+1$ conformal points that lie on it.

$$s^* = \bigwedge_{i=1}^{n+1} \mathbb{P}(x_i). \tag{10}$$

A hyper-plane Θ in $\mathbb{R}^n$ is represented by:

$$\Theta(n,\delta) = n + \delta e_\infty, \tag{11}$$

where n is the normal vector to the hyper-plane in $\mathbb{R}^n$ and $\delta > 0$ is its oriented distance (with respect to n) to the origin. A hyper-plane in $\mathbb{R}^n$ can be defined by n points belonging to it, thus, in the conformal model is represented by the outer product of the n image conformal points plus the vector e_∞.

$$\Theta^* = \left(\bigwedge_{i=1}^{n} P_i \right) + e_\infty. \tag{12}$$

Distance between Points

The inner product between the points $\mathbb{P}(x)$ and $\mathbb{P}(y)$ is directly proportional to the square of the Euclidean distance of the points $\{x, y\}$ multiplied by -2,

$$\mathbb{P}(x) \bullet \mathbb{P}(y) = x \cdot y - \frac{1}{2}y^2 - \frac{1}{2}x^2 = -\frac{1}{2}(x-y)^2 \tag{13}$$

$$(x-y)^2 = -2(\mathbb{P}(x) \bullet \mathbb{P}(y)) \tag{14}$$

The conformal points $\mathbb{P}(x)$ and $\mathbb{P}(y)$ are null and as a consequence, $\mathbb{P}(x) \bullet \mathbb{P}(y) = 0$ if and only if $x = y$.

Distance between Point and Hyper-plane

Similarly, it is possible to calculate the signed distance between a point and a hyper-plane. For a point $\mathbb{P}(x)$ and hyper-plane $\Theta(n, \delta)$, the signed distance of the point to the hyper-plane is given by:

$$\mathbb{P}(x) \bullet \Theta(n, \delta) = x \cdot n - \delta. \tag{15}$$

The distance is zero if the point x lies in the hyper-plane. The distance is positive if the point x is on the same side of the plane as the normal vector n and negative if it is on the opposite side [10].

Point Inside or Outside a Hyper-sphere

The inner product between a point $\mathbb{P}(x)$ and a hyper-sphere $\mathbb{S}(c, \rho)$ can be used to decide whether a point is inside or outside of the hyper-sphere,

$$\mathbb{P}(x) \bullet \mathbb{S}(c, \rho) = \frac{1}{2}\rho^2 - \frac{1}{2}(c - x)^2. \tag{16}$$

The sign of the result indicates the location of the point x with respect to the hyper-sphere [10].

- $\text{Sign}(\mathbb{P}(x) \bullet \mathbb{S}(c, \rho)) > 0$: x is inside the hyper-sphere.
- $\text{Sign}(\mathbb{P}(x) \bullet \mathbb{S}(c, \rho)) < 0$: x is outside the hyper-sphere.
- $\mathbb{P}(x) \bullet \mathbb{S}(c, \rho) = 0$: x is on the hyper-sphere.

2.3   Inversion on a Hyper-sphere

Figure 1 shows the inversion of a point with respect to circle $\mathbb{S}$ with center c and radius ρ, where $A^\wedge$ is the inverse point of A, when A is inside of the circle. In the calculation the following constraints must be satisfied: $CA \times CA^\wedge = \rho^2$ and $\angle CAB$ and $\angle CBA^\wedge$ are equal to 90°. Likewise, the point A can be the inverse point of the point $A^\wedge$ with respect to circle if, and only if, the next condition is satisfied:

$$\frac{\overline{CA}}{\overline{CB}} = \frac{\overline{CB}}{\overline{CA^\wedge}} \quad \Leftrightarrow \quad \overline{CB}^2 = \overline{CA} \times \overline{CA^\wedge} \tag{17}$$

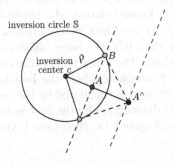

Fig. 1. Inversion of a point A with respect to a circle

In CGA, the Euclidean transformations are representable by versors as follows [5,18]:

$$\sigma \mathbb{P}(x)^{\wedge} = G\mathbb{P}(x)G^{-1} \tag{18}$$

where $(\mathbb{P}(x)^{\wedge})^2 = \mathbb{P}(x)^2 = 0$, G is a versor which can be defined as any element which can be factored into a geometric product of unit vectors v_i as follows [11]:

$$G = v_1 v_2 \cdots v_k, \tag{19}$$

The scalar σ is applied to control the weight. In CGA, the inversion of point x with respect to a hyper-sphere $\mathbb{S}(c, \rho)$ can be obtained as follows:

$$\mathbb{P}(x)^{\wedge} = (\mathbb{S}(c, \rho))\mathbb{P}(x)\mathbb{S}(c, \rho)^{-1}, \tag{20}$$

where $(\mathbb{S}(c, \rho))^{-1} = \frac{1}{\rho^2}\mathbb{S}(c, \rho)$, and

$$\sigma \mathbb{P}(x)^{\wedge} = \left(\frac{(x-c)^2}{\rho}\right)^2 [g(x) + \frac{1}{2}[g(x)]^2 e_{\infty} + e_0], \tag{21}$$

where

$$g(x) = \frac{\rho^2}{(x-c)^2}(x-c) + c \tag{22}$$

is the formula of the inversion of a point with respect to a hyper-sphere in Euclidean notation. Figure 2 (b) shows several properties of the inversion transformation on a circle in $\mathbb{R}^2$.

3   Conformal Geometric Optimization Algorithm

In the CGEOA, the Euclidean points are encoded in the conformal space. The inner product is used to obtain distances between pairs of points and to identify when a point is inside or outside of a sphere. The center of the hyper-sphere is estimated as the mean of the selected population and the radius is the maximum distance that exists between the center and the points of the selected population. The hyper-sphere represents a potential area in which to search for a global minimum. All of the points located outside the hyper-sphere are inverted with respect to the hyper-sphere so that their images are now inside the hyper-sphere with a probability higher than 0.5. We have called this process the inversion with respect to a hyper-sphere. The other points (the ones already in the hyper-sphere) are moved to the hemisphere where the best two points of population are located. A direction vector x^d is calculated by the means of the lineal combination between the vectors determined by the best point x^b with respect to the center of the hyper-sphere. Figure 2 (a) shows the trajectory of a single point.

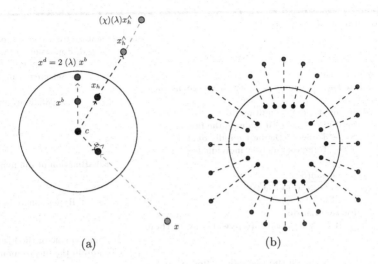

Fig. 2. Geometric Interpretation from CGEOA. (a) Trajectory for a single point in the CGEOA. (b) An illustration of the inversion of points which lie either on lines or on circles.

A candidate point $x_h^\wedge$ is calculated as follows:

1. The direction vector x^d is calculated as follows:

$$x^d = 2(\lambda)\overline{x^b} \tag{23}$$

where

$$\overline{x^b} = x^b - c \tag{24}$$

2. The candidate point x_h is calculated as follows:

$$x_h = c + [(x^\wedge - c) + (1 - \lambda)(x^d - c) \tag{25}$$

where $x^\wedge$ can either be a point which previously was outside the hyper-sphere, and it was inverted with respect to the hyper-sphere, or it is a point which is inside the hyper-sphere. λ is a random component.

3. Finally, the point x_h can be the final candidate point. However, given a probability higher than 0.5, this point can be inverted again. The new inverse point $x_h^\wedge$ can be calculated as follows:

$$x_h^\wedge = c + \chi(\lambda)\left(\frac{\overline{x}}{\|\overline{x}\|}\right) \tag{26}$$

where χ is a scalar, and,

$$\overline{x} = x^{\text{new inverse point}} - c \tag{27}$$

The algorithm determines whether the new point is better than the current point. The current point is updated when the new point is better than the current point. Figure 3 shows the pseudo code of the CGEOA.

```
1: procedure CGEOA(ps,d)
2:                                                          ▷ ps: percentage from selected population
3:                                                          ▷ d: dimension of the problem
4:    n ← 100                                               ▷ Population size
5:    η ← n × ps                                            ▷ Selected population size
6:    Create and initialize a d-dimensional population x
7:    xᵗ ← x                                                ▷ Candidate Population
8:    while not stopping condition do
9:        Aptitude ← f(x) To evaluate the fitness.
10:       Sₓ ← To Select the best η individuals
11:       xᵇ ← To Select the best individual
12:                                                         ▷ Estimation of the hyper-sphere
13:       c ← mean(Sₓ)                                      ▷ Center
14:       ρ ← √(−2(ℙ(c) • ℙ(xᵇ)))                           ▷ Radio
15:       S ← 𝕊(c, ρ)                                       ▷ Hyper-sphere in the CGA
16:       for i ← 1, n do
17:           if (λ > 0.5) and (𝕊(c, ρ) • ℙ(xᵢ) < 0) then
18:                                                         ▷ Point outside of the hyper-sphere
19:                                                         ▷ To obtain the inverse point from x
20:               xᵢᵗ ← GetInverse(𝕊(c, ρ), ℙ(xᵢ))
21:           else
22:               xᵢᵗ ← xᵢ
23:           end if
24:           x⃗ ← c + (xᵇ − c)                              ▷ Direction vector
25:           for j ← 1, d do
26:               xᵢ,ⱼᵗ = cⱼ + (xᵢ,ⱼ − cⱼ) + (1 − λ)(x⃗ⱼ − cⱼ)
27:           end for
28:           if λ < 0.5 then
29:               tmp^ = GetInverse(𝕊(c, ρ), ℙ(xᵢᵗ))
30:               x̄ = tmp^ − c
31:               xᵢᵗ = c + χ(λ)(x̄/‖x̄‖)
32:           end if
33:       end for
34:       for i ← 1, n do                                   ▷ Update population
35:           if f(xᵢᵗ) < f(xᵢ) then
36:               xᵢ = xᵢᵗ
37:           end if
38:       end for
39:   end while
40:   return x^{FB}                                         ▷ The best point solution for the problem
41: end procedure

42: procedure GetInverse(𝕊(c, ρ), ℙ(x))
43:                                                         ▷ 𝕊(c, ρ): Hyper-sphere in the conformal space
44:                                                         ▷ ℙ(x): Point x in the conformal space
45:    if (𝕊(c, ρ) • (𝕊(c, ρ)) == ((𝕊(c, ρ) • ℙ(x)) then
46:        return x                                         ▷ c is equal to x
47:    else
48:                                                         ▷ Inversion: 𝕊(c, ρ)ℙ(x)𝕊(c, ρ)⁻¹
49:        return (𝕊(c,ρ)•𝕊(c,ρ))/((x−c)²) (x − c) + c
50:
51:    end if
52: end procedure
```

Fig. 3. Conformal Geometric Optimization Evolutionary Algorithm Pseudocode

4   Implementation

The standard Perceptron creates decision regions by the composition of hyperplane functions. Figure 4 shows the topology of the MLP. The input signals $x = [x_1, x_2, ..., x_n]$ are propagated through the network from left to right. The symbols $o_1, ...o_k, ..o_q$ denote the output vector o of the hidden layer to the output layer. The outputs of the output layer are described by the vector $y = [y_1, y_2, ..., y_k, ..., y_q]$. The model is implemented with the sigmoid function. The output of the neuron k of the hidden layer and the output layer can be written as a function composition between the linear associator and the sigmoid function $o_k = g \circ f(\overrightarrow{x}, w, \delta)$. The function $f(\overrightarrow{x}, w, \delta)$ is computed as follows:

$$f(\overrightarrow{x}, w, \delta) = \mathbb{P}(\overrightarrow{x}) \bullet \Theta(w, \delta) = \overrightarrow{x} \cdot w - \delta \qquad (28)$$

where x denotes the input vector, w is the weight vector, and δ is the bias. The ouput s of a neuron is computed as follws:

$$s = g \circ f(\overrightarrow{x}, w, \delta)$$
$$= \frac{1}{1 + exp(-\overrightarrow{x} \cdot w - \delta))} \qquad (29)$$

The training by using the CGEOA algorithm starts with the parameters (weights and bias) encoding. The individual particle structure is presented in Figure 5. The parameter $w_{1,k}^i$ describes the weight value for the connection of the input vector in its position 1 and the k neuron in the hidden layer. The parameter $w_{i,j}^o$ describes the weight value for the connection of the i-th neuron in the hidden layer to the j-th neuron in the output layer. Also, the parameter δ_k^i, δ_k^o describes the bias value of the k-th neuron in the input layer and output layer respectively.

5   Experimental Setup

To evaluate the performance of the MLP using the CGEOA, several experiments were conducted on 3 data sets listed in Table 1. The data sets were obtained from the UCI machine learning benchmark repository [15]. The data set of each experiment was normalized by the min-max normalization method which maps into a new interval the original data, i.e. for a value $x_{(i,j)}$ of the original data set is mapped into value $x_{(i,j)}^{new}$ by the equation (30).

$$x_{(i,j)}^{new} = \frac{x_{(i,j)} - \min_{x_{(j)}}}{\max_{x_{(j)}} - \min_{x_{(j)}}}(\max^{new} - \min^{new}) + \min^{new} \qquad (30)$$

where $\min_{x_{(j)}}$ and $max_{x_{(j)}}$ are the minimum and maximum value of the column j of the original data set, $\max^{new}$ and $\min^{new}$ is the new maximum and minimum value for the new interval. The attributes of all data sets used in our experiments were scaled linearly in the interval (-1,+1).

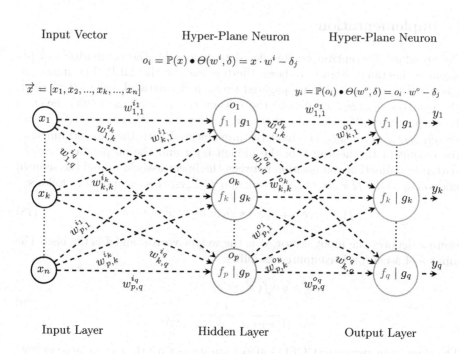

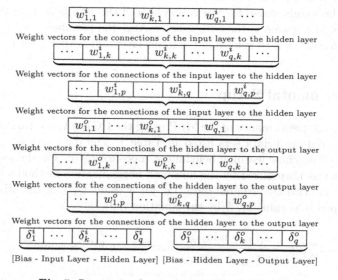

Fig. 4. Multilayer Perceptron Model

Fig. 5. Structure of a single particle of the PSO

Table 1. Description of data sets used in the experiments

Data set	Attributes	Classes	Instances	Subsets		
				Training	Validation	Testing
Breast cancer	9	2	699	350	175	174
Diabetes	8	2	768	384	192	192
Heart disease	13	2	303	149	74	74

Table 2 shows the data set of Table 1 partioned in 3 subsets (training, validation and test stage). The training data columns made up of a training set and a validation set, where the 50% of the data set is for the training set, 25% for the validation set and the remaining part for the testing set [17]. The classification problem of the breast cancer data set is the identification of breast lumps as benign or malignant. The data set, collected at the University of Wisconsin, Madison by Dr. William Wolberg which is called breast-cancer-wisconsin.data in the repository. Diabetes data set is analyzed to predict diabetes among Pima Indians. The data set is called pima-indians-diabetes.data in the repository. All patients reported are females of at least 21 years of age. There are 768 samples in the data set which 500 are classified as diabetes negative and 268 as diabetes positive. For the heart disease data set, the goal is to predict the presence/absence of a heart disease. The initial Heart disease data set was created at Cleveland Clinic Foundation by Dr. Robert Detrano, which is called as processed.cleveland.data in the repository. It contains 303 samples but 6 of them contain missing data, so they are discarded.

Table 2. Description of the subsets used in the experiments

Data set			Subsets		
			Training Data		Test Data
	Clases	Instances	Training set	Validation set	Testing set
Breast cancer	Benign	458	230	115	113
	Malignant	241	120	60	61
Diabetes	Negative	500	250	125	125
	Positive	268	134	67	67
Heart disease	Negative	160	80	40	40
	Positive	137	69	34	34

The objective function is based on the mean square error (MSE). The MLP using the CGEOA was trained 30 times on 6 different topologies. For each topology the number of neurons in the hidden layer was varied from 1 up to 6. The topology which contains a minimum mean error rate using the testing data sets on 30 independent runs was chosen as the winner. To describe the Artificial Neural Network (ANN) topology we use a sequence: number of input neurons - number of hidden neurons - number of output neurons. For instance, the sequence 9-1-1 refers to a network with 9 input neurons, 1 hidden and 1 output neurons.

The outputs are encoded by 1-of-c representation for c classes. The class is set according to winner-takes-all methodology. The population size is 100 points initialized in the interval (-1,+1). The CGEOA uses a value for $\chi = 1.5$ and a 50% of the best individuals are selected. The stop criteria is given by the number of generations in the interval (1,100) or whether the MSE on the validation data set presents an increment during the training.

6   Experimental Results

Experimental details, results and comparisons with other works are presented in this section. Table 3 presents the results of the MLP using the CGEOA over 30 independent runs on the breast cancer, diabetes, and heart disease data sets. The Mean, Median, SD, Min, and Max indicate the mean, median, standard deviation, minimum and maximum values, respectively.

Table 3. Results of HC-MLP for the data sets

Data set		Error Rate - Percentage of wrong classification			Topology
		Training set	Validation set	Testing set	Average of Number of Generations
	Mean	3.3333	5.0190	1.7050	
	Median	2.8571	4.8571	1.7241	9-4-2
Breast cancer	SD	0.9501	1.1297	0.7314	
	Min	1.7143	3.1429	0.5747	32
	Max	5.7143	7.4286	3.4483	
	Mean	23.2378	20.3125	21.0417	
	Median	23.3073	20.3125	20.8333	8-6-2
Diabetes	SD	1.1396	1.0051	1.6218	
	Min	21.3542	18.7500	18.7500	23
	Max	26.3021	22.9167	27.0833	
	Mean	11.8345	24.4595	18.0180	
	Median	10.7383	23.6486	17.5676	13-3-2
Heart disease	SD	2.5265	3.4104	3.1208	
	Min	9.3960	20.2703	10.8108	22
	Max	20.1342	33.7838	28.3784	

For comparison purposes, Table 4 shows the classification error rate statistics, over test data sets of 8 algorithms from [9]. The algorithms are the following:

1. MD PSO [9] : Evolutionary artificial neural networks by multi-dimensional particle swarm optimization.
2. PSO-PSO [25] : An ANN evolved by a new evolutionary system and its application.
3. PSO-PSO:WD [2] : Particle swarm optimization of neural network architectures and weights.
4. IPSONet [24] : An improved particle swarm optimization for evolving feedforward artificial neural networks.
5. EPNet [23]: A new evolutionary system for evolving artificial neural networks.

6. GA (Sexton & Dorsey, 2000) [21]: Reliable classification using neural networks: a genetic algorithm and back-propagation comparison.
7. GA (Cantu-Paz & Kamath, 2005) [3]: An empirical comparison of evolutionary algorithms and neural networks for classification problems.
8. BP [9] : Back-propagation algorithm.

Proposals 1, 2, 3 and 4, are based on a PSO algorithm. Proposal 5 applies simulated annealing whereas proposals 6 and 7 apply genetic algorithms. Finally, a classical back-propagation algorithm is listed as number 8. $\mu_{<rank>}$ and σ represent the mean and standard deviation of the classification on breast cancer, diabetes, and heart disease data set.

Table 4. Mean (μ) and standard deviation (σ) of classification error rates (%) over test data sets

Algorithm	Data set Breast cancer		Diabetes		Heart disease	
	$\mu_{<rank>}$	σ	$\mu_{<rank>}$	σ	$\mu_{<rank>}$	σ
1. MD PSO	$0.39_{<1>}$	0.31	$20.55_{<1>}$	1.22	$19.53_{<5>}$	1.71
2. PSO-PSO	$4.83_{<9>}$	3.25	$24.36_{<6>}$	3.57	$19.89_{<6>}$	2.15
3. PSO-PSO:WD	$4.14_{<8>}$	1.51	$23.54_{<5>}$	3.16	$18.1_{<3>}$	3.06
4. IPSONet	$1.27_{<2>}$	0.57	$21.02_{<2>}$	1.23	$18.14_{<4>}$	3.42
5. EPNet	$1.38_{<3>}$	0.94	$22.38_{<4>}$	1.40	$16.76_{<2>}$	2.03
6. GA (Sexton & Dorsey, 2000)	$2.27_{<5>}$	0.34	$26.23_{<8>}$	1.28	$20.44_{<7>}$	1.97
7. GA (Cantu-Paz & Kamath, 2005)	$3.23_{<7>}$	1.1	$24.56_{<7>}$	1.65	$23.22_{<8>}$	7.87
8. BP	$3.01_{<6>}$	1.2	$29.62_{<9>}$	2.2	$24.89_{<9>}$	1.71
9. MLP - CGEOA	$1.70_{<4>}$	0.57	$21.04_{<3>}$	1.62	$18.01_{<2>}$	3.41
Minimum value error MLP-CGEOA	0.7314		18.75		10.8108	
Maximum value error MLP-CGEOA	3.4483		27.0833		28.3784	

MLP - CGEOA obtains a mean error rate of 1.70 with a standard deviation of 0.57 on the breast cancer data set which places the proposed approach as the fourth best of the list. However, out of the 30 runs there is one solution with error rate of 0.7314 which ranks MLP - CGEOA as the second best classifier for the breast cancer data set. Notice that the maximum error rate is still better than the mean value of approaches 2 and 3. Likewise, MLP - CGEOA obtains a mean error rate of 21.04 with a standard deviation of 1.62 on the diabetes data set which places the proposed approach as the third of the list. However, out of the 30 runs there is one solution with error rate of 18.75 which ranks MLP - CGEOA as the best classifier for the diabetes data set. Notice that the maximum error rate is still better than the mean value of approach 8. Finally, MLP - CGEOA obtains a mean error rate of 18.01 with a standard deviation of 3.41 on the heart disease data set which places the proposed approach as the second best of the list. However, out of the 30 runs there is one solution with error rate of 10.8108 which ranks MLP - CGEOA as the best classifier for the heart disease data set.

7   Conclusions

In this paper, we propose a novel algorithm called Conformal Geometric Evolutionary Optimization Algorithm. According to numerical experiments the inverse operator helps in the exploration of the search space. The use of CGA makes easier the computation of geometric features of the population, Also, operations in linear algebra can be extended in the conformal space. The new reproduction operator defines non linear search directions. Training a MLP using CGEOA gives competitive results for the classifying data sets used in the experiments, however, more experiments are being performed with a benchmark of functions of several kinds as to fully measure the potential of the operator. The contribution of this paper is the novel application of conformal geometric algebra in evolutionary algorithms for the first time.

Acknowledgments. The authors gratefully acknowledge the financial support from National Council for Science and Technology of Mexico (CONACYT) and from the Center for Research in Mathematics (CIMAT).

References

1. Bayro-Corrochano, E., Reyes-Lozano, L., Zamora-Esquivel, J.: Conformal geometric algebra for robotic vision. Journal of Mathematical Imaging and Vision 24(1), 55–81 (2006)
2. Carvalho, M., Ludermir, T.B.: Particle Swarm Optimization of Neural Network Architectures and weights. In: Proceedings of the 7th International Conference on Hybrid Intelligent Systems, HIS 2007, pp. 336–339. IEEE Computer Society, Washington, DC (2007)
3. Cantu-Paz, E., Kamath, C.: An empirical comparison of combinations of evolution algorithms and neural networks for classification problems. IEEE Transactions on Systems, Man, and Cybernetics, Part B 35(5), 915–927 (2005)
4. Clifford, W., Tucker, R., Smith, H.: Mathematical papers. Macmillan and Co. (1882)
5. Dorst, L., Fontijne, D., Mann, S.: Geometric Algebra for Computer Science: An Object-Oriented Approach to Geometry. The Morgan Kaufmann Series in Computer Graphics, pp. 23–57, 355–389. Morgan Kaufmann Publishers Inc., San Francisco (2007)
6. Hestenes, D., Li, H., Rockwood, A.: A unified algebraic framework for classical geometry. In: Geometric Computing with Clifford Algebra. Springer (1999)
7. Hestenes, D., Sobczyk, G.: Clifford Algebra to Geometric Calculus: A Unified Language for Mathematics and Physics, Fundamental Theories of Physics, vol. 5. Kluwer Academic Publishers, Dordrecht (1987)
8. Kennedy, J., Eberhart, R.C.: Swarm Intelligence. Morgan Kauffmann Publishers Inc., San Francisco (2001)
9. Kiranyaz, S., Ince, T., Yildirim, A., Gabbouj, M.: Evolutionary artificial neural networks by multi-dimensional particle swarm optimization. Neural Networks 22(10), 1448–1462 (2009)

10. Hildenbrand, D., Fontijne, D., Perwass, C., Dorst, L.: Geometric Algebra and its Application to Computer Graphics. In: Proceedings of the 25th Annual Conference of the European Association for Computer Graphics (Interacting with Virtual Worlds). INRIA and the Eurographics Association, Grenoble (2004)
11. Li, H., Hestenes, D., Rockwood, A.: Generalized homogeneous coordinates for computational geometry. In: Sommer, G. (ed.) Geometric Computing with Clifford Algebra, pp. 27–52. Springer (2001)
12. Moraglio, A., Poli, R.: Product geometric crossover. In: Runarsson, T.P., Beyer, H.-G., Burke, E.K., Merelo-Guervós, J.J., Whitley, L.D., Yao, X. (eds.) PPSN 2006. LNCS, vol. 4193, pp. 1018–1027. Springer, Heidelberg (2006)
13. Moraglio, A., Di Chio, C., Poli, R.: Geometric particle swarm optimization. In: Ebner, M., O'Neill, M., Ekárt, A., Vanneschi, L., Esparcia-Alcázar, A.I. (eds.) EuroGP 2007. LNCS, vol. 4445, pp. 125–136. Springer, Heidelberg (2007)
14. Moraglio, A., Togelius, J.: Geometric differential evolution. In: Proceedings of the 11th Annual Conference on Genetic and Evolutionary Computation, pp. 1705–1712 (2009)
15. Murphy, P., Aha, D.: UCI Repository of machine learning databases. Tech. Rep. (1994)
16. Perwass, C., Banarer, V., Sommer, G.: Spherical decision surfaces using conformal modelling. In: Michaelis, B., Krell, G. (eds.) DAGM 2003. LNCS, vol. 2781, pp. 9–16. Springer, Heidelberg (2003)
17. Prechelt, L.: PROBEN1 - A Set of Benchmarks and Benchmarking Rules for Neural Network Training Algorithms. Tech. Rep. 21–94, Fakultat fur Informatik, Universitat Karlsruhe, D-76128 Karlsruhe, Germany (1994)
18. Rosenhahn, B., Sommer, G.: Pose estimation in Conformal Geometric Algebra Part I: The stratification of mathematical spaces. J. Math. Imaging Vision 22(1), 27–48 (2005)
19. Rosenhahn, B., Sommer, G.: Pose Estimation in Conformal Geometric Algebra. Part II: Real-time pose estimation using extended feature concepts. Journal of Mathematical Imaging and Vision 22(1), 49–70 (2005)
20. Rubio, J.P.S., Aguirre, A.H., Guzmán, R.H.: A conic higher order neuron based on geometric algebra and its implementation. In: Batyrshin, I., Mendoza, M.G. (eds.) MICAI 2012, Part II. LNCS, vol. 7630, pp. 223–235. Springer, Heidelberg (2013)
21. Sexton, R.S., Dorsey, R.E.: Reliable classification using neural networks: A genetic algorithm and back propagation comparison. Decision Support Systems 30(1), 11–22 (2000)
22. Vince, J.A.: Geometric Algebra - An Algebraic System for Computer Games and Animation, pp. I–XVIII, 1–195. Springer (2009) ISBN 978-1-84882-378-5
23. Yao, X., Liu, Y.: A new evolutionary system for evolving artificial neural networks. IEEE Transactions on Neural Networks 8(3), 694–713 (1996)
24. Yu, J., Xi, L., Wang, S.: An Improved Particle Swarm Optimization for Evolving Feedforward Artificial Neural Networks. Neural Processing Letters 26(3), 217–231 (2007)
25. Zhang, C., Shao, H.: An ANN's evolved by a new evolutionary system and its application. In: Procceddings of the 39th IEEE Conference on Decision and Control, vol. 4, pp. 3562–3563 (2000)

RotaSVM: A New Ensemble Classifier

Shib Sankar Bhowmick[1,2,*], Indrajit Saha[1,*],
Luis Rato[2], and Debotosh Bhattacharjee[1]

[1] Department of Computer Science and Engineering,
Jadavpur University,
Kolkata 700032, West Bengal, India
[2] Department of Informatics,
University of Evora,
Evora 7004-516, Portugal
shibsankar.ece@gmail.com, indra@icm.edu.pl,
lmr@di.uevora.pt, debotosh@ieee.org

Abstract. In this paper, an ensemble classifier, namely RotaSVM, is proposed that uses recently developed rotational feature selection approach and Support Vector Machine classifier cohesively. The RotaSVM generates the number of predefined outputs of Support Vector Machines. For each Support Vector Machine, the training data is generated by splitting the feature set randomly into S subsets. Subsequently, principal component analysis is used for each subset to create new feature sets and all the principal components are retained to preserve the variability information in the training data. Thereafter, such features are used to train a Support Vector Machine. During the testing phase of RotaSVM, first the rotation specific Support Vector Machines are used to test and then average posterior probability is computed to classify sample data. The effectiveness of the RotaSVM is demonstrated quantitatively by comparing it with other widely used ensemble based classifiers such as Bagging, AdaBoost, MultiBoost and Rotation Forest for 10 real-life data sets. Finally, a statistical test has been conducted to establish the superiority of the result produced by proposed RotaSVM.

Keywords: Principal component analysis, rotational feature selection, statistical test, support vector machine.

1 Introduction

Integration of classifiers nowadays is drawing much attention of the machine learning and pattern recognition communities and growing rapidly [1–10]. In integrated classification techniques, an ensemble of classifiers is generated by giving similar or different permutated training data sets. Thereafter, class label of the test sample is assigned by either majority voting or averaging the output probabilities of the ensemble. Recent research shows that ensemble based classifiers, such as Bagging [11], AdaBoost [12, 13], Random Forest [14] and Rotation

---

* Both the authors contributed equally.

M. Emmerich et al. (eds.), *EVOLVE - A Bridge between Probability, Set Oriented Numerics,* 47
and Evolutionary Computation IV, Advances in Intelligent Systems and Computing 227,
DOI: 10.1007/978-3-319-01128-8_4, © Springer International Publishing Switzerland 2013

Forest [15], are used more often to increase the prediction accuracy of learning systems [9, 16–20].

Among these ensemble classifiers, Rotation Forest [15] performs much better than other ensemble methods. It uses rotational feature sets for decision tree classifiers. In Rotation Forest, rotational feature sets subsequently undergo Principal Component Analysis (PCA) to preserve the variability information of the training data. Here the main idea is to simultaneously increase diversity and individual accuracy within the decision tree classifiers. Diversity is achieved by using PCA, which is used to extract the principal components of rotational features for each classifier and accuracy is sought by keeping all principal components [15, 21].

The Support Vector Machine (SVM) is a state-of-the-art classification method introduced in 1992 by Boser *et al.* [22]. The basic idea of SVM is to find a hyperplane which separates the d-dimensional data perfectly into two classes. However, since classification data is often not linearly separable, SVM introduced the notion of a "kernel induced feature space" which embed the data into a higher dimensional feature space where the data is linearly separable. For this purpose, first the hyperplane is found, which separates the largest possible fraction of points such that points on the same side belong to the same class, while the distance of each class from the hyperplane is maximized.

As both Rotation forest and Support Vector Machine are successfully used in classification, thus their integration may achieve even higher prediction accuracy than either of them. Hence, in this paper, an ensemble classifier, named as RotaSVM, is proposed by integrating rotational feature selection scheme with SVM. The RotaSVM produces the number of predefine outputs of SVMs. For each SVM, the training data is generated by splitting the feature set randomly into S subsets. Subsequently, principal component analysis is used for each subset to create new feature sets and all the principal components are retained to preserve the variability information of the training data. Thereafter, such features are used to train a Support Vector Machine. During the testing phase of RotaSVM, the sample data are the input to the rotation specific Support Vector Machines. Subsequently, it is classified by computing average posterior probability. The experimental studies were conducted with available 10 real-life data sets[1]. The results show that RotaSVM can produce significantly lower prediction error more often than Rotation Forest and other ensemble based classifiers such as Bagging, AdaBoost and MultiBoost for all the data sets. Finally, t-test [24] has been conducted to establish the statistical significance of the result produced by RotaSVM.

The rest of this paper is organized as follows: Section 2 briefly describes the Support Vector Machine classifier. The proposed RotaSVM is discussed in Section 3. Section 4 shows the empirical results. Finally, Section 5 concludes this paper with an additional note of future work.

---

[1] UCI repository [23].

2   Brief Description of Support Vector Machine

In this section, we briefly discuss Support Vector Machine classifier which is inspired by statistical learning theory. It performs structural risk minimization on a nested set structure of separating hyperplanes [25]. Viewing the input data as two sets of vectors in a d-dimensional space, an SVM constructs a separating hyperplane in that space, one which maximizes the margin between the two classes of points. To compute the margin, two parallel hyperplanes are constructed on each side of the separating one. Intuitively, a good separation is achieved by the hyperplane that has the largest distance to the neighboring data points of both classes. The larger margins or distances between these parallel hyperplanes indicate better generalization error of the classifier. Fundamentally, the SVM classifier is designed for two-class problems. It can be extended to handle multiclass problems by designing a number of one-against-all or one-against-one twoclass SVMs [26]. For example, a K-class problem is handled with K two-class SVMs [27]. For linearly nonseparable problems, SVM transforms the input data into a very high-dimensional feature space and then employs a linear hyperplane for classification. Introduction of a feature space creates a computationally intractable problem. SVM handles this by defining appropriate kernels so that the problem can be solved in the input space itself. The problem of maximizing the margin can be reduced to the solution of a convex quadratic optimization problem, which has a unique global minimum.

For a binary classification training data problem, suppose a data set consists of N feature vectors (x_i, y_i), where $y_i \in \{+1, -1\}$, denotes the class label for the data point x_i. The problem of finding the weight vector ν can be formulated as minimizing the following function:

$$L(\nu) = \frac{1}{2} \|\nu\|^2 \tag{1}$$

subject to

$$y_i[\nu \cdot \phi(x_i) + b] \geq 1, i = 1, \ldots, N \tag{2}$$

Here, b is the bias and the function $\phi(x)$ maps the input vector to the feature vector. The SVM classifier for the case on linearly inseparable data is given by

$$f(x) = \sum_{i=1}^{N} y_i \beta_i K(x_i, x) + b \tag{3}$$

where K is the kernel matrix, and N is the number of input patterns having nonzero values of the Langrangian multipliers β_i. These N input patterns are called support vectors, and hence the name SVMs. The Langrangian multipliers β_i can be obtained by maximizing the following:

$$Q(\beta) = \sum_{i=1}^{N} \beta_i - \frac{1}{2} \sum_{i=1}^{N} \sum_{j=1}^{N} y_i y_j \beta_i \beta_j K(x_i, x_j) \tag{4}$$

subject to

$$\sum_{i=1}^{N} y_i \beta_i = 0 \quad 0 \leq \beta_i \geq C, \quad i = 1, ..., N \tag{5}$$

where C is the cost parameter, which controls the number of non separable points. Increasing C will increase the number of support vectors thus allowing fewer errors, but making the boundary separating the two classes more complex. On the other hand, a low value of C allows more non separable points, and therefore, has a simpler boundary. Only a small fraction of the β_i coefficients are nonzero. The corresponding pairs of x_i entries are known as support vectors and they fully define the decision function. Geometrically, the support vectors are the points lying near the separating hyperplane. $K(x_i, x_j) = \phi(x_i).\phi(x_j)$ is called the kernel function. The kernel function may be linear or nonlinear, like polynomial, sigmoidal, radial basis functions (RBF), etc. RBF kernels are of the following form:

$$K(x_i, x_j) = e^{-\gamma |x_i - x_j|^2} \tag{6}$$

where x_i denotes the ith data point and γ is the weight. In this paper, the above mentioned RBF kernel is used. In addition, the extended version of the two-class SVM that deals with multiclass classification problem by designing a number of one against all two-class SVMs, is used here.

3   Proposed RotaSVM

Consider a training set $\mathcal{L} = \{(x_i, y_i)\}_{i=1}^{N}$ consisting of N independent instances, in which each (x_i, y_i) is described by an input attribute vector $x_i = (x_{i1}, x_{i2}, ..., x_{id}) \in \mathbb{R}^d$ and a class label y_i. In a classification task, the goal is to use the information only from $\mathcal{L}$ to construct a classifier which performs well on unseen data. For simplicity of the notations, let X be a $N \times d$ data matrix composed with the values of d input attributes for each training instance and Y be a column vector of size N, containing the outputs of each training instance in $\mathcal{L}$. Moreover, $\mathcal{L}$ can also be expressed by concatenating X and Y vertically, that is, $\mathcal{L} = [XY]$. Also let $F = \{X_1, X_2, ..., X_d\}^T$ be the attribute or feature set composed of d input attributes or features and ω be the set of class labels $\{\omega_1, \omega_2, ..., \omega_c\}$, from which Y takes values.

In RotaSVM, SVM runs T number of times with different rotational feature set. During the training of each SVM, the feature set F is randomly split into S (S is an input parameter of RotaSVM) subsets, which may be disjoint or intersecting. To maximize the chance of high diversity, disjoint subsets are chosen. Subsequently, a submatrix $X_{t,s}$, where t is the timestamp of the SVM classifier runs and s is the subset number, is created with the attributes in $F_{t,s}$. From this submatrix, $X_{t,s}$, a new bootstrap sample $X'_{t,s}$ of size 75% is selected. Thereafter, PCA technique is applied to each subset to obtain a matrix $D_{t,s}$ where all principal components are retained in order to preserve the variability information in the data. Thus, S axis rotations take place to form the new attributes for SVM classifier. Subsequently, the matrix $D_{t,s}$ is arranged into a block diagonal matrix R_t. To construct the training set for classifier SVM_t the rows of R_t are rearranged, so that they correspond to the original attributes in F. The rearranged rotation matrix is R_t^a and training set for classifier SVM_t is $[XR_t^a, Y]$. Note that the reason behind selecting 75% is to avoid getting identical coefficients of the principal components if the same attribute subset is chosen for different classifiers and to increase the diversity among the ensemble classifiers. Details of RotaSVM are mentioned in Algorithm 1.

In the testing phase, given a test sample $\mathcal{I}$, let $SVM_{t,i}(\mathcal{I}R_t^a)$ be the posterior probability produced by the classifier SVM$_t$ on the hypothesis that $\mathcal{I}$ belongs to class ω_i. Then the confidence for a class is calculated by the average posterior probability of combination SVMs:

$$\psi_i(\mathcal{I}) = \frac{1}{T} \sum_{t=1}^{T} SVM_{t,i}(\mathcal{I}R_t^a), (i = 1, 2, \ldots, c) \tag{7}$$

Thereafter, $\mathcal{I}$ is assigned to the class with the largest confidence. Note that while running the RotaSVM algorithm to solve a classification task, some parameters like T and S are needed to be specified in advance.

Algorithm 1. RotaSVM

Require: For Training
  X, Data Set
  Y, Class Label
  T, Number of time SVM runs
  S, Number of Feature Sets
Require: For Testing
  $\mathcal{I}$, A data object to classify
Ensure:
  Class lable of $\mathcal{I}$
  Prediction Error of RotaSVM

1: **for** $(t = 1, 2, \ldots, T)$ **do**
2:     Randomly split the attribute set F into S subsets, $F_{t,s}$ where $(s = 1, 2, \ldots, S)$.
3:     **for** $(s = 1, 2, \ldots, S)$ **do**
4:         Create submatrix $X_{t,s}$ using X and $F_{t,s}$.
5:         Create a new bootstrap sample $X'_{t,s}$ of size 75% form $X_{t,s}$.
6:         Apply PCA on $X'_{t,s}$ to obtain the coefficient matrix $D_{t,s}$.
7:     **end for**
8:     Arrange the matrices $D_{t,s}(s = 1, 2, \ldots, S)$ into a block diagonal matrix R_t.
9:     Construct the rotation matrix R_t^a by rearranging the rows of R_t.
10:     Train the classifier SVM_t using $[XR_t^a Y]$ as the training set.
11: **end for**
12: Test the sample $\mathcal{I}$ using different SVM_t and compute average posterior probability to assign class label.
13: **return** *Class label of $\mathcal{I}$ and Prediction Error of RotaSVM.*

4   Empirical Results

The effectiveness of the RotaSVM is quantitatively measured by comparing it with Bagging, AdaBoost, MultiBoost and Rotation Forest for 10 real-life data sets. In this section, details of data sets, performance metrics and results are discussed.

4.1   Data Sets

Table 1 gives the information about data sets, where their different characteristics and variety of fields are described in first three columns by giving the name, sample size and number of classes of each data set, respectively. The last column summarize the information of total number of input attributes in the data sets. All the data sets taken in this experiment, have only numerical attributes. During the pre-processing of data, instances that contain missing values are deleted from the data sets.

Table 1. Summery of the data sets

Data Set	Size	Classes	Total number of input attributes
Balance	625	3	4
BCW	691	2	9
Dermatology	366	6	34
Ecoli	366	8	7
Glass	214	6	10
Ionosphere	351	2	34
Iris	150	3	4
Sonar	208	2	60
Vehicle	94	4	18
Wine	178	3	13

4.2   Experimental Setup

The experimental settings are kept same as used in [11–15] for other ensemble based classifiers such as Bagging, AdaBoost, MultiBoost and Rotation Forest. The classification tree for all these algorithms is created by the "Treefit" algorithm in Matlab. The implementations of these methods are done in Matlab software with version 7.1. For RotaSVM and Rotaion Forest, we have fixed the ensemble size or the number of times SVM and Decision Tree runs arbitrarily at $T = 6$. The value of S is not fixed for each data set, thus we have adjusted it manually. Here, each method run 20 times and their prediction errors are summarized by computing mean, standard deviation as well as the *kappa* index [28]. Finally, statistical test has been conducted to show the superiority of the results produced by RotaSVM. Note that RBF (Radial Basis Function) kernel is used for SVM in our experiment. Here the parameters of SVM such as γ for kernel function and the soft margin C (cost parameter), are set to be 0.5 and 2.0, respectively.

Table 2. Mean and standard deviation of prediction errors (expressed in %) for 20 runs of each method on real-life data sets

Data Set	RotaSVM	Bagging	AdaBoost	MultiBoost	Rotation Forest
Balance	8.32 ±0.91	15.03 ±0.93•	21.62 ±0.75•	18.03 ±0.83•	9.62 ±0.72•
BCW	3.47 ±0.29	3.45 ±0.34	3.15 ±0.35	3.13 ±0.30	2.89 ±0.33○
Dermatology	2.33 ±0.78	3.6 ±0.78•	3.14 ±0.80•	3.02 ±0.79•	2.62 ±0.68
Ecoli	15.80 ±2.11	17.23 ±1.36•	15.7 ±1.16	14.99 ±1.16○	16.37 ±1.33•
Glass	1.25 ±0.71	24.53 ±0.92•	24.56 ±0.88•	24.37 ±0.83•	24.39 ±0.56•
Ionosphere	1.33 ±0.59	8.46 ±0.74•	6.3 ±0.77•	6.27 ±0.80•	5.53 ±0.72•
Iris	2.67 ±0.94	4.7 ±1.23•	5.33 ±1.08•	5.8 ±1.15•	4.37 ±1.00•
Sonar	12.98 ±0.68	23.08 ±1.82•	17.98 ±2.14•	18.63 ±2.23•	17.26 ±2.24•
Vehicle	11.71 ±1.51	25.79 ±0.97•	23.68 ±1.06•	23.65 ±0.89•	21.82 ±1.00•
Wine	1.49 ±0.38	4.55 ±1.77•	3.31 ±0.81•	3.23 ±1.01•	5.11 ±1.77•

"•" indicates that RotaSVM is significantly better and "○" denotes that RotaSVM is significantly worse at the significance level $\alpha = 0.05$.

Table 3. One-tailed paired t-test results of RotaSVM in comparison with other methods for real-life data sets

Algorithm		Bagging	AdaBoost	MultiBoost	Rotation Forest
	t-test Result				
	Win	9	8	8	8
RotaSVM	Tie	1	2	1	1
	Loss	0	0	1	1

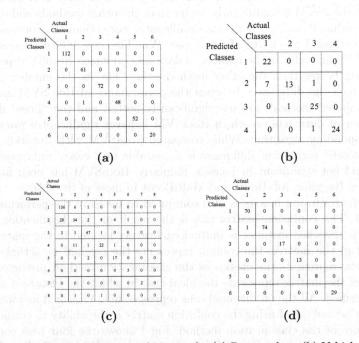

Fig. 1. Best Confusion matrix out of 20 runs for (a) Dermatology (b) Vehicle (c) Ecoli and (d) Glass data sets

Table 4. Average values of *Kappa* Index for different data sets

Data Set	RotaSVM	Bagging	AdaBoost	MultiBoost	Rotation Forest
Balance	0.87	0.65	0.59	0.62	0.71
BCW	0.91	0.91	0.92	0.92	0.93
Dermatology	0.93	0.90	0.91	0.92	0.93
Ecoli	0.64	0.62	0.64	0.65	0.63
Glass	0.95	0.55	0.55	0.56	0.56
Ionosphere	0.94	0.86	0.87	0.87	0.88
Iris	0.92	0.88	0.87	0.86	0.87
Sonar	0.78	0.55	0.61	0.60	0.61
Vehicle	0.80	0.53	0.55	0.55	0.57
Wine	0.93	0.89	0.90	0.91	0.89

4.3   Results

In Table 2, the mean and standard deviation of the prediction errors (expressed in %) for each method on 10 date sets are reported, where the values of standard deviation are followed after "$\pm$". RotaSVM gives consistent results for all data sets. Moreover, the minimum error is achieved by RotaSVM for "Glass" data set. In order to see whether RotaSVM is significantly better or worse than other methods from the statistical viewpoint, a one-tailed paired *t*-test [24] is performed at $\alpha = 0.05$ significance level. The results for which a significant difference of RotaSVM with other methods are found and marked with a bullet or an open circle next to the values of standard deviation in Table 2. A bullet indicates that RotaSVM is significantly better than the other methods and an open circle gives that RotaSVM performs significantly worse than the corresponding method. As it can be seen from Table 3, the "Win-Tie-Loss" information is given, where the "Win" value is the number of data sets on which RotaSVM performs significantly better than the other methods, the "Tie" is the number of data sets on which the differences between the performance of RotaSVM and that of the compared methods are not significantly better, and the "Loss" denotes the number of data sets on which RotaSVM behaves significantly worse than the corresponding algorithm. While compared RotaSVM with Rotation forest, the statistically significant difference is favourable in 8 cases, unfavourable in 1 cases and not significant in 1 cases. Similarly, RotaSVM has been found to outperform Bagging, AdaBoost and MultiBoost in most of the cases.

   Here, the confusion matrix is also computed to measure the performance of RotaSVM. The confusion matrix [29] is the result of the classification phase where each classified instance is mutual exclusively located in the matrix. For every cell in the matrix, the column represents the original or actual classes and the row represents the classes as the classification method predicted. The diagonal of the matrix represents the ideal case in which the instances are correctly classified. All the off diagonal cells represent miss classified instances. An important advantage of using the confusion matrix is the ability to consider the performance of the classification method. Fig 1 shows the four best confusion matrices produced by the RotaSVM for Dermatology, Vehicle, Ecoli and Glass data sets, respectively. The accuracy assessment of different methods have also been justified by measuring *kappa* index, and reported in Table 4. The higher value of *kappa* (close to 1) indicates better accuracy. For most of the cases, it is found from the Table 4 that the *kappa* values are also better for RotaSVM.

5   Conclusions

In this paper, RotaSVM ensemble classifier is developed that uses rotational feature selection approach with the integration of Support Vector Machine classifier. To generate the number of predefined outputs of Support Vector Machines, the training data is prepared by splitting the feature set randomly into S number of subsets. Subsequently, principal component analysis is used for each subset to generate new feature sets, which are reassembled to preserve the variability information in the data. Such features are later used to train Support Vector Machines. Finally, the classification is done by computing average posterior probability. The results, demonstrate the superiority of the RotaSVM quantitatively by comparing it with Bagging, AdaBoost and MultiBoost for 10 real-life data sets taken from UCI Machine Learning Repository. Statistical test like one-tailed paired t-test has been performed to show the superiority of the result produced by RotaSVM.

There are still some interesting issues in RotaSVM that are needed to investigate in future. In this regards, trade-off between the parameters like T and S can be archived automatically using multiobjective optimization techniques [30, 31]. Moreover, RotaSVM can be applied for pixel classification of satellite imagery [20, 32, 33], microarray classification [34, 35], protein translational modification site prediction [36, 37], human leukocyte antigen class II binding peptide prediction [38] ect. The authors are currently working in this direction.

Acknowledgments. This work is partially supported by Erasmus Mundus Mobility with Asia (EMMA) fellowship grant 2012 from European Union (EU) and University with Potential for Excellence (UPE) - Phase II project grant from University Grants Commission (UGC) in India. Moreover, one of the authors, Mr. I. Saha, would also like to express sincere thanks to the All India Council for Technical Education (AICTE) for providing National Doctoral Fellowship (NDF).

References

1. Benediktsson, J.A., Kittler, J., Roli, F. (eds.): MCS 2009. LNCS, vol. 5519. Springer, Heidelberg (2009)
2. Haindl, M., Kittler, J., Roli, F. (eds.): MCS 2007. LNCS, vol. 4472. Springer, Heidelberg (2007)
3. Oza, N.C., Polikar, R., Kittler, J., Roli, F. (eds.): MCS 2005. LNCS, vol. 3541. Springer, Heidelberg (2005)
4. Roli, F., Kittler, J., Windeatt, T. (eds.): MCS 2004. LNCS, vol. 3077. Springer, Heidelberg (2004)

5. Windeatt, T., Roli, F. (eds.): MCS 2003. LNCS, vol. 2709. Springer, Heidelberg (2003)
6. Roli, F., Kittler, J. (eds.): MCS 2002. LNCS, vol. 2364. Springer, Heidelberg (2002)
7. Kittler, J., Roli, F. (eds.): MCS 2001. LNCS, vol. 2096. Springer, Heidelberg (2001)
8. Kittler, J., Roli, F. (eds.): MCS 2000. LNCS, vol. 1857. Springer, Heidelberg (2000)
9. Maulik, U., Bandyopadhyay, S., Saha, I.: Integrating clustering and supervised learning for categorical data analysis. IEEE Transactions on Systems, Man and Cybernetics, Part A 40(4), 664–675 (2010)
10. Saha, I., Maulik, U., Bandyopadhyay, S., Plewczynski, D.: Unsupervised and supervised learning approaches together for microarray analysis. Fundamenta Informaticae 106(1), 45–73 (2011)
11. Breiman, L.: Bagging predictors. Machine Learning 24(2), 123–140 (1996)
12. Freund, Y., Schapire, R.E.: Experiments with a new boosting algorithm. In: Proceedings of the Thirteenth International Conference on Machine Learning, pp. 148–156 (1996)
13. Freund, Y., Schapire, R.E.: A decision-theoretic generalization of on-line learning and an application to boosting. Journal of Computer and System Sciences 55(1), 119–139 (1997)
14. Breiman, L.: Random forests. Machine Learning 45(1), 5–32 (2001)
15. Rodriguez, J.J., Kuncheva, L.I., Alonso, C.J.: Rotation Forest: A new classifier ensemble method. IEEE Transactions on Pattern Analysis and Machine Intelligence 28(10), 1619–1630 (2006)
16. Giacinto, G., Roli, F.: An approach to the automatic design of multiple classifier systems. Pattern Recognition Letters 22(1), 25–33 (2001)
17. Perdisci, R., Ariu, D., Fogla, P., Giacinto, G., Lee, W.: McPAD: A multiple classifier system for accurate payload-based anomaly detection. Computer Networks 53, 864–881 (2009)
18. Giacinto, G., Perdisci, R., Rio, M.D., Roli, F.: Intrusion detection in computer networks by a modular ensemble of one-class classifiers. Information Fusion 9(1), 69–82 (2008)
19. Giacinto, G., Roli, F., Didaci, L.: Fusion of multiple classifiers for intrusion detection in computer networks. Pattern Recognition Letters 24(12), 1795–1803 (2003)
20. Saha, I., Maulik, U., Bandyopadhyay, S., Plewczynski, D.: SVMeFC: SVM ensemble fuzzy clustering for satellite image segmentation. IEEE Geoscience and Remote Sensing Letters 9(1), 52–55 (2011)
21. Krogh, A., Vedelsby, J.: Neural network ensembles, cross validation, and active learning. In: Advances in Neural Information Processing Systems 7, pp. 231–238 (1995)
22. Boser, B.E., Guyon, I.M., Vapnik, V.N.: A training algorithm for optimal margin classifiers. In: Proceedings of the Fifth Annual Workshop on Computational Learning Theory, pp. 144–152 (1992)
23. Blake, C.L., Merz, C.J.: UCI repository of machine learning databases (1998), http://www.ics.uci.edu/~mlearn/mlrepository.html
24. Ferguson, G.A., Takane, Y.: Statistical analysis in psychology and education (2005)
25. Vapnik, V.: Statistical Learning Theory. Wiley, New York (1998)
26. Goh, K.S., Chang, E.Y., Li, B.: Using one-class and two-class SVMs for multiclass image annotation. IEEE Transactions on Knowledge and Data Engineering 17(10), 1333–1346 (2005)

27. Juang, C.F., Chiu, S.H., Shiu, S.J.: Fuzzy system learned through fuzzy clustering and support vector machine for human skin color segmentation. IEEE Transactions on Systems, Man and Cybernetics, Part A: Systems and Humans 37(6), 1077–1087 (2007)

28. Cohen, J.A.: Coefficient of agreement for nominal scales. Educational and Psychological Measurement 20(1), 37–46 (1960)

29. Marom, N.D., Rokach, L., Shmilovici, A.: Using the confusion matrix for improving ensemble classifiers. In: Proceedings of the IEEE Twenty Sixth Convention of Electrical and Electronics Engineers, pp. 555–559 (2010)

30. Deb, K.: Multi-objective Optimization Using Evolutionary Algorithms. Wiley, Chichester (2001)

31. Saha, I., Maulik, U., Plewczynski, D.: A new multi-objective technique for differential fuzzy clustering. Applied Soft Computing 11(2), 2765–2776 (2011)

32. Maulik, U., Saha, I.: Modified differential evolution based fuzzy clustering for pixel classification in remote sensing imagery. Pattern Recognition 42(9), 2035–2149 (2009)

33. Maulik, U., Saha, I.: Automatic fuzzy clustering using modified differential evolution for image classification. IEEE Transactions on Geoscience and Remote Sensing 48(9), 3503–3510 (2010)

34. Saha, I., Maulik, U., Bandyopadhyay, S., Plewczynski, D.: Improvement of new automatic differential fuzzy clustering using svm classifier for microarray analysis. Expert Systems with Applications 38(12), 15122–15133 (2011)

35. Saha, I., Plewczynski, D., Maulik, U., Bandyopadhyay, S.: Improved differential evolution for microarray analysis. International Journal of Data Mining and Bioinformatics 6(1), 86–103 (2012)

36. Plewczynski, D., Basu, S., Saha, I.: AMS 4.0: Consensus prediction of post-translational modifications in protein sequences. Amino Acids 43(2), 573–582 (2012)

37. Saha, I., Maulik, U., Bandyopadhyay, S., Plewczynski, D.: Fuzzy clustering of physicochemical and biochemical properties of amino acids. Amino Acids 43(2), 583–594 (2012)

38. Saha, I., Mazzocco, G., Plewczynski, D.: Consensus classification of human leukocyte antigens class II proteins. Immunogenetics 65, 97–105 (2013)

Finding Biologically Plausible Complex Network Topologies with a New Evolutionary Approach for Network Generation

Gordon Govan, Jakub Chlanda, David Corne, Alex Xenos, and Pierluigi Frisco

School of Mathematical and Computer Sciences
Heriot-Watt University
Edinburgh, UK
gmg8@hw.ac.uk, {j.chalanda,Pierluigi.Frisco}@gmail.com,
dwcorne@macs.hw.ac.uk, xenosa@yahoo.gr

Abstract. We explore the recently introduced Structured Nodes (SN) network model, for which earlier work has shown its capability in matching several topological properties of complex networks. We consider a diverse set of empirical biological complex networks as targets and we use an evolutionary algorithm (EA) approach to identify input for the SN model allowing it to generate networks similar to these targets.

We find that by using the EA the structural fit between SN networks and the targets is improved.

The combined SN/EA approach is a promising direction to further investigate the growth, properties and behaviour of biological networks.

1 Introduction

Networks have recently been used in several fields of biology providing important advances in the study of process and phenomena when these are modelled as networks [13,14]. Network models have been used to bring insight into the development and dynamics of these empirical networks [6,15], but these models are limited in capturing several topological features of empirical networks [2]. The present study has been fuelled by the need to have an algorithm creating networks matching several topological features present in empirical networks [3,7,11].

The ability to generate networks similar to empirical ones is paramount: it allows to test theories that cannot be tested in empirical networks, it allows to understand how the empirical network could have developed under different stimuli, etc.

Some network models, such as the Erdős-Rényi model, are very limited in replicating empirical networks [10,16]. Others, as the Watts-Strogatz model [18,5], are able to capture a few global topographical measures of some empirical networks. The Barabási-Albert network model is able to generate networks having a scale-free degree distribution. This feature is found in many empirical networks. Unfortunately, the Barabási-Albert network model in not able to capture other network topological measures [4,2].

M. Emmerich et al. (eds.), *EVOLVE - A Bridge between Probability, Set Oriented Numerics,* 59
and Evolutionary Computation IV, Advances in Intelligent Systems and Computing 227,
DOI: 10.1007/978-3-319-01128-8_5, © Springer International Publishing Switzerland 2013

Recently, the structured nodes (SN) network model has been presented [11]. It has been proven that this model is able to replicate several topological features of empirical networks better than other network models. This increased flexibility is due to the complex input needed by the SN model. Due to its complexity, it is difficult to find the appropriate input allowing the SN model to generate networks with desired properties.

In this paper we present the results obtained using an evolutionary algorithm (EA) [8] in finding the input allowing the SN model to generate networks with specific properties. As we aimed to use the SN model to create complex networks having (three) different but correlated topological features, our study falls in the realm of multi-objective optimisation.

The target empirical networks we considered are the:

Escherichia coli protein-protein interaction network,
Saccharomyces cerevisiae (yeast) gene network,
Methicillin-resistant Staphylococcus aureus (MRSA) gene network.

The SN model has been previously employed [8] to create networks similar to the *E. coli* and *S. cerevisiae* networks, but there the input for the SN model was found by trial and error.

The rest of the paper is organised as follows. Sect. 2.1 gives a background on networks and their topological properties, Sect. 2.2 provides an overview of the SN and other network models, Sect. 3 gives a description of the EA, Sect. 4 describes the methodology and the obtained results while in Sect. 5 we draw conclusions on this study.

2 Networks

2.1 Topological Features

In this section we introduce the network terminology we employ together with short definitions of the topological features we consider. Further details on network topological features can be found in [13].

Networks are composed of a set of *nodes* connected by *edges*. In this paper the networks have at most one edge between a pair of nodes, and edges are bi-directional, that is they can be travelled in either direction. The networks are also connected, meaning that it is possible starting from any node and travelling along edges, to reach any other node in the network.

The *degree* of a node is given by the number of edges connecting it to other nodes, the *path length* between two nodes in a network is given by the minimum number of edges that have to be traversed in order to go from one node to the other, while the *clustering coefficient (clust. coef.)* of a node with k neighbours having e edges between themselves is $\frac{2e}{k(k-1)}$.

The network topological features considered in this study are:

average degree: the sum of the degree of each node divided by the number of nodes;

average path length: the sum of the shortest path lengths between all pairs of
    different nodes divided by the number of pairs of different nodes;
average clustering coefficient: the sum of the clustering coefficient for each node
    divided by the number of nodes;
degree distribution: the probability distribution indicating the probability to find
    in a network a node with a given degree;
path length distribution: the probability distribution indicating the probability
    to find in a network a given path length between two nodes.

The average degree, average path length, and average clustering coefficient are
the target measurements in the EA used in this study. Clearly, these three measurements are dependent from each other: attempts to change one of them will
affect the others. A good example of this is given by the Watts-Strogatz network
model. If in this model the edges are rewired with the intention of reducing the
average path length of the network then the average clustering coefficient will
become smaller [18]. As these three measurements are not independent from each
other the present study is a suitable task to be performed by a multi-objective
optimisation technique.

2.2   Network Models

Given an input, a network model is an algorithm able to generate networks.
The topological features of the generated networks depend on the input and the
model itself. Two well known networks models which are often used in modelling
biological networks are the Erdős-Rényi model and the Watt-Strogatz model.

The Erdős-Rényi (ER) model [10] (the networks generated by it are also
known as *random networks*) starts from a fixed number of nodes and it adds
edges with a fixed probability p_{ER}. Networks created by the ER model are likely
to have a low average path length but they fail to account for the local clustering
that characterises many empirical networks.

The Watts-Strogatz (WS) model [18] starts from a regular lattice in which
each node has degree n_{WS}, then it rewires each edge with probability p_{WS}.
In our implementation nodes are arranged in a ring and each node has out-
going edges to its $n_{WS}/2$ clockwise and $n_{WS}/2$ counter-clockwise neighbours.
Networks created by the WS model are likely to have the *small-world* property (the minimum path length between any pair of nodes is approximately
equivalent to a comparable random network but the nodes of the network have
greater local inter-connectivity than a random network) found in many empirical
networks [6,12,17].

Both these network models have an input with few parameters but they are
very limited in the range of network topological features they can replicate.

The Structured Nodes Model. The Structure Nodes (SN) model is a network
model which aims to generate networks by using methods inspired by biological
processes [11]. It does so by giving nodes a structure and then connecting pairs
of nodes depending on their structure.

The best way to describe how the SN model works is with an example. Let us consider the *alphabet* $\alpha = \{A, B, C, D\}$. This model generates networks in which each node has a unique *structure* or word over the letters in the given alphabet, $S \in \alpha^+$. The algorithm implementing the SN model starts with a network having only one initial node. Let us assume that in this example the *initial node* is ABCDAB (that is, the structure of this initial node is ABCDAB). A new node and edges are added to the network following the steps:

1. randomly select a node in the network;
2. copy the structure of the chosen node and then change it (according to some possible *operations*, see below) in order to get a new structure (the chosen node remains in the network);
3. if the new structure is already present in the network, then do not add it to the network and go to step 1, otherwise go on to step 4;
4. add edges connecting the new node to the other nodes in the network according to a *distance measure* that operates on the node structures (explained below);
5. if at least one edge has been added, then the new structure and all the new edges are added to the network, otherwise the new structure is not added to the network.

These steps are repeated until the network reaches a given number of nodes.

For this example there is not much choice in implementing step 1: as there is only one node (with structure ABCDAB) this is the only node that can be chosen. Step 2 copies and changes the chosen structure. Let us assume that we change this structure replacing the D with a C. In this way the structure ABCCAB is obtained and, as this structure is not present in the network, we move to step 4.

In order to implement step 4 in this example a distance measure between node structures must be defined. In this example a *Hamming distance* is used to measure the number of mismatches between two structures. The Hamming distance between ABCDAB and ABCCAB is 1 (because there is only 1 mismatch in the fourth character). We decide that an edge between two nodes is added only if the Hamming distance is at most 1. This is indeed the case in our example, so an edge is added between the two nodes. Now the network is composed by two nodes and one edge between them.

The Hamming distance is configurable with values for Max_distance and Unit_distance. Max_distance specifies the maximum Hamming distance that is allowed between two nodes, above which they will not be connected by an edge. Unit_distance controls whether the node structure is treated as a list of single elements, or as a list of tuples. When treated as a list of tuples, the Hamming distance measures the number of tuples that do not match (multiple mismatches in a single tuple are treated the same as a single mismatch). The use of tuples simulates the use of *codons* in the genetic code, where a codon is a triplet (3-tuple) of nucleotides [1]. In this example we employed a Max_distance and Unit_distance of 1.

The network in this example will grow further starting again from step 1 until it reaches a predefined size (number of nodes).

The operations node structures can undergo are:

mutation: one symbol in the chosen structure is changed to another symbol of the alphabet. For instance (as in our example), the structure ABCDAB can mutate into ABCCAB;

deletion: one symbol in the chosen structure is removed. For instance, the structure ABCDAB can have its D removed to become ABCAB;

addition: a symbol from the alphabet is added anywhere in the chosen structure. For instance, the structure ABCDAB can have a A added after its D and become ABCDAAB;

duplication : a sub-word of the chosen structure is replicated. For instance, the structure ABCDAB can have CDA replicated and become ABCDACDAB.

In the algorithm implementing the SN model each of these operations has a probability to occur. We refer the reader to [11] for a complete explanation of the SN model.

It should be clear that the SN model requires lots of input parameters (alphabet, structure of initial node, types of operations, type of distance measure, etc.) in order to run. It should be also clear that the algorithm implementing the SN model is stochastic: random numbers are employed to chose the operation to implement, the node to be selected, etc. The number of parameters that the SN model requires is much greater than the other network models mentioned above. Generating a set of input parameters for the SN model that will produce a network with desired features is a difficult task, and could take weeks of effort if changing the parameters through trial and error in order to obtain specific networks. The development of the EA described in the following was motivated by the need of having a more efficient way to find the input allowing the SN model to generate networks with desired topological features.

3   An Evolutionary Algorithm for the SN model

We implemented an EA taking taking as input:

a sample set of input parameters to the SN model;

a set of EA parameters (number of generations to run, the population size to keep between generations, probabilities to crossover or mutate, etc.);

the topological network features.

An initial population is created mutating the sample set of input parameters for the SN model. The initial population is composed of different input parameters for the SN model. Each member of that population is run through the SN model such that several (a user specified number) networks are generated for each individual. The fitness of the ensemble of networks deriving from the same individual is computed.

In each generation a new population (with a user specified number of individuals) is obtained from the existing population through mutation and crossover.

New networks are generated through the SN model for each individual in the new population. The fitness of the ensemble of networks deriving from each individual in the new population is computed. The individuals of the two populations ('existing' and 'new') are sorted according to their fitness. The non-dominated individuals are kept and the rest of the individuals are discarded.

The output of the EA is the best individual. It is returned when the program has run through the specified number of generations.

3.1 Individuals

In the program an *individual* is a single set of input parameters to the SN model. The fitness of the ensemble of networks generated by running the SN model on these input parameters (a user specified number of times) are associated to each individual. Each network generated from each individual derives from a different sequence of random numbers (i.e., n different random sequences are used to generate the n networks associated to each individual).

The average degree, clustering coefficient, and path length are the measurements determining the fitness of each individual.

3.2 Mutation and Crossover

A new generation of individuals is created by performing mutation and crossover operations on an existing *parent* population. This is done until the population in the new generation reaches a user defined size.

The implemented crossover and mutation operators are:

crossover: two individuals (first and second parent) are chosen from the parent population. The initial node and alphabet of the new individual are set to be the same as the first parent. The probabilities to add, mutate, remove, and delete elements are set to be the same as the second parent. If the crossover operation fails to produce a unique individual for the new population then a new pair of parents is chosen and the crossover operation is tried again. This process is repeated and if no unique individual is found in a set number of attempts (set to 50 for the tests in Sect. 4), then one of the mutation operators is used to create a new individual instead;

mutate initial node: an individual is chosen from the parent population and then the initial node is mutated by either adding or removing a character;

mutate distances: an individual is chosen from the parent population and then either the Max_distance or Unit_distance fields of the individual are either incremented or decremented by 1;

mutate alphabet: an individual is chosen from the parent population and then either a new character is added or removed from the alphabet of that individual. This may require the initial node to be adapted to match the new alphabet;

mutate probabilities: an individual is chosen from the parent population and then the probabilities that are used to mutate a node are changed. One of the probabilities is changed by an amount taken from a Gaussian distribution, and the rest of the probabilities are modified so that they all add up to 1.

If all of the mutation operators fail to produce a new unique individual then the process is repeated until they do. Each mutation and crossover operation produces one new individual.

3.3 Parallel Evaluation of Individuals

Evaluating the fitness of an individual in our EA is a computationally expensive task both in time and memory. To reduce the time taken for each generation to be evaluated, the evaluation of individuals is done in parallel. Each individual is evaluated independently from all others. This has been implemented using a software architecture called a *farm* [9]. A farm is used when there is a set of data all of which is to be operated on by the same algorithm. The farm will *farm out* (partition and distribute) the data to a set of processing elements such that every piece of data is processed once, and then all the results are brought back together.

The amount of data in the input parameters for the SN model and the result of the evaluation (measurements from the generated networks) is small in comparison to the amount of work that is needed to be performed for the evaluation. This means that this algorithm is well suited to not just symmetric multiprocessing, but distributed computing running on a cluster of computers.

A lightweight framework was created to evaluate the individuals in parallel, either on a multi-core computer or a cluster of computers (on which multi-core capabilities would still be taken advantage of). The framework acts a task supervisor and will re-distribute any processes that fail on remote machines.

3.4 Sorting the Population

The 'new' population is merged to the 'existing' population once each individual in both populations have had their fitness evaluated. The non-dominated individuals of the merged population (those which are most Pareto optimal [8]) are then taken to be the 'existing' population for the next generation (i.e. the dominated solutions are removed from the merged population). An individual a is a member of the non-dominated solutions set if there exists no other individual b that has values no worse than those of a and at least one value better than one in a. In this context, values refer to the distances between the average of each measurement and the measurement from the empirical network.

This manner of choosing individuals creates a weak ordering between them. This means that there are individuals x and y, where $x \neq y$ and neither $x < y$ nor $y < x$ are true. This is the case for all individuals in the non-dominated set. If the full number of non-dominated individuals after each generation were kept, then the population could increase.

3.5 Fitness Function

The fitness function is based on the average degree, average path length, and average clustering coefficient of each of the n networks generated by each individual. Target values for each of these three topological features are given as input to the EA.

To calculate the fitness, the average of each of the three features is calculated. Each of the measurements is normalised: the difference between the average and the target value is divided by the target value. The reason for this division is to give a similar weight to each of the measurements (for instance, the target average degree may be over 100, but the target average clustering coefficient can never be greater than 1). Normalising the distances from the targeted values means that the clustering coefficient is going to affect the fitness much more than if it wasn't normalised.

The equation used for the fitness function is (where deg is average degree, $p.l.$ is average path length, and $c.c.$ is average clustering coefficient):

$$\text{fitness} = \frac{|\text{avg deg} - \text{target deg}|}{\text{target deg}} + \frac{|\text{avg p.l.} - \text{target p.l.}|}{\text{target p.l.}} + \frac{|\text{avg c.c.} - \text{target c.c.}|}{\text{target c.c.}}$$

3.6 Hardware and Software Setup

The EA and SN programs were written in Java and can be downloaded from `http://www.macs.hw.ac.uk/~gg32/projects/modelnetwork` along with the properties files used for the tests in this paper.

The tests were performed using 4 computers in a cluster. The computers were each equipped with two Intel Xeon E5506 processors for a total of 8 cores running at 2.13GHz, and had 12 GB of main memory per computer. The computers used a 64 bit install of CentOS GNU/Linux operating system and the Java HotSpot VM.

4   Results

Tables 1, 2 and 3 show the comparison of the measurements between the:

empirical networks (i.e. the target measurements from the EA);
ER model;
WS model;
SN model employed in [11] (where the results were generated through trial and error);
present study (using an EA).

In these tables the measurements are given alongside their standard deviation indicating how many of these runs had measurements deviating by at most 10% or at most 20% from the average.

The degree and path length distributions are shown in Figs. 1, 2, and 3. In these figures the distributions of the empirical networks are compared with examples networks produced from the EA, ER and WS models.

The probabilities used for crossover and mutation in the EA were:

```
crossover: 0.2
mutate initial node: 0.2
mutate distances: 0.2
mutate alphabet: 0.2
mutate probabilities: 0.2
```

4.1 *E. coli*

We considered the largest connected component of the protein-protein interaction network of *E. coli*. It consists of 230 nodes and 1390 edges. The EA ran for 200 generations, creating a new population of 20 individuals in each generation and took almost an hour to execute. Neither the ER or WS models are able to capture the average path length and average clustering coefficient of the empirical network. The results from the *E. coli* network are displayed in Table 1 and plots of the distributions are shown in Fig. 1.

Table 1. Results for *E. coli*

	Avg. Degree	Avg. Path Length	Avg. Clust. Coef.	10%	20%
Empirical	6.04	3.78	0.22	-	-
Erdős-Rényi	6.02	3.22	0.025	30	68
Watts-Strogatz	6.00	3.53	0.22	78	99
Trial and error [11]	6.03	3.85	0.26	12	34
EA	6.60	3.77	0.20	88	100

The networks produced from the EA are more accurate than those generated through trial and error when looking at the average path lengths and clustering coefficients but they are less accurate when looking at the average degree. The major difference between the networks obtained by trial and error and those produced from the EA is the consistency of the results. The trial and error input parameters, only 34% of the networks had measurements that deviated by at most 20% from the average value. The results from the EA create networks that are much more consistent; 88% fall into the 10% range, and all fall within the 20%.

The consistency of these measurements within the networks shows that the EA and the SN model are able to produce a large number of networks to meet a specification.

The networks generated by the EA fail to accurately model the average degree distribution of the empirical network. The *E. coli* network has a long tail of degrees which the networks from the SN model do not feature. The *E. coli*

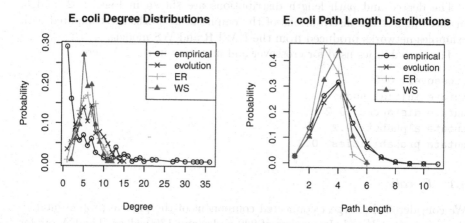

Fig. 1. Degree and path length distributions of the empirical *E. coli* and generated networks

network has many nodes which have only one or two connecting edges, which again is not captured by the networks from the SN model. Both the ER and WS models also fail to capture this property.

The path length distribution of the networks generated from the EA is very close to that of the *E. coli* network. Both distributions probabilities peak at the same path length and fall away at the same rate. The ER and WS models both have a smaller range of path lengths and do not capture the distribution of the empirical network.

The input parameters for the SN model resulting from the EA are:

```
Prob_to_add = 0.0
Prob_to_delete = 0.0
Prob_to_duplicate = 0.0
Prob_to_mutate = 1.0
Max_distance = 1
Unit_distance = 1
Initial_node = CCBAC
Alphabet = A,B,C,D
```

The `Prob_to_mutate` being set at 1 means that the length of the structure of a node cannot be different from that of the `Initial_node`. As the alphabet size is 4 and the length of the initial node is 5, then there is a maximum of 1024 (4^5) node structures that the SN model can produce.

4.2 *S. cerevisiae*

We considered the gene network of *S. cerevisiae*. It consists of 3186 nodes and 130234 edges. The EA ran for 200 generations, creating a new population of 20 individuals in each generation and took 52 hours to execute. The network has a very high clustering coefficient, which could be considered its distinguishing feature. This high clustering coefficient is higher than both the ER and even the WS models are able to produce. The results from the *S. cerevisiae* network are displayed in Table 2, and plots of the distributions are shown in Fig. 2.

Table 2. Results for *S. cerevisiae*

	Avg. Degree	Avg. Path Length	Avg. Clust. Coef.	10%	20%
Empirical	40.87	3.564	0.85	-	-
Erdős-Rényi	40.70	2.57	0.013	100	100
Watts-Strogatz	40.00	3.67	0.69	100	100
Trial and error [11]	41.5	3.04	0.44	15	66
EA	48.50	3.31	0.80	20	50

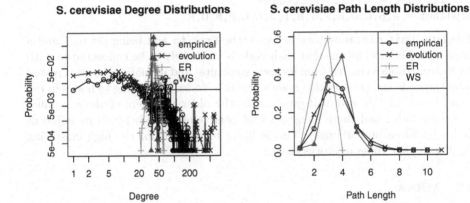

Fig. 2. Degree and path length distributions of the empirical *S. cerevisiae* and generated networks

As we found for the *E. coli* network, the networks generated by the EA do not match the average degree of the *S. cerevisiae* network as well as the networks from the trial and error input parameters, but they do have a far better value for the average clustering coefficient and a better value for the average path length. The average clustering coefficient of 0.44 from the trial and error networks was nowhere near the value of 0.85 that was measured in the empirical network. The value of 0.80 from the networks generated from the output of the EA is much closer, and better captures the locality of clusters in the network. This improved accuracy comes at a loss of precision; only 50% of the generated networks fall within the 20% deviation range.

Both the degree distribution and the path length distribution of a network generated by the EA bear a close resemblance to the matching distributions from the empirical network. The degree distribution from the SN model captures the slope seen in the *S. cerevisiae* when they are plotted on a log-log plot. The high clustering coefficient of the networks is likely to limit the variations that are possible in the internal structures of the networks. The ER and WS models produce a degree distribution completely unlike that of the empirical network, as they fail to produce a long-tailed distribution.

The input parameters for the SN model resulting from the EA are:

```
Prob_to_add = 1.0
Prob_to_duplicate = 0.0
Prob_to_delete = 0.0
Prob_to_mutate = 0.0
Max_distance = 1
Unit_distance = 5
Initial_node = ABCBACDDADCCADBAA
Alphabet = A,B,C,D,E,F,G,H,I,J,K,L,M,N,O,P
```

The large `Unit_distance` together with the `Prob_to_add` being set to 1 and a large alphabet size, means that each node is very likely to be connected to both its *children* (nodes created from its node structure) and its *siblings* (nodes created from the node structure of this node's parent). Nodes connecting to nodes in the same *family* in this way will give rise to the older members of those families becoming hubs, and there being a lot of interconnections between its adjacent nodes (children and siblings). This is likely to account for the high clustering coefficient that these networks have.

4.3   MRSA

We considered the gene network of *Methicillin-resistant Staphylococcus aureus*. It consists of 2321 nodes and 37864 edges. The EA ran for 200 generations, creating a new population of 20 individuals in each generation and took 21 hours to execute. The WS model is able to produce networks with the same clustering coefficient as the empirical network, but not with a similar average path length. This network was not studied in [11]. The results from the MRSA network are displayed in Table 3 and plots of the distributions are shown in Fig. 3.

Table 3. Results for MRSA

	Avg. Degree	Avg. Path Length	Avg. Clust. Coef.	10%	20%
Empirical	16.31	3.95	0.33	-	-
Erdős-Rényi	16.30	3.05	0.007	99	100
Watts-Strogatz	16.00	3.51	0.33	100	100
EA	18.51	4.07	0.47	88	100

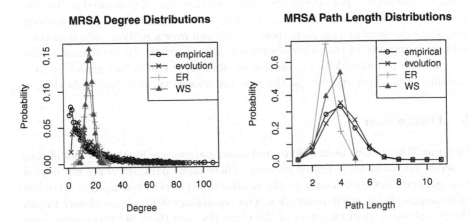

Fig. 3. Degree and path length distributions of the empirical MRSA and generated networks

The networks produced by the SN model closely match the empirical network on all 3 of the global topological features. The deviations from the mean measurements show that the networks are very consistent in their structures with all of them deviating from the mean by less than 20%. The WS model is also accurately captures the average degree and clustering coefficient of the empirical network and is not far off capturing the average path-length.

When looking at the distributions, the networks generated from the results of the EA very closely match the degree distribution and path length distribution of the empirical network. Both of the degree distributions show an early peak, and then a long tail that run down along side each other. Likewise the path length distributions rise and fall at the same rate and peak on the same path length. The WS model which had done well on the global measurements fails to match either the degree of path length distributions of the empirical network. The WS model has a short range of both degrees and path lengths whereas the empirical network has a much wider range.

The input parameters for the SN model resulting from the EA are:

```
Prob_to_add = 0.002
Prob_to_delete = 0.002
Prob_to_duplicate = 0.002
Prob_to_mutate = 0.994
Max_distance = 2
Unit_distance = 1
Initial_node = ABCABABAABBCA
Alphabet = A,B,C
```

These input parameters show that the overwhelming majority of the operations that will be perform to change node structures will be to mutate the chosen node structure. We had said that for matching the *E. coli* network that only mutating limits the number of possible nodes structures that can be generated in the network. The longer initial node that is used here means that this is not a problem as despite the smaller alphabet, there can be well over a million node structures without using any of the other operators. The increased `Max_distance` means that nodes are more likely to connect to siblings, children, and grandchildren, creating hubs which account for the long tail seen in the degree distribution.

5   Discussion

Using an EA we have been able to find input parameters for the SN model that generate networks with desired features. These input parameters are not always able generate networks closer to the empirical networks than using a trial and error approach for input properties. The results are still able to closely match the global topological measures of the networks, and these networks often have degree and path length distributions that are very similar to those from the empirical networks.

The clustering coefficient of the *S. cerevisiae* network is extremely high and could be though of as the most distinguishing feature of the network. It may have been the hardest feature to match, and there may not be a great number of differences between networks with such an extreme feature. That would explain the similarities in the degree distributions and path length distributions.

In addition to giving us input parameters which are able to generate networks with desired features, the nature of the SN model allows us to reason on the input parameters and how these parameters affect the running of the SN model. This gives insight into how the SN model produces networks with desired features, and this insight may reflect the natural processes that created the empirical networks.

There is further work that can be done to enable the EA to target more specific features of networks. The degree distribution of the *E. coli* network is not matched by networks generated from the EA, if the EA could be changed to allow it to target the distributions then it would be much more capable of matching all the features of networks that we desire.

References

1. Alberts, B., Johnson, A., Lewis, J., Raff, M., Roberts, K., Walter, P.: Molecular Biology of the Cell, 4th edn. Garland Science (2002)
2. Amaral, L.A.N., Scala, A., Barthélémy, M., Stanley, H.E.: Classes of small-world networks. Proceedings of the National Academy of Sciences 97(21), 11149–11152 (2000)
3. Barabàsi, A.L., Albert, R.: Emergence of scaling in random networks. Science 286, 509–512 (1999)
4. Barabási, A.L., Albert, R.: Emergence of scaling in random networks. Science 286(5439), 509–512 (1999)
5. Barrat, A., Weigt, M.: On the properties of small-world network models. Eur. Phy. Jour. B - Condensed Matter and Complex Systems 13(3), 547–560 (2000)

6. Bassett, D.S., Bullmore, E.: Small-world brain networks. The Neuroscientist 12(6), 512–523 (2006)
7. Chung, F., Lu, L., Dewey, T.G., Galas, D.J.: Duplication models for biological networks. Journal of Computational Biology 10(5), 677–687 (2003)
8. Coello, C.A.C., Lamont, G.B., Veldhuizen, D.A.V.: Evolutionary Algorithms for Solving Multi-Objective Problems. In: Genetic and Evolutionary Computation, Springer-Verlag New York, Inc. (2006)
9. Cole, M.: Bringing skeletons out of the closet: a pragmatic manifesto for skeletal parallel programming. Parallel Computing 30(3), 389–406 (2004)
10. Erdos, P., Renyi, A.: On random graphs I. Publicationes Mathematicae 6, 290–297 (1959)
11. Frisco, P.: Network model with structured nodes. Phys. Rev. E 84, 021931 (2011)
12. He, Y., Chen, Z.J., Evans, A.C.: Small-world anatomical networks in the human brain revealed by cortical thickness from MRI. Cereb. Cortex 17(10), 2407–2419 (2007)
13. Junker, B.H., Schreiber, F. (eds.): Analysis of Biological Networks. Wiley-Blackwell (2008)
14. Kleinberg, J., Easley, D.: Networks, Crowds, and Markets: Reasoning About a Highly Connected World. Cambridge University Press (2011)
15. Maslov, S., Sneppen, K.: Specificity and stability in topology of protein networks. Science 296(5569), 910–913 (2002)
16. Milo, R., Shen-Orr, S., Itzkovitz, S., Kashtan, N., Chklovskii, D., Alon, U.: Network motifs: Simple building blocks of complex networks. Science 298(5594), 824–827 (2002)
17. Sporns, O., Zwi, J.D.: The small world of the cerebral cortex. Neuroinformatics 2(2), 145–162 (2004)
18. Watts, D.J., Strogatz, S.H.: Collective dynamics of 'small-world' networks. Nature 393(6684), 440–442 (1998)

Fitness Landscape Analysis of NK Landscapes and Vehicle Routing Problems by Expanded Barrier Trees

Bas van Stein, Michael Emmerich, and Zhiwei Yang

LIACS, Leiden University,
Niels Bohrweg 1, 2333-CA Leiden,
The Netherlands
{bvstein,emmerich,zyang}@liacs.nl

Abstract. Combinatorial landscape analysis (CLA) is an essential tool for understanding problem difficulty in combinatorial optimization and to get a more fundamental understanding for the behavior of search heuristics. Within CLA, Barrier trees are an efficient tool to visualize essential topographical features of a landscape. They capture the fitness of local optima and how they are separated by fitness barriers from other local optima. The contribution of this study is two-fold: Firstly, the Barrier tree will be extended by a visualization of the size of fitness basins (valleys below saddle points) using expandable node sizes for saddle points and a graded dual-color scheme will be used to distinguish between penalized infeasible and non-penalized feasible solutions of different fitness. Secondly, fitness landscapes of two important NP hard problems with practical relevance will be studied: These are the NK landscapes and Vehicle Routing Problems (with time window constraints). Here the goal is to use EBT to study the influence of problem parameters on the landscape structure: for NK landscapes the number of interacting genes K and for Vehicle Routing Problems the influence of the number of vehicles, the capacity and time window constraints.

1 Introduction

Much research has been done to combinatorial optimization problems, though we might not even have reached the top of the iceberg yet. Before designing algorithms for a problem, an important task is to understand the properties of the search landscape. Classical calculus has focused mainly on the characterization of continuous search landscapes, whereas landscape analysis tools for discrete search spaces are only discussed more recently also because their application requires fast computing machinery. Our aim is to improve tools for landscape analysis and to study two discrete landscapes with relevance for science and technology, namely the NK landscape problems and the Vehicle Routing problem with time windows (VRPTW). These problems have in common that for a given problem size their difficulty can be scaled by problem parameters, for

M. Emmerich et al. (eds.), *EVOLVE - A Bridge between Probability, Set Oriented Numerics,* 75
and Evolutionary Computation IV, Advances in Intelligent Systems and Computing 227,
DOI: 10.1007/978-3-319-01128-8_6, © Springer International Publishing Switzerland 2013

instance the level of gene interaction K in NK landscapes and the number of vehicles and capacity in VRPTW.

To understand the difficulty of a landscape for local search procedures, the number of local optima is an important characteristic. Beyond this, one may also ask how the attractor basins of local optima are connected with each other. Local Optima Network Analysis [8] and Barrier Tree Analysis [11] provide interesting tools for doing so. In this work we focus on the latter. The idea in Barrier Tree Analysis is to compute a tree that characterizes the topographical structure of a combinatorial landscape. Essentially it provides for each pair of local optima the information on the height of the barrier that separates them from each other. The turning point on an optimal path across this 'fitness barrier' is called a saddle point. For a given landscape this information can be condensed in a hierarchical structure – a barrier tree of the landscape – for which the leaf nodes are local optima and the branching nodes are saddle points.

In Section 2 we provide definition of classical barrier trees and related concepts and refer to some related work. In Section 3 we introduce a new type of barrier trees, we will term *expanded barrier trees* (EBT) and outline an algorithm to compute these. Then, we use the EBT to analyse two problem spaces: Section 4 deals with NK landscapes that are models of evolution of genotypes with interacting genes. Section 5 applies EBT to Vehicle Routing Problems (with Time Windows). Section 6 concludes with a summarizing discussion.

2   Barrier Trees and Related Work

Abstractly, a combinatorial landscape [11] can be defined as a triple $(X, f, \mathcal{N})$, where X denotes a finite, but possibly large, search space, $f : X \to \mathbb{R}$ denotes a fitness function (or height function) that assigns a fitness value to each point in the landscape, and $\mathcal{N} : X \to \wp(X)$ denotes a neighborhood function which declares a neighborhood on X by assigning the set of direct neighbors to points in X. Note, that $\wp(X)$ denotes the set of all subsets (or power set) of X. The neighborhood function is often related to the search heuristic that is used to find an optimal solution. For instance, the set of neighbors can be given by the set of solutions that can be reached from a given point in X by a single mutation or step.

A few formal definitions are required to precisely define barrier trees:

Definition 1 (Path). *Let $N : X \to \wp(X)$ be a neighborhood function. A sequence $p_1, \ldots, p_\ell$ for some $\ell \in \mathbb{N}$ and $p_1, \ldots, p_\ell \in X$ is called a* path *connecting x_1 and x_2, iff $p_1 = x_1$, $p_{i+1} \in N(p_i)$, for $i = 1, \ldots, \ell - 1$, and $p_\ell = x_2$. We say that ℓ is the length of the path.*

Definition 2. *Let $\mathbb{P}_{x_1, x_2}$ denote the set of all paths between x_1 and x_2.*

Definition 3 (Saddle point). *Let $\hat{f}(x_1, x_2) = \min_{\mathbf{p} \in \mathbb{P}_{x_1,x_2}} (\max_{x_3 \in \mathbf{p}} f(x_3))$. A point s on some path $\mathbf{p} \in \mathbb{P}_{x_1,x_2}$ for which $f(s) = \hat{f}(x_1, x_2)$ is called a saddle point between x_1 and x_2.*

Definition 4 (Basin). *The basin $B(s)$ of a point s is defined as*

$$B(s) = \{x \in X | \exists \mathbf{p} \in \mathbb{P}_{s,x} : \max_{z \in \mathbf{p}} f(z) \leq f(s)\}.$$

Theorem 1 ([3]). *Suppose for two points x_1 and x_2 that $f(x_1) \leq f(x_2)$. Then, either $B(x_1) \subseteq B(x_2)$ or $B(x_1) \cap B(x_2) = \emptyset$.*

Theorem 1 implies that the barrier structure of the landscape can be represented as a tree [10] where the saddle points are branching points and the local optima are the leaves.

Example 1. An example of a barrier tree for the search space of a 3-D binary cube is provided in Fig. 1. The search space X is given by the binary numbers of length 3. The hamming neighborhood is applied. Height values are indicated by numbers in the upper part of the nodes.

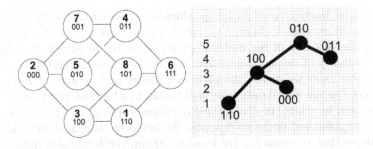

Fig. 1. A binary landscape (left) and its barrier tree (right)

3   Expanded Barrier Trees

The barrier trees describe the notions of saddle points, barriers and basins which give us a vivid landscape visualization of different problems. However, the information of the basin size is not presented yet in barrier trees. Here, we extend the barrier trees to expanded barrier trees, which consist of more details of the landscapes. The addition of basin sizes are attached to each node in the barrier tree. The size of the basin will be represented by the size of the node, which is the logarithm of the number of configurations that belong to the basin. The edge length represents distance in fitness values. After the modification of barrier trees, the expanded barrier trees can be defined as follows:

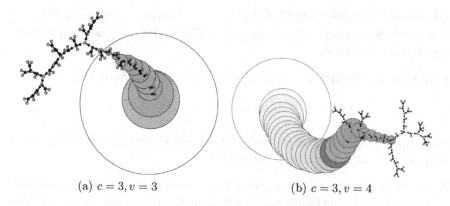

(a) $c = 3, v = 3$ (b) $c = 3, v = 4$

Fig. 2. Example of Landscapes of VRPTW

Definition 5 (Expanded Barrier Trees). *An expanded barrier tree is a barrier tree with labeled nodes and edges. To each saddle point s in the barrier tree is assigned a natural number size(s) = |B(s)|. And to each edge is given a value length(e) = distance(x_1, x_2), which denotes the distance in height between the saddle points or local optimum x_1 and x_2.*

The expanded barrier tree can be graphically represented: Saddle points are represented by means of disks. The radius of the disk shows the size of the basin and the distance of the nodes is expressed as the length of the edge. For better readability of the graphs we also use a coloring of the nodes. Depending on the fitness value we depict the node in a darker or brighter color. The best fitness value (so the global optimum) will be black, while the worst fitness value (the root) will be white. In constrained optimization (e.g. in VRP) a two color scheme is applied: red nodes to describe the infeasible solutions and the green ones indicate that the solution are feasible. Again, the brightness indicates the height of the function value at the node. See Figure 2, for two examples.

In order to generate a barrier tree the *flooding algorithm* (see Algorithm 1) is commonly used [3]. It constructs a barrier tree in discrete landscapes with finite search space $\mathcal{X}$ and linearly ordered search points (e.g. by means of the objective function values). The flooding algorithm builds up the barrier tree in the following way. First, the elements of the search space are sorted in ascending order and send to a queue. Then, the search points are removed one by one from the queue in ascending order and for each point x the following cases are processed:

1. if x_i has no neighbor, it is a local minimum.
2. if x_i has neighbors in only one basin say $B(x_{i_1})$, it also belongs to $B(x_{i_1})$
3. if x_i has neighbors in $n > 1$ basins $B(x_{i_1}, x_{i_2}, ...x_{i_n})$, it is a saddle point. These basins are combined to one $B(x_i)$ and $B(x_{i_1}, x_{i_2}, ...x_{i_n})$ are removed.

After this process, the barrier tree is generated. Note, that if the height function is not injective the flooding algorithm can still be used but the barrier tree may not be uniquely defined. A detailed description of the flooding algorithm is provided with Fig. 1. We indicated with blue color font the parts of the algorithm that have been added to the canonical flooding algorithm, in order to compute expanded barrier trees.

Algorithm 1. Flooding algorithm

1: Let $x^{(1)}, \ldots, x^{(N)}$ denote the elements of the search space sorted in ascending order.

2: Let $size(1) \leftarrow 1, \ldots, size(n) \leftarrow 1$.
3: $i \to 1; \mathcal{B} = \emptyset$
4: **while** $i \leq N$ **do**
5:     **if** $N(x_i) \cap \{x^{(1)}, \ldots, x^{(i-1)}\} = \emptyset$ [i. e., $x^{(i)}$ has no neighbor that has been processed.] **then**
6:         $\{x^{(i)}$ is local minimum$\}$
7:         **Draw** $x^{(i)}$ as a new leaf representing basin $B(x^{(i)})$ located at the height of f in the 2-D diagram
8:         $\mathcal{B} \leftarrow \mathcal{B} \cup \{B(x^{(i)})\}$ {Update set of basins}
9:     **else**
10:        Let $\mathcal{T}(x^{(i)}) = \{B(x^{(i_1)}), \ldots, B(x^{(i_N)})\}$ be the set of basins $B \in \mathcal{B}$ with $N(x^{(i)}) \cap B \neq \emptyset$.
11:        **if** $|\mathcal{T}(x^{(i)})| = 1$ **then**
12:            Let $size(i_1) \leftarrow size(i_1) + 1$.
13:            $B(x^{(i_1)}) \leftarrow B(x^{(i_1)}) \cup \{x^{(i)}\}$
14:        **else**
15:            $\{x^{(i)}$ is a saddle point$\}$
16:            Let $size(i) \leftarrow size(i_1) + \ldots + size(i_N) + 1$.
17:            **Draw** $x^{(i)}$ as a new branching point with edges to the nodes $B(x^{(i_1)}), \ldots, B(x^{(i_N)})$. The length of the edges is given by $length((i, i_1)) = f(x^{(i)}) - f(x^{i_1}), \ldots, length((i, i_N)) = f(x^{(i)}) - f(x^{i_N})$, respectively.
18:            {Update set of basins}
19:            $B(x^{(i)}) = B(x^{(i_1)}) \cup \cdots \cup B(x^{(i_N)}) \cup \{x^{(i)}\}$
20:            Remove $B(x^{(i_1)}), \ldots, B(x^{(i_N)})$ from $\mathcal{B}$
21:            $\mathcal{B} \leftarrow \mathcal{B} \cup \{B(x^{(i)})\}$
22:        **end if**
23:     **end if**
24: **end while**

---

4   Studies on NK Landscapes

NK Landscapes were introduced by [6] as abstract models for fitness functions based on interacting genes. In NK Landscapes N genes are represented by variables from a finite alphabet, typically of size two. The degree of epistasis (gene interaction) is given by the parameter K. With increasing values of K the ruggedness of an adaptive landscape grows. Besides theoretical biology, NK landscapes have been used as test problem generators for Genetic Algorithms (GAs)[7].

The standard NK Landscapes are fitness functions $F : \{0, 1\}^N \to \mathbb{R}^+$ that are generated by an stochastic algorithm. Gene interaction data is stored in a randomly generated *epistasis matrix* E, and used to generate a fitness function [1]. The genotype's fitness F is the average of N fitness components F_i, $i = 1, \ldots, N$. Each gene's fitness component F_i is determined by the allele x_i, and also by K alleles at K epistatic genes distinct from i:

$$F(\mathbf{x}) = \frac{1}{N} \sum_{i=1}^{N} F_i(x_i; x_{i_1}, \ldots, x_{i_k}), \ \mathbf{x} \in \{0,1\}^N \tag{1}$$

where $\{i_1, \ldots, i_k\} \subset \{1, \ldots, N\} - \{i\}$. There are two ways for choosing K other genes: '**adjacent neighborhoods**', where the K genes nearest to position i on the vector are chosen; and '**random neighborhoods**', where these positions are chosen randomly on the vector. The components $F_i : \{0,1\}^K \to [0,1)$, $i = 1, \ldots, N$ are computed based on the *fitness matrix* F. For any i and for each of the 2^{K+1} bit combinations a random number is drawn independently from a uniform distribution over $[0,1)$. The algorithm also creates an epistasis matrix E which for each gene i contains references to its K epistatic genes.

Expanded barrier trees of NK landscapes were computed and displayed for $N = 10$ and varying K, in Figure 5 for adjacent neighborhoods and in Figure 6 for random neighborhoods.

In order to describe our results we will use terminology of axial trees[12,9]. For these trees, that are used as data structures to describe 'natural' trees in geomorphology and biology, branches of different degree are defined. The main branch has degree zero and branches of degree one are side branches of the main branch, and and so on. In the context of EBT we will call 'thick' branches consisting of chains of big nodes to have a higher degree than branches consisting of smaller nodes. However, we will use this definition in a rather informal, descriptive way based on visual appearance. The following observations were found interesting:

1. As expected, the complexity of the NK landscape and thus of the barrier trees grow with K.
2. Landscapes with adjacent neighborhood look slightly more complex than those with random neighborhood.
3. All expanded barrier trees have a relatively large basin for the highest saddle point (white disk). This tendency is getting less, however, as K grows.
4. For each expanded barrier tree, there exists one 'main' branch with small lateral branches which have only one or two local optima. This means a randomized local research cannot be trapped easily, since the basins of these local optima are relatively small and the barriers are not too high.

Figure 3 displays the numbers of leaves, average basin size and average edge length of the NK landscape expanded barrier trees with adjacent neighborhoods and random neighborhoods. The numbers of the leaves linearly increase with the number of neighborhoods(K), which means the number of the local optima will grow with K. The average basin size drops dramatically and the average edge length increases exponentially. We can attain that the algorithms are much more

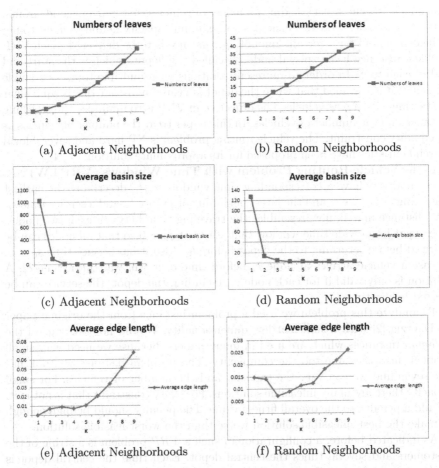

Fig. 3. The number of leaves , average basin size and average edge length of NK landscape with adjacent neighborhoods and random neighborhoods

easily to trap in the local optima and the energy needed to escape from the local optima increase rapidly. Comparing NK landscape with adjacent neighborhoods and those with random neighborhoods, for the latter the number of leaves of NK landscape with random neighborhoods are lower and the average basin size is smaller and edge lengths are shorter. This means the NK landscape with adjacent neighborhoods present more traps to local search heuristics than those with random neighborhoods.

5   Studies on Vehicle Routing Problems

In this section we will apply expanded barrier tree analysis to a constrained real world optimization problem from logistics. The **Vehicle Routing Problem (VRP)** is a generalization of the Traveling Salesperson Problem (TSP).

The goal is to generate a schedule for a fleet of vehicles that delivers goods from a depot to customers. Each vehicle has a maximum capacity Q and each customer v_i has a demand for a certain amount, q_i, that needs to be delivered. Each VRP problem instance has a special node v_0, called the depot, which is the start and end of each tour. In order to represent solutions with multiple tours as a single sequence, it is assumed that when the depot is visited mid-tour this is equivalent to starting with a new vehicle. The objective in VRP is to minimize the traveling distance and the number of vehicles. In this paper treat the number of vehicles as an equality constraint. The vehicle routing problem is NP hard and several local search heuristics have been proposed for its approximate solution; cf. [4].

In the **Vehicle Routing Problem with Time Windows (VRPTW)** customer nodes v_i have a corresponding time window $[e_i, l_i]$ describing the earliest beginning of service e_i and the latest beginning of service l_i, and a service time s_i. The distance matrix now resembles the traveling time between each two nodes. The depot has a wide time window that starts at $e_0 = 0$ so that all nodes can be serviced before returning to the depot before l_0. Also it has service time $s_0 = 0$. In case a vehicle arrives at node v_i before time e_i it will wait until time e_i. A solution is only valid if for each node v_i, including the depot, the service can be started before time l_i.

To analyze this problem we used a problem instance from the original paper by Dantzig [2] with a sufficient low dimensionality. We did not take one of the *Solomon* instances which are used in many papers, because we need very small problem instances due too the complexity. The problem instance we took has size seven and we can vary the capacity and the number of vehicles. For small capacity there are many infeasible solutions. For every constraint that is not met, we add a penalty to the overall fitness value. The penalty should be high enough to make the best infeasible solution worse than the worst feasible solution.

As indicated before, a configuration of such a *VRP* problem is a string of the customers and several times the central depot. Each time the central depot is in the string means that we use the next vehicle. We define a neighbor of such a configuration, a configuration where two cities are swapped (i.e. two configurations with Cayley distance of 1). Swapping seems a elementary operation to modify traveling salesman like problems such as this VRP problem.

The problem, that was taken, has six customers and a central depot. For simplicity we number the depot and the customers where the depot is defined as number one. Each customer has a distance to each other customer and to the central depot. The distance matrix T is given below:

$$
T = \begin{pmatrix}
 & 1 & 2 & 3 & 4 & 5 & 6 & 7 \\
1 & -1 & 10 & 20 & 25 & 12 & 20 & 2 \\
2 & 10 & -1 & 25 & 20 & 20 & 10 & 11 \\
3 & 20 & 25 & -1 & 10 & 25 & 11 & 25 \\
4 & 25 & 20 & 10 & -1 & 30 & 22 & 10 \\
5 & 12 & 20 & 25 & 30 & -1 & 30 & 20 \\
6 & 20 & 10 & 11 & 22 & 30 & -1 & 12 \\
7 & 2 & 11 & 25 & 10 & 20 & 12 & -1
\end{pmatrix}
\tag{2}
$$

Extra constraints will now be added to our problem by using Time Windows [5]. We took the instance above and added the following time window constraints: $e_1 = 0$, $l_1 = 999$, $e_2 = 0$, $l_2 = 20$, $e_3 = 15$, $l_3 = 40$, $e_4 = 10$, $l_4 = 30$, $e_5 = 8$, $l_5 = 20$, $e_6 = 20$, $l_6 = 40$, $e_7 = 0$, $l_7 = 5$. These time windows are chosen keeping in mind that there should still exist some feasible solutions.

Four galleries of expanded barrier trees have been computed. Figure 7 shows the results of expanded barrier trees on VRP. And those in Figure 8 shows results of VRPTW. The capacity c and the number of vehicles v are changed from 1 to 4 and from 1 to 5, respectively. Recall that green colors indicate feasible solutions and red colors infeasible ones. Note, that the edge length is not displayed to make the graphics more readable.

The results look very different to the trees of the NK landscape problem. Some interesting findings are:

- As expected, the number of feasible solutions grows with c and v. More surprisingly, for the VRP a single increment of a parameter can cause a transition from a completely feasible search space to a completely infeasible search space or vice versa. This can be observed in the transition from ($c = 2$, $v = 3$) to ($c = 2$, $v = 4$) in Figure 7 or in the transition from ($c = 2$, $v = 2$) to ($c = 3$, $v = 2$).
- This fast transition is not observed the problem with time windows. Rather, in the problems with time windows feasible and infeasible solutions coexist in the search space. An implication of this can be the isolation of feasible components from each other in the landscape. This can be observed in Figure 8 for ($c = 2$, $v = 3$). These isolated local optima can be potential traps for optimization algorithms which do not accept moves to infeasible solutions.
- Again we can characterize the expanded barrier trees as axial trees with a wide zero order branch. Branches of higher order occur typically only in deeper regions of the landscape. This implies that it is relatively unlikely to get trapped in local optima in early stages of a local search. However, as the differences in depth between the small side branches and the main axis are large, it is difficult to escape from these side branches, once they have been entered, even if worsening of the function value would be accepted (as in simulated annealing).

The trees were generated on a Intel Core i72675QM CPU, with two times 2.2 GHz. Only one core was used. The installed working memory is 4.00 GB and the compiler is DEV C++. According to the results, there is no significant difference between the time consumption of VRP trees and VRPTW trees, so we take the VRPTW as an example here. Figure 4 illustrates the time consumption of constructing the expanded barrier trees of VRPTW. From which we can discover the following things:

- As we can see, the time consumption increases rapidly with the number of vehicles (v). And based on this figure, we can easily predict that it will be time-consuming to generating an expanded barrier tree when the number of vehicle grows larger.

– However, what surprises us most is that the time consumption hardly changes when the capacity of the vehicle (c) changes from 1 to 6. This reveals that (c) is not a sensitive parameter we suspected in the process of generating a expanded barrier tree on VRPTW.

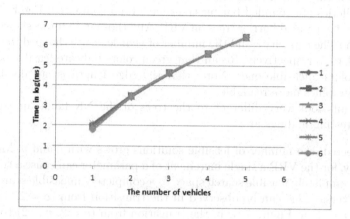

Fig. 4. The time consumption of constructing VRPTW trees

6   Summary and Outlook

Expanded barrier trees were proposed as a visualization tool for landscape analysis of combinatorial landscapes. As compared to standard barrier trees also the size of the saddle points was defined and a color code for distinguishing infeasible from feasible subspaces.

Expanded barrier trees were computed for two problem classes - NK landscapes and VRP/VRPTW) constraints. For both classes it was shown that the problem difficulty depends crucially on the choice of some parameters.

For NK landscapes the main observation was that for a small value of K there are fewer local optima and branches are only of low order and the highest saddle point has a basin of significant size. For larger values of K the trees get more highly branched and the differences in the size of the saddle points tend to align.

In case of VRP transitions between problems with many feasible solutions and a high fraction of infeasible solutions in the search space were rapid. In case of time window constraints also isolated regions of infeasible solutions occured for parameter values in these critical transitions – a problem that needs to be adressed by optimization algorithm design, e.g. by allowing to tunnel infeasible subspaces or relax constraint penalties. The analysis captured some interesting features of these landscapes, such as disconnected feasible regions.

So far, the studies were confined to small instances, and future work needs to clarify whether or not the insights gained from small models generalize to problems larger input sizes. An interesting observation is also that *expanded*

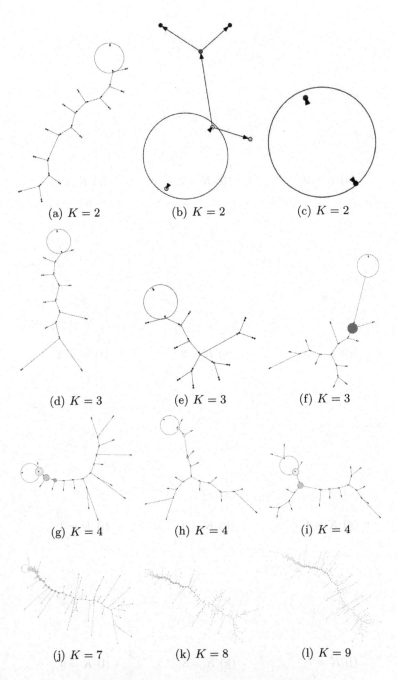

(a) $K = 2$ (b) $K = 2$ (c) $K = 2$

(d) $K = 3$ (e) $K = 3$ (f) $K = 3$

(g) $K = 4$ (h) $K = 4$ (i) $K = 4$

(j) $K = 7$ (k) $K = 8$ (l) $K = 9$

Fig. 5. Landscapes of NK Adjacent Neighbours

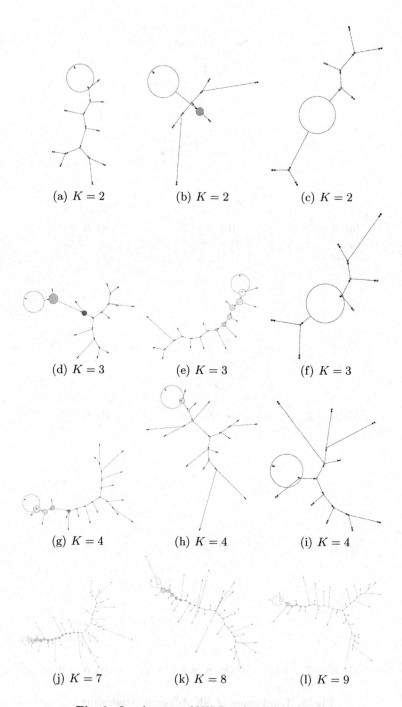

(a) $K = 2$ (b) $K = 2$ (c) $K = 2$

(d) $K = 3$ (e) $K = 3$ (f) $K = 3$

(g) $K = 4$ (h) $K = 4$ (i) $K = 4$

(j) $K = 7$ (k) $K = 8$ (l) $K = 9$

Fig. 6. Landscapes of NK Random Neighbours

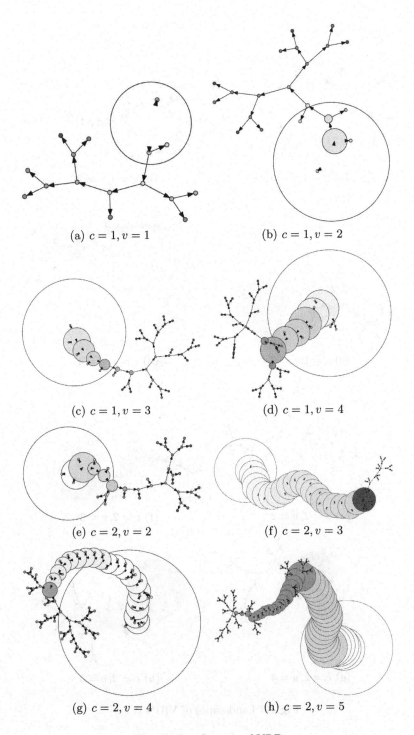

(a) $c = 1, v = 1$

(b) $c = 1, v = 2$

(c) $c = 1, v = 3$

(d) $c = 1, v = 4$

(e) $c = 2, v = 2$

(f) $c = 2, v = 3$

(g) $c = 2, v = 4$

(h) $c = 2, v = 5$

Fig. 7. Landscapes of VRP

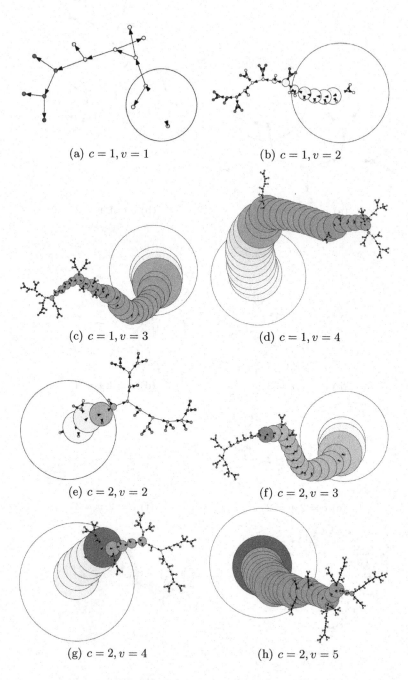

(a) $c = 1, v = 1$

(b) $c = 1, v = 2$

(c) $c = 1, v = 3$

(d) $c = 1, v = 4$

(e) $c = 2, v = 2$

(f) $c = 2, v = 3$

(g) $c = 2, v = 4$

(h) $c = 2, v = 5$

Fig. 8. Landscapes of VRPTW

barrier trees can be characterized in the terminology of axial trees, an aspect that could be elaborated further. Moreover, alternative visualization techniques such as local optima network should be studied for the same landscapes (cf. [13]) in future work.

Acknowledgements. The authors acknowledge support of Agentschap NL, The Netherlands within the DELIVER project for their kind support. The software used for generating the tree visualizations is based on the program GraphViz. The source code will be made available under http://natcomp. liacs.nl.

References

1. Altenberg, L.: Nk-fitness landscapes. In: Bäck, T., et al. (eds.) Handbook of Evolutionary Computation. Oxford University Press (1997)
2. Dantzig, G.B., Ramser, J.H.: The truck dispatching problem. Management Science 6(1), 80–91 (1959)
3. Flamm, C., Hofacker, I.L., Stadler, P.F., Wolfinger, M.T.: Barrier trees of degenerate landscapes. Zeitschrift für Physikalische Chemie 216, 155 (2002)
4. Gambardella, L.M., Taillard, É., Agazzi, G.: MACS-VRPTW: A multiple ant colony system for vehicle routing problems with time windows. In: New Ideas in Optimization, ch. 5, pp. 63–76. McGraw-Hill (1999)
5. Kallehauge, B., Larsen, J., Madsen, O.B.G., Solomon, M.M.: Vehicle routing problem with time windows. Column Generation, 67–98 (2005)
6. Kauffman, S.A.: The origins of order: Self-organization and selection in evolution. Oxford University Press, USA (1993)
7. Li, R., Emmerich, M.T.M., Eggermont, J., Bovenkamp, E.G.P., Bäck, T., Dijkstra, J., Reiber, J.H.C.: Mixed-integer nk landscapes. In: Runarsson, T.P., Beyer, H.-G., Burke, E.K., Merelo-Guervós, J.J., Whitley, L.D., Yao, X. (eds.) PPSN 2006. LNCS, vol. 4193, pp. 42–51. Springer, Heidelberg (2006)
8. Ochoa, G., Tomassini, M., Vérel, S., Darabos, C.: A study of nk landscapes' basins and local optima networks. In: Proceedings of the 10th Annual Conference on Genetic and Evolutionary Computation, pp. 555–562. ACM (2008)
9. Prusinkiewicz, P., Lindenmayer, A.: The algorithmic beauty of plants. Springer (1996)
10. Rammal, R., Toulouse, G., Virasoro, M.A.: Ultrametricity for physicists. Reviews of Modern Physics 58(3), 765 (1986)
11. Stadler, P.: Towards a theory of landscapes. In: Complex Systems and Binary Networks. Lecture Notes in Physics, vol. 461, pp. 78–163. Springer, Heidelberg (1995)
12. Strahler, A.N.: Quantitative analysis of watershed geomorphology. Transactions of the American Geophysical Union 38(6), 913–920 (1957)
13. Tomassini, M., Daolio, F.: A complex-networks view of hard combinatorial search spaces. In: Tantar, E., Tantar, A.-A., Bouvry, P., Del Moral, P., Legrand, P., Coello Coello, C.A., Schütze, O. (eds.) EVOLVE- A Bridge between Probability, Set Oriented Numerics and Evolutionary Computation. SCI, vol. 447, pp. 221–244. Springer, Heidelberg (2013)

Sewer Network Design Optimization Problem Using Ant Colony Optimization Algorithm and Tree Growing Algorithm

Ramtin Moeini[1] and Mohammed Hadi Afshar[2]

[1] School of Civil Engineering, Iran University of Science and Technology,
P.O. 16765-163, Narmak, Tehran, Iran
rmoeini@iust.ac.ir
[2] School of Civil Engineering & Enviro-hydroinformatic Center of Excellence, Iran
University of Science and Technology, P.O. 16765-163, Narmak, Tehran, Iran
mhafshar@iust.ac.ir

Abstract. This paper presents an adaptation of the Ant Colony Optimization Algorithm (ACOA) for the efficient layout and pipe size optimization of sewer network. ACOA has a unique feature namely incremental solution building mechanism which is used here for this problem. Layout and pipe size optimization of sewer network is a highly constrained Mixed-Integer Nonlinear Programming (MINLP) problem presenting a challenge even to the modern heuristic search methods. ACOA equipped with a Tree Growing Algorithm (TGA) is proposed in this paper for the simultaneous layout and pipe size determination of sewer networks. The TGA is used in an incremental manner to construct feasible tree-like layouts out of the base layout, while the ACOA is used to optimally determine the cover depths of the constructed layout. Proposed formulation is used to solve three test examples of different scales and the results are presented and compared with other existing methods. The results indicate the efficiency of the proposed method to optimally solve the problem of layout and size determination of sewer network.

Keywords: Ant Colony Optimization Algorithm, Tree Growing Algorithm, sewer network, layout, pipe sizing.

1 Introduction

In the last 20 years, a new kind of approximate algorithm has emerged which basically tries to combine basic heuristic methods in higher level frameworks aimed at efficiently and effectively exploring a search space. These methods are nowadays commonly called meta-heuristics. This class of algorithms includes, but is not restricted to, Ant Colony Optimization Algorithms (ACOA), Evolutionary Computation (EC) including Genetic Algorithms (GA), Simulated Annealing (SA) and Tabu Search (TS). Nowadays, meta-heuristic algorithms are being used more and more to solve optimization problems other than those which were originally developed.

M. Emmerich et al. (eds.), *EVOLVE - A Bridge between Probability, Set Oriented Numerics,* 91
and Evolutionary Computation IV, Advances in Intelligent Systems and Computing 227,
DOI: 10.1007/978-3-319-01128-8_7, © Springer International Publishing Switzerland 2013

Sewer network is essential structures of any urban area which is designed to protect human and environmental health. Construction of a sewer network, however, is quite an expensive task. Therefore, designing the minimum cost sewer network is one of the key issues in any society which can be achieved by formulating and solving the underlying design problem as an optimization problem. The sewer network design optimization problem includes two sub-problems of optimal layout determination and optimal sizing of sewer network components. These sub-problems are strongly inter-connected and should be handled simultaneously if an optimal solution to the whole problem is required. Simultaneous layout and pipe size optimization of sewer network is a highly constrained Mixed-Integer Nonlinear Programming (MINLP) problem presenting a challenge even to the modern heuristic search methods.

Many researches have been applied for sewer network design optimization problem. Due to the complexity of the problem, however, most of the existing researches are restricted to considering some simplified form of the problem. Some researchers have focused on the component sizing and neglected the influence of the layout on the sizing of the components. In contrast, others have focused on the layout determination and neglected the influence of the component sizing on the final solution. Haestad [1] and Guo et al. [2] reviewed a significant amount of research works in the field of sewerage and drainage systems developed in the last 40 years. Most of the researchers have addressed the optimal component sizing problem while only a few researchers have addressed the problem of layout and in particular the joint problem of layout and size optimization of sewer network. For example, in the field of optimal component sizing, the methods such as enumeration approaches [3,4], Linear Programming [5,6], Nonlinear Programming [7,8], Dynamic Programming [9,10,11,12], Evolutionary Algorithms (EA) such as GA [13,14], ACOA [15,16], Particle Swarm Optimization [17] and Cellular Automata [18,19] and hybrid methods [20,21] have been proposed to solve optimal component sizing of sewer network. Furthermore, some researchers focused on the layout and some other on the more interesting problem of joint layout and size optimization of the sewer network. For example, Dynamic Programming [22, 23, 24], hybrid method [25] and EA such as GA [26], SA [27] and ACOA [28] have been proposed to solve optimal layout and joint layout and size of sewer network.

In this paper, the ACOA equipped with a Tree Growing Algorithm (TGA) is used to efficiently solve the sewer network joint layout and size optimization problem and its performance is compared with other existing methods. The incremental solution building mechanism inherent in the ACOA is used to explicitly enforce a set of most important constraints present in the layout and size optimization of sewer network design. A TGA is used along with the ACOA to construct feasible tree-like layouts out of the base layout defined for the sewer network, while ACOA is used to optimally determine the cover depths of the constructed feasible layout. Proposed formulations are used to solve three hypothetical test examples and the results are presented and compared. The results indicate the efficiency and effectiveness of the proposed methods to simultaneously determine the layout and pipe size of the sewer network.

The layout of the paper is as follows. In section 2, the sewer network design optimization problem is formulated presenting the objective function and the constraints. The ACOA and TGA are explained briefly in sections 3 and 4, respectively. In section 5, a new formulation named ACOA-TGA is proposed for the layout and size optimization of sewer networks. Performance of proposed ACOA-TGA formulation is assessed in section 6 by solving three test examples and comparing the results with other available results. Finally, some concluding remarks are addressed in Section 7.

2 Sewer Network Design Problem

Sewer network is essential structures of modern cities in which it is a collection of sewer pipes, manholes, pumping stations, and other appurtenances. The sewer network is designed to collect wastewater and transmit them by gravity to treatment plants. Effective sewer network should be designed to avoid many problems, such as public health threats and environmental damages. Designing an effective sewer network with minimum cost is a quite complex task which can be achieved by formulating and solving the underlying design problem as an optimization problem.

The design process of sewer network may be divided into two phases: (1) Optimal layout determination, (2) Optimal hydraulic design of the network components. These two phases, however, are not independent and should be handled simultaneously. Cost-effective sewer network design is normally interpreted as finding the solution for this problem that minimizes infrastructural cost, without violating operational requirements. The sewer network design optimization problem is a complex highly constrained MINLP optimization problem consisting of topological and hydraulic constraints. The problem of joint optimal layout and component size determination of a sewer system, here without any pump and drop, can be formulated as:

$$Minimize \ C = \sum_{l=1}^{N} L_l K_p(d_l, E_l) + \sum_{m=1}^{M} K_m(h_m) \qquad (1)$$

Where, C is defined as cost function of sewer network; N is defined as total number of sewer pipes; M is defined as total number of manholes; L_l is defined as the length of pipe l ($l=1,\ldots, N$); K_p is defined as the unit cost of sewer pipe provision and installation defined as a function of its diameter (d_1) and average cover depth (E_l); and K_m is defined as the cost of manhole construction as a function of manhole height (h_m).

This problem is subjected to the topological constraint requiring tree-like layout and hydraulic constraints regarding cover depths, sewer flow velocity, minimum sewer pipe slope, relative flow depth, commercial pipe diameters, partially-full pipe condition, and progressive pipe diameters [28]. The constraints can be defined as:

$$E_{\min} \leq E_l \leq E_{\max} \qquad \forall l = 1,....,N \qquad (2)$$

$$S_l \geq S_{\min} \qquad \forall l = 1,....,N \qquad (3)$$

$$\begin{aligned} V_l \leq V_{\max} \\ V_l^* \geq V_{clean} \end{aligned} \qquad \forall l = 1,....,N \qquad (4)$$

$$\beta_{\min} \leq \left(\frac{y}{d}\right)_l \leq \beta_{\max} \qquad \forall l = 1,....,N \qquad (5)$$

$$Q_l = \frac{1}{n} A_l R_l^{2/3} S_l^{1/2} \qquad \forall l = 1,....,N \qquad (6)$$

$$d_l \in D \qquad \forall l = 1,....,N \qquad (7)$$

$$d_l \geq d_l \qquad \forall l = 1,....,N \qquad (8)$$

$$X_{ij} + X_{ji} = 1 \qquad \forall l = 1,......,N \qquad (9)$$

$$\sum_{j=1}^{N_i} X_{ij} = 1 \qquad \forall i = 1,....,K \qquad (10)$$

$$\sum_{j=1}^{N_i} X_{ji} Q_l - \sum_{j=1}^{N_i} X_{ij} Q_l = 0 \qquad \forall i = 1,....,K \qquad (11)$$

Where, E_{min} is defined as minimum cover depth of sewer pipe; E_{max} is defined as maximum cover depth of sewer pipe; E_l is defined as average cover depth of pipe l; V_l is defined as flow velocity of pipe l at the design flow; V_l^* is defined as the maximum flow velocity of pipe l at the beginning of the operation; V_{clean} is defined as self-cleaning flow velocity of sewer; V_{max} is defined as maximum allowable flow velocity of sewer; S_l is defined as slope of the sewer pipe l; $S_{\min}$ is defined as minimum sewer pipe slope; d_l is defined as diameter of sewer pipe l; y_l is defined as sewer flow depth in pipe l; $\beta_{\max}$ is defined as maximum allowable relative flow depth; $\beta_{\min}$ is defined as minimum allowable relative flow depth; D is defined as discrete set of commercially available sewer pipe diameters; A_l is defined as wetted cross section area of sewer pipe l; R_l is defined as hydraulic radius of the sewer pipe l; n is defined as Manning constant; d_l is defined as set of upstream pipe diameters of pipe l; X_{ij} is defined as a binary variable with a value of 1 for pipe l with a flow direction from node i to node j and zero otherwise; N_i is defined as the number of pipes connected to node i; K is defined as the total number of nodes and Q_l is defined as the discharge of pipe l considered between node i and node j.

A note should be regarded that, as a topological constraint, the sewer network is designed such that the flow of the sewer in the system is due to gravity requiring that the network has a tree-like layout. For this, information regarding the topography of the areas considered for design of the network are collected and used to define the generic locations of the sewer network components leading to a connected undirected graph often referred to as the base layout. The graph vertices represent the fixed position manholes, outlet locations or wastewater treatment plants while graph edges represent the sewer pipes excited between manholes or manholes and outlets or wastewater treatment plants. In the form of a tree-like structure, the sewer pipes configuration should be branched defined by a set of root nodes and a number of non-root nodes and edges located between nodes. The root nodes represent the outlets or the wastewater treatment plants, the non-root nodes represent the manholes, and the edges define the pipes of the sewer network with the condition that only one edge leaves from each node [27]. The sewer network design problem formulated above is clearly a MINLP problem which can not be solved using conventional methods. The complexity of the problem is mostly due to the topological constraints requiring a systematic and efficient method to enforce them if an optimal solution is required.

3   Ant Colony Optimization Algorithm (ACOA)

Based up on ant behavior, Colorni et al. [29] developed ACOA. In the ACOA, a colony of artificial ants cooperates in finding good solutions to difficult discrete optimization problems. Ant system (AS) is the original and simplest ACOA used to solve many difficult combinatorial optimization problems. So far, other version of ACOAs were proposed to overcome the limitations of AS.

Application of the ACOA to an arbitrary combinatorial optimization problem requires that the problem can be projected on a graph [30]. For this, consider a graph $G = (DP, OP, CO)$ in which $DP=\{dp_1, dp_2, ..., dp_I\}$ is the set of decision points at which some decisions are to be made, $OP=\{op_{ij}\}$ is the set of options j ($j=1,2,......,J$) at each decision point i ($i=1,2,......,I$) and finally $CO =\{co_{ij}\}$ is the set of costs associated with options $OP =\{op_{ij}\}$.

The basic steps of the ant algorithms may be defined as follows [30]:

1. The total number of ants m in the colony is chosen and the amount of pheromone trail on all options, $OP =\{op_{ij}\}$, are initialized to some proper value.
2. Ant number k is placed on the ith decision point of the problem. The starting decision point can be chosen randomly or following a specific rule out of n total decision points of the problem.
3. A transition rule is used for the ant k currently placed at decision point i to decide which option to select. The transition rule used here is defined as follows:

$$P_{ij}(k,t) = \frac{[\tau_{ij}(t)]^\alpha [\eta_{ij}]^\beta}{\sum_{j=1}^{J} [\tau_{ij}(t)]^\alpha [\eta_{ij}]^\beta} \tag{12}$$

Where $p_{ij}(k,t)$ is defined as the probability that the ant k selects option op_{ij} from the ith decision point at iteration t; $\tau_{ij}(t)$ is defined as the concentration of pheromone on option op_{ij} at iteration t; η_{ij} is defined as the heuristic value representing the local cost of choosing option j at decision point i, and α and β are two parameters that control the relative weight of pheromone trail and heuristic value, respectively.

4. The cost $f(\varphi)$ of the trial solution is calculated. The generation of a complete trial solution and calculation of the corresponding cost is called a cycle (k).
5. Steps 2 to 4 are repeated for all ants leading to the generation of m trial solutions and the calculation of their corresponding cost, referred to as iteration (t).
6. Each iteration is followed, by the process of pheromone updating. The following form of the pheromone updating is used here as suggested in [30]:

$$\tau_{ij}(t+1) = \rho\tau_{ij}(t) + \Delta\tau_{ij} \tag{13}$$

Where $\tau_{ij}(t+1)$ is defined as the amount of pheromone trail on option j at ith decision point, op_{ij}, at iteration $t+1$; $\tau_{ij}(t)$ is defined as the concentration of pheromone on option op_{ij} at iteration t; ρ is defined as the coefficient representing the pheromone evaporation; $\Delta\tau_{ij}$ is defined as the change in pheromone concentration associated with option op_{ij}.

7. The process defined by steps 2 to 6 is continued until the iteration counter reaches its maximum value defined by the user or some other convergence criterion is met.

Premature convergence to suboptimal solutions is an issue that can be experienced by all ant algorithms. To overcome the problem of premature convergence whilst still allowing for exploitation, the Max-Min Ant system (MMAS) was developed. The basis of MMAS is the provision of dynamically evolving bounds on the pheromone trail intensities such that the pheromone intensity on all paths is always within a specified lower bound, $\tau_{min}(t)$, of a theoretically asymptotic upper limit, $\tau_{max}(t)$ [30]. The upper and lower pheromone bound at iteration t is given, respectively, by:

$$\tau_{max}(t) = \frac{1}{1-\rho}\frac{R}{f\left(s^{gb}(t)\right)} \tag{14}$$

$$\tau_{min}(t) = \frac{\tau_{max}(t)\left(1 - \sqrt[n]{P_{best}}\right)}{(NO_{avg} - 1)\sqrt[n]{P_{best}}} \tag{15}$$

Where $\tau_{max}(t)$ and $\tau_{min}(t)$ are defined as the upper and lower bound of pheromone trail at iteration t, respectively; R is defined as the pheromone reward factor; $f\left(s^{gb}(t)\right)$ is defined as the cost of global best solution at iteration t; P_{best} is defined as a specified probability that the current global-best solution will be created again and NO_{avg} is the average number of options at each decision points.

4   Tree Growing Algorithm (TGA)

Sewer network has a tree-like layout. Here, therefore, the TGA is used in an incremental manner to construct feasible tree-like layouts out of the base layout. The use of TGA for tree network construction has already been attempted for the layout optimization of tree pipe network [31]. The TGA is easily implemented using three vectors **C**, **A** and **CC** in which **C**= the set of nodes contained within the growing tree; **A**= the set of pipes within the growing tree and **CC**= the set of nodes adjacent to the growing tree. Starting from the *Root-Node*.

1. Initialize **C**= [*Root-Node*], **A**= [ ], and **CC**= [nodes in the base graph connected to the root node].
2. Chose a node, i, at random from **CC** and set **C**=**C**+ [i].
3. Of all the pipes connected to node i in the base graph, identify those with both end nodes within the growing tree **C** as potential candidates to be included in the growing tree and choose one of the pipe, a, randomly as the newly connected pipe and set **A**=**A**+ [a].
4. Identify nodes connected to i in the base graph and update **CC** by removing node i and adding any of the nodes connected to i that are feasible candidates to form a tree layout. **CC** now contains all feasible choices for the next node of the tree. The feasibility of this approach is guaranteed by the condition that each node of the network is visited once and only once.
5. Repeat from step (2) by considering i as the current decision point until **CC** is empty.

The process defined above leads to the construction of a spanning tree layout out of base layout defined. Here, this process equipping with ACOA is used to construct tree-like layout out of base layout defined for sewer network.

5   Proposed ACOA-TGA for Layout and Size Optimization of Sewer Network

Here, a new formulation named ACOA-TGA is proposed for layout and size optimization of sewer network. ACOAs enjoy a unique feature, namely incremental solution building capability, which is not observed in other evolutionary search methods. This capability is reflected in the process of solution building by ants in which each ant is required to choose an option out of the available options present at a decision point of the problem in turn. In ACOA-TGA, the incremental solution building capability of ACOA equipped with TGA is used to construct tree-like feasible layouts out of the base layout during the solution construction. In this approach, the TGA is responsible to keep the ants' options limited to those forming tree layouts while the ACOA determines the nodal cover depths.

The use of TGA for tree network construction has already been attempted by Moeini and Afshar [28] for joint layout and size optimization of sewer networks. However, Moeini and Afshar [28] used pipe diameters as decision variables of the problem in which an assumption of sewer flow at maximum allowable relative depth is made allowing for the calculation of the optimal pipe slopes, while the formulation proposed here does not consider such limiting assumptions. This will be shown to lead to improved convergence characteristics and better results in the test example section.

Solution of the sewer network design as an optimization problem by the ACOA requires that the problem is defined as a graph. Construction of the graph for the application of the ACOA requires that decision points, options available at each decision point, and the costs associated with each of these options are defined. Here nodal elevations of the base layout are considered as the decision variables of the problem leading to an easy definition of the problem graph. The layout side of the problem is resolved by augmenting the ACOA with a TGA with a role to guide the ants to create the required tree structure of the network when deciding on the pipe sizing aspect of the problem. Here, the nodes of the base layout are taken as decision points of the graph and the options available at each decision point are defined by the aggregation of the those cover depths of the nodes which can contribute to a tree-like layout.

Starting from the root node, the ACOA-TGA is easily implemented as follow.

1. Initialize $\mathbf{C}=$ [*Root-Node*], $\mathbf{A}=$ [ ], and $\mathbf{CC}=$ [nodes in the base graph connected to root node].
2. Form the tabu list, $\mathbf{T}$, as the aggregations of the cover depths of the components of $\mathbf{CC}$. An assumption that the nodes contained in $\mathbf{CC}$ are the upstream nodes of pipes within growing tree will lead to easier application of proposed formulation.
3. Let the ant chose a discrete cover depth, c, at random from $\mathbf{T}$.
4. Determine to which node, i, of $\mathbf{CC}$ the chosen discrete cover depth, c, belong and set $\mathbf{C}=\mathbf{C}+$ [i].
5. Of all the pipes connected to node i in the base graph, identify those with both end nodes within the growing tree $\mathbf{C}$ as potential candidates to be included in the growing tree and choose the pipe with the steepest slope, a, as the newly connected pipe and set $\mathbf{A}=\mathbf{A}+$ [a].
6. Identify nodes connected to i in the base graph and update $\mathbf{CC}$ by removing node i and adding any of the nodes connected to i that are feasible candidates to form a tree layout.
7. Repeat from (2) by considering i as the current decision point until $\mathbf{CC}$ is empty.

Fig. 1(b) shows the graphical representation of the proposed ACOA-TGA for the problem of Fig. 1(a), in which the circles represent the network nodes, the bold circles represent the decision points, the numbers in the circles represent the node numbers, C_{jn}^{in} represents the set of discrete cover depths of node jn such that the pipe with the downstream and upstream nodes in and jn , respectively, C_{jn}^{in-kn} represents the set of discrete cover depths of node jn present in both C_{jn}^{in} and C_{jn}^{kn}, and the brackets represent the aggregation of all these feasible cover depths as options available to the ant at the current decision point to form

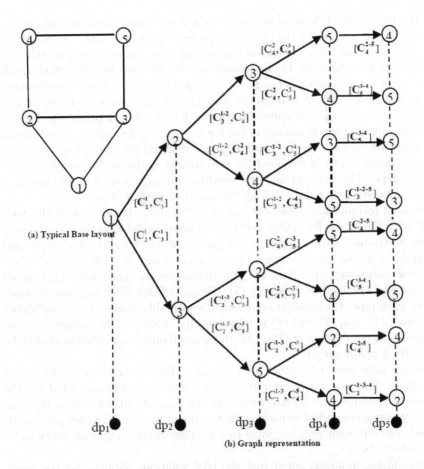

(a) Typical Base layout

(b) Graph representation

Fig. 1. Graph representation of a typical problem for ACOA-TGA

a tree layout, and finally the bold C_{jn}^{in} indicates a decision by the ant leading to the inclusion of node jn in the tree under construction. This would automatically define the node jn as the next decision point to move to.

The process defined above leads to the construction of a spanning tree network with known nodal covers and hence nodal elevations. The network so constructed, however, will not contain some of the pipes present in the looped base layout. To complete the construction of a tree-like network containing all pipes of the base layout, the connection of the pipes which are absent in the constructed layout are cut at either end and added to the constructed spanning tree to form the final network layout. A dummy node is used to define the cut end of the pipe. A note has to be made here regarding the way the pipes are cut and the nodal covers of the resulting dummy nodes are computed. Three different methods are used here for comparison purposes.

In the first method denoted by suffix 1, the cover depth of the dummy nodes are taken equal to those of the adjacency nodes requiring that the pipes are cut at the upstream end, the end with higher nodal elevation, to restore the tree layout of the final network. In the second method, denoted by suffix 2, the rank of each node of the network defined as the minimum number of pipes between the root node and the current node in the constructed tree layout is calculated and the pipes are cut at the end node of higher rank. The cover depth of the dummy node introduced at the adjacency of the cut is then taken equal to the minimum allowable cover depth. In the third method, denoted by suffix 3, a minimum allowable cover depth is assumed for the dummy nodes to be introduced for each cut pipe. The cut end and the resulting dummy node location are then decided upon such that the resulting cut pipe has steeper slope.

Proposed formulations generate a directed tree-like layout out of the base layout with known nodal covers and hence known nodal elevations as defined. Having determined the average cover depths, nodal elevations, pipe slopes and the layout of a trial sewer network, the remaining task is to determine the diameter of each sewer pipe. Here the pipe diameters are calculated such that all the constraints defined by Eqs. (4), (5), (7) and (8) are fully satisfied, if possible. For each pipe, the smallest commercially available diameter fully satisfying constraints (4), (5), (7) and (8) is taken as pipe diameter. To enforce the progressive diameter constraint, Eq. (8), the pipe diameter calculation should be started from upstream pipes.

Another note should be made here regarding the construction of the layout when more than one wastewater treatment plant is to be considered for the network. A dummy node is introduced into the base layout network as the base graph root node connected to each of the treatment plant via dummy paths with no discharge but with very low local cost leading to very high possibility to be selected by ants.

It should be, however, noted that the trial solutions obtained by the application of the proposed formulations may violate some of the constraints of the problem. To encourage the ants to make decision leading to feasible solutions, a higher cost is associated to the solutions that violate the problem constraints defined by Eqs. (2) to (11). This may be done via the use of a penalty method as:

$$F_p = F + \alpha_p \times \sum_{g=1}^{G} CSV_g \qquad (16)$$

Where, F_P is defined as penalized objective function; F is defined as original objective function defined by Eq.1; CSV_g is defined as a measure of the violation of constraint g; G is defined as total number of constraints and α_p is defined as the penalty constant assumed large enough so that any infeasible solution has a total cost greater than that of any feasible solution.

6   Results of Test Examples and Discussions

Performance of the proposed algorithms is now tested against three test examples of sewer network design citied in Moeini and Afshar [28]. The geometry of the area along with the ground elevation of benchmark point of three test examples are shown in Moeini and Afshar [28]. The set of diameters ranging from 100 mm up to 1500 mm with an interval of 50 mm from 100 mm to 1000 mm and an interval of 100 mm from 1000 mm to 1500 mm is used as the set of commercially available pipe diameters for all the pipes. Other parameters used to solve this problem are listed in Table 1. Here the following explicit relation is used for the pipe installation and manhole costs:

$$K_p = 10.93e^{3.43d_l} + 0.012E_l^{1.53} + 0.437E_l^{1.47}d_l$$
$$K_m = 41.46h_m$$

(17)

In this section, the results obtained using proposed formulations for the test examples are presented and compared. All the results presented hereafter are based on a uniform discretization of the allowable range of cover depths into 30 intervals for all proposed formulations and test examples. A MMAS is used as the ACOA with the parameter values of $\alpha = 1$, $\beta = 0.2$, $P_{best} = 0.2$ and $\rho = 0.95$ for all test examples which is obtained by some preliminary runs for the best performance of the algorithm. The results are obtained using colony sizes of 50, 100, 200 and the maximum of 500, 1000 and 2000 iterations for test examples I, II and III, respectively.

Table 1. Design parameters for test examples

Parameters	Test example	Parameter values
Area size	I	200 (m) * 200 (m)
	II	400 (m) * 400 (m)
	III	800 (m) * 800 (m)
All pipe length	I, II, III	100 (m)
Uniform population distribution at end of the design period	I, II, III	2500
Uniform population distribution at beginning of the design period	I, II, III	4000
Return factor	I, II, III	0.8
water consumption per person per day	I, II, III	$250 \; l/(d*c)$
Maximum Velocity	I, II, III	6 (m/s)
Self-cleaning Velocity	I, II, III	0.75 (m/s)
Minimum slope	I, II, III	0.0005
Maximum allowable relative flow	I, II, III	0.83
Minimum allowable relative flow	I, II, III	0.1
Maximum cover depth	I, II, III	10 (m)
Minimum cover depth	I, II, III	2.5 (m)
Manning coefficients	I, II, III	0.015

Table 2. Maximum, minimum and average solution costs obtained over 10 runs

Formulation	test example	Cost value			Scaled Standard deviation	No. of runs with final feasible solution
		Minimum	Maximum	Average		
ACOA-TGA1	I	23467.8	23467.8	23467.8	0	10
ACOA-TGA2		23467.8	23467.8	23467.8	0	10
ACOA-TGA3		23467.8	23467.8	23467.8	0	10
ACOA-TGA1	II	86531.4	88607.0	87667.5	0.0081	10
ACOA-TGA2		86425.0	87958.5	87294.5	0.0061	10
ACOA-TGA3		86411.0	87581.3	87138.8	0.0041	10
ACOA-TGA1	III	--------	--------	--------	--------	0
ACOA-TGA2		395246	425028	410020	0.0201	10
ACOA-TGA3		386995	413864	404345	0.0181	10

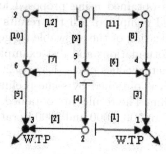

Fig. 2. Optimal tree-like layout of test example I obtained using ACOA-TGA3

Table 2 shows the results of 10 runs carried out using randomly generated initial guesses for the test examples along with the scaled standard deviation and the number of feasible final solutions. It is clearly seen from Table 2 that all proposed formulations have been able to produce near-optimal solutions for all three test examples. However, the quality of the final solutions in all measures such as the minimum, maximum, average cost and the scaled standard deviation are improved when using the proposed ACOA-TGA3. In fact, ACOA-TGA3 has been able to outperform other formulations due to the fact that solutions constructed by the ACOA-TGA3 will be less prone to violate the problem constraint in particular for the third test example.

These problems were also solved by Moeini and Afshar [28] using three different formulations named ACOA1, ACOA2 and ACOA-TGA in which pipe diameters were taken as decision variables. Comparison of the results obtained by Moeini and Afshar [28] with those produced here indicates that proposed ACOA-TGA formulations have superior performance for the test examples considered. Moeini and Afshar [28] reported optimal solutions of 24514.9, 89568.3 and 397563 unit cost for test examples I, II and III, respectively, obtained by the ACOA-TGA. These can be compared with the costs of 23467.8, 86411 and 386995 obtained using proposed ACOA-TGA3 for the same test examples

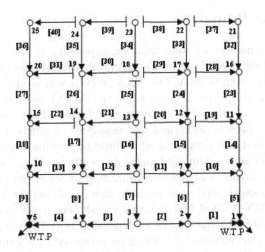

Fig. 3. Optimal tree-like layout of test example II obtained using ACOA-TGA3

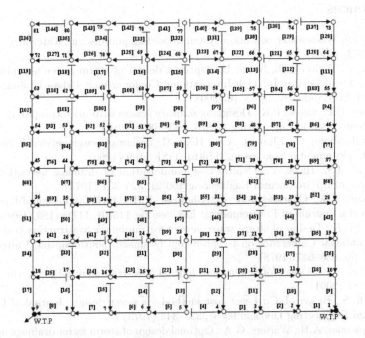

Fig. 4. Optimal tree-like layout of test example III obtained using ACOA-TGA3

indicating superiority of the proposed method. In addition, all the runs carried out for the test examples produced final feasible solutions using ACOA-TGA3 formulations presented here while ACOA-TGA of Moeini and Afshar [28] produced 10, 10 and 9 final feasible solutions for test example I, II and III, respectively. Furthermore, Figs. 2, 3 and 4 show the optimal tree-like layout for the test examples I, II and III, respectively, obtained using ACOA-TGA3.

7 Concluding Remarks

In this paper, incremental solution building capability of the ACOA was exploited for the efficient layout and pipe size optimization of sewer network. In the proposed method referred to as ACOA-TGA, a TGA was used to guide the ants to construct only feasible tree layouts while the ACOA was used to optimally determine the cover depths of the constructed layout. Proposed formulations were used to solve three hypothetical test examples of different scales and the results were presented and compared with those of existing results. The results indicated the ability of the proposed methods to efficiently and optimally solve the problem of the layout and size determination of sewer networks. While all proposed formulations showed good performance in solving the problems under consideration the ACOA-TGA3 formulation was shown to produce better results and to be less sensitive to the randomly generated initial guess represented by the scaled standard deviation of the solutions produced in ten different runs.

References

1. Haestad: Wastewater collection system modeling and design. Waterbury, Haestad methods (2004)
2. Guo, Y., Walters, G., Savic, D.: Optimal design of storm sewer networks: Past, Present and Future. In: 11th International Conference on Urban Drainage, Edinburgh, Scotland, UK, pp. 1–10 (2008)
3. Desher, D.P., Davis, P.K.: Designing sanitary sewers with microcomputers. Journal of Environmental Engineering 112(6), 993–1007 (1986)
4. Charalambous, C., Elimam, A.A.: Heuristic design of sewer networks. Journal of the Environmental Engineering 116(6), 1181–1199 (1990)
5. Dajani, J.S., Hasit, Y.: Capital cost minimization of drainage networks. ASCE Journal of Environmental Engineering 100(2), 325–337 (1974)
6. Elimam, A.A., Charalambous, C., Ghobrial, F.H.: Optimum design of large sewer networks. Journal of Environmental Engineering 115(6), 1171–1189 (1989)
7. Price, R.K.: Design of storm water sewers for minimum construction cost. In: 1st International Conference on Urban Strom Drainage, Southampton, United Kingdom, pp. 636–647 (1978)
8. Swamee, P.K.: Design of Sewer Line. Journal of Environmental Engineering 127(9), 776–781 (2001)
9. Walsh, S., Brown, L.C.: Least cost method for sewer design. Journal of Environmental Engineering Division 99(3), 333–345 (1973)
10. Templeman, A.B., Walters, G.A.: Optimal design of storm water drainage networks for roads. In: Inst. of Civil Engineers, London, pp. 573–587 (1979)
11. Yen, B.C., Cheng, S.T., Jun, B.H., Voohees, M.L., Wenzel, H.G.: Illinois least cost sewer system design model. User's guide, Department of Civil Engineering, University of Texas at Austin (1984)
12. Kulkarni, V.S., Khanna, P.: Pumped wastewater collection systems optimization. ASCE Journal of Environmental Engineering 111(5), 589–601 (1985)
13. Liang, L.Y., Thompson, R.G., Young, D.M.: Optimising the design of sewer networks using genetic algorithms and tabu search. Engineering, Construction and Architectural Management 11(2), 101–112 (2004)

14. Afshar, M.H., Afshar, A., Marino, M.A., Darbandi, A.A.S.: Hydrograph-based storm sewer design optimization by genetic algorithm. Canadian Journal Civil Engineering 33(3), 310–325 (2006)
15. Afshar, M.H.: Partially constrained ant colony optimization algorithm for the solution of constrained optimization problems: Application to storm water network design. Advances in Water Resources 30(4), 954–965 (2007)
16. Afshar, M.H.: A parameter free Continuous Ant Colony Optimization Algorithm for the optimal design of storm sewer networks: Constrained and unconstrained approach. Advances in Engineering Software 41, 188–195 (2010)
17. Izquierdo, J., Montalvo, I., Perez, R., Fuertes, V.S.: Design optimization of wastewater collection networks by PSO. Computers and Mathematics with Applications 56(3), 777–784 (2008)
18. Guo, Y.: Sewer Network Optimal Design Based on Cellular Automata Principles. In: Proceeding of 2005 XXXI IAHR Congress, Seoul, Korea, pp. 6582–6593 (2005)
19. Guo, Y., Walters, G.A., Khu, S.T., Keedwell, E.C.: A novel cellular automata based approach to storm sewer design. Engineering Optimization 39(3), 345–364 (2007)
20. Guo, Y., Keedwell, E.C., Walters, G.A., Khu, S.T.: Hybridizing cellular automata principles and NSGA for multi-objective design of urban water networks. Computer Science and Mathematics, 546–559 (2007)
21. Pan, T.C., Kao, J.J.: GA-QP Model to Optimize Sewer System Design. ASCE Journal of Environmental Engineering 135(1), 17–24 (2009)
22. Argaman, Y., Shamir, U., Spivak, E.: Design of Optimal Sewerage Systems. Journal of the Environmental Engineering Division 99(5), 703–716 (1973)
23. Mays, L.W., Wenzel, H.G.: Optimal Design of Multilevel Branching Sewer Systems. Water Resources Research 12(5), 913–917 (1976)
24. Walters, G.A.: The design of the optimal layout for a sewer network. Engineering Optimization 9, 37–50 (1985)
25. Li, G., Matthew, R.G.S.: New approaches for optimization of urban drainage system. Journal of Environmental Engineering 116(5), 927–944 (1990)
26. Weng, H.T., Liaw, S.L., Huang, W.C.: Establishing an Optimization Model for Sewer System Layout with Applied Genetic Algorithm. Environmental Informatics Archives 2, 781–790 (2004)
27. Diogo, A.F., Graveto, V.M.: Optimal Layout of Sewer Systems: A Deterministic versus a Stochastic Model. ASCE Journal of Hydraulic Engineering 132(9), 927–943 (2006)
28. Moeini, R., Afshar, M.H.: Layout and size optimization of Sanitary Sewer network using intelligent ants. Advances in Engineering Software 51, 49–62 (2012)
29. Colorni, A., Dorigo, M., Maniezzo, V.: The Ant System: Ant Autocatalytic Optimization Process. Technical Report, TR 91-016, Politiecnico Di Milano (1991)
30. Afshar, M.H., Moeini, R.: partially and Fully Constrained Ant Algorithms for the Optimal Solution of Large Scale Reservoir Operation problems. Journal Water Resource Management 22(1), 1835–1857 (2008)
31. Afshar, M.H., Marino, M.A.: Application of an ant algorithm for layout optimization of tree networks. Engineering Optimization 38(3), 353–369 (2006)

An Ant-Based Optimization Approach
for Inventory Routing

Vasileios A. Tatsis[1], Konstantinos E. Parsopoulos[1],
Konstantina Skouri[2], and Ioannis Konstantaras[3]

[1] Department of Computer Science, University of Ioannina,
GR-45110 Ioannina, Greece
{vtatsis,kostasp}@cs.uoi.gr
[2] Department of Mathematics, University of Ioannina, GR-45110 Ioannina, Greece
kskouri@uoi.gr
[3] Department of Business Administration, University of Macedonia,
GR-54006 Thessaloniki, Greece
ikonst@uom.gr

Abstract. The inventory routing problem (IRP) is a major concern in
operation management of a supply chain because it integrates transportation activities with inventory management. Such problems are
usually tackled by independently solving the underlying inventory and
vehicle routing sub-problems. The present study introduces an ant-based
solution framework by modeling the IRP problem as a vehicle routing
task. In this context, a mixed-integer mathematical model for the IRP
is developed, where a fleet of capacitated homogeneous vehicles transport different products from multiple suppliers to a retailer to meet the
demand for each period over a finite planning horizon. In our model,
shortages are allowed while unsatisfied demand is backlogged and can
be met in future periods. The mathematical model is used to find the
best compromise among transportation, holding, and backlogging costs.
The corresponding vehicle routing problem is solved using an ant-based
optimization algorithm. Preliminary results on randomly generated test
problems are reported and assessed with respect to the optimal solutions
found by established linear solvers such as CPLEX.

1 Introduction

The main target of Supply Chain Management (SCM) is to align the various
stages of the supply chain. Integrating the decisions in planning the different
activities, has shown to produce improved global performance. An example of
integrating and coordinating decisions can be found in Vendor Managed Inventory (VMI), where customers make their inventory data available to their
suppliers (distributors), who then take the responsibility of deciding when to re-
plenish which customers. Thus, the supplier has to choose how often, when, and
in what quantities the different customers are replenished. This integrated inventory and distribution management offers more freedom for designing efficient
vehicle routes, while optimizing inventory across the supply chain.

The underlying optimization problem that has to be addressed by the supplier,
namely the simultaneous decision on the replenishment quantities and the vehicle

M. Emmerich et al. (eds.), *EVOLVE - A Bridge between Probability, Set Oriented Numerics,* 107
and Evolutionary Computation IV, Advances in Intelligent Systems and Computing 227,
DOI: 10.1007/978-3-319-01128-8_8, © Springer International Publishing Switzerland 2013

routes to visit all customers, is known as the *Inventory Routing Problem* (IRP). IRP is one of the most interesting and challenging issues in the field of supply chain management and typically considers a distribution firm that operates a central vehicle station, supplying a number of geographically scattered customers by a fleet of homogeneous vehicles, over a period of time [6].

The IRP literature is extensive and includes several variants of the problem, mainly depending on the nature of the demand at customers (deterministic, stochastic etc.) as well as on the length of the planning horizon (finite, infinite etc.) We can indicatively mention several relevant works considering single-period IRPs with stochastic demand [13] or deterministic demand [9]; multi-period finite horizon IRPs with constant or dynamic demand [1, 2, 7, 17]; as well as infinite horizon IRPs with deterministic or stochastic demand [8, 14]. For recent detailed reviews of the IRP field, the reader is referred to [3, 16].

In the present work, we propose a constructive meta-heuristic approach for solving a two-echelon supply chain problem, where one retailer is served multi-products by different suppliers, using a fleet of capacitated homogeneous vehicles. This forms a multi-product, multi-period, finite horizon IRP where the retailer's demands are assumed to be known for all periods. The IRP problem is first modeled as an equivalent vehicle routing (VR) task. Then, we use an ant-based algorithm that combines elements from some of the most successful Ant Colony Optimization (ACO) variants, namely (Elitist) Ant System ((E)AS) [5, 10, 12] and Max-Min Ant System ($\mathcal{MMAS}$) [19], to solve the corresponding VR problem. Preliminary experimental results on a set of randomly generated test problems are reported. The results are compared with ones obtained using the CPLEX software, offering preliminary evidence regarding the competitiveness and weaknesses of the proposed approach.

The rest of the paper is organized as follows: Section 2 contains the mathematical formulation of the problem, while Section 3 describes the algorithm, in detail. Preliminary experimental results are reported in Section 4. Finally, the paper concludes in Section 5.

2   Problem Formulation

We consider a many-to-one, part-supply network that is similar to the one proposed in [17]. The network consists of one retailer, N suppliers, and a vehicle station. Each supplier provides a distinct product to the retailer. Henceforth, we will denote each supplier (and his product) with the corresponding index $i = 1, 2, \ldots, N$, while the index 0 will denote the station, and $N + 1$ will denote the retailer. A fleet of homogeneous capacitated vehicles housed at the vehicle station, transports products from the suppliers to meet the demand specified by the retailer over a finite horizon, while backlogging is allowed. The vehicles return to the vehicle station at the end of each trip. If the demand for more than one period is collected, the inventory is carried forward subject to product-specific holding cost. The unsatisfied demand for a specific product leads the related product-specific shortage cost.

The main objective of the problem is the minimization of the total transportation, inventory, and shortages costs over the planning horizon. Putting it formally, let the sets of indices:

$$\text{Suppliers:} \quad I_s = \{1, 2, \ldots, N\},$$
$$\text{Vehicles:} \quad I_v = \{1, 2, \ldots, M\}, \tag{1}$$
$$\text{Time periods:} \quad I_p = \{1, 2, \ldots, T\},$$

and $I'_s = I_s \cup \{N + 1\}$. We use the following notation, which is similar to [17], although our model assumes a finite fleet size instead of the unlimited number of vehicles in [17]:

C: capacity of each vehicle.

F: fixed vehicle cost per trip (same for all periods).

V: travel cost per unit distance.

M: size of the vehicle fleet.

d_{it}: retailer's demand for product from supplier i in period t.

c_{ij}: travel distance between supplier i and j where $c_{ij} = c_{ji}$ and the triangle inequality, $c_{ik} + c_{kj} \geqslant c_{ij}$, holds for all i, j, k with $i \neq j$, $k \neq i$, and $k \neq j$.

h_i: holding cost at the retailer for product i per unit product per unit time.

s_i: backlogging cost at the retailer for product i per unit product per unit time.

I_{i0}: inventory level of product i at the retailer, at the beginning of period 1.

a_{it}: total amount to be picked up at supplier i in period t.

I_{it}: inventory level of product from supplier i at the retailer, at end of period t.

q_{ijt}: quantity transported through the directed arc (i, j) in period t.

x_{ijt}: number of times that the directed arc (i, j) is visited by vehicles in period t.

Then, the mathematical formulation of the problem is defined as follows:

$$\text{minimize} \quad Z_1 + Z_2 + Z_3 + Z_4 + Z_5, \tag{2}$$

where:

$$Z_1 = \sum_{i=1}^{N} h_i \sum_{t=1}^{T} I_{it}^+,$$

$$Z_2 = \sum_{i=1}^{N} s_i \sum_{t=1}^{T} (-I_{it})^+,$$

$$Z_3 = V \left(\sum_{\substack{j=1 \\ j \neq i}}^{N} \sum_{i=0}^{N} c_{ij} \left(\sum_{t=1}^{T} x_{ijt} \right) \right), \tag{3}$$

$$Z_4 = V \left(\sum_{i=1}^{N} c_{i,N+1} \left(\sum_{t=1}^{T} x_{i,N+1,t} \right) \right),$$

$$Z_5 = (F + c_{N+1,0}) \sum_{i=1}^{N} \sum_{t=1}^{T} x_{0it},$$

where $x^+ = \max\{x, 0\}$, subject to the constraints:

$$\text{(C1):} \quad I_{it} = I_{it-1} + a_{it} - d_{it}, \quad i \in I_s, t \in I_p, \tag{4}$$

$$\text{(C2):} \quad \sum_{\substack{i=0 \\ i \neq j}}^{N} q_{ijt} + a_{jt} = \sum_{\substack{i=1 \\ i \neq j}}^{N+1} q_{jit}, \quad j \in I_s, t \in I_p, \tag{5}$$

$$\text{(C3):} \quad \sum_{i=1}^{N} q_{i,N+1,t} = \sum_{i=1}^{N} a_{it}, \quad t \in I_p, \tag{6}$$

$$\text{(C4):} \quad \sum_{\substack{i=0 \\ i \neq j}}^{N} x_{ijt} = \sum_{\substack{i=1 \\ i \neq j}}^{N+1} x_{jit}, \quad j \in I_s, t \in I_p, \tag{7}$$

$$\text{(C5):} \quad \sum_{j=1}^{N} x_{0jt} = \sum_{j=1}^{N} x_{j,N+1,t}, \quad t \in I_p, \tag{8}$$

$$\text{(C6):} \quad q_{ijt} \leqslant C\, x_{ijt}, \quad i \in I_s, j \in I'_s, i \neq j, t \in I_p, \tag{9}$$

$$\text{(C7):} \quad \sum_{i=1}^{N} x_{0it} \leqslant M, \quad t \in I_p, \tag{10}$$

$$\text{(C8):} \quad \sum_{t=1}^{T} a_{it} = \sum_{t=1}^{T} d_{it}, \quad i \in I_s, \tag{11}$$

$$\text{(C9):} \quad C\, x_{ijt} - q_{ijt} \leqslant C - 1, \quad i \in I_s, j \in I'_s, t \in I_p, \tag{12}$$

$$\text{(C10):} \quad a_{jt} \leqslant \sum_{\substack{i=1 \\ i \neq j}}^{N} C\, x_{ijt}, \quad j \in I_s, t \in I_p, \tag{13}$$

$$\text{(C11):} \quad x_{ijt} \leqslant \sum_{\substack{k=0 \\ k \neq i,j}}^{N} x_{kit} \quad i \in I_s, j \in I'_s, t \in I_p \tag{14}$$

$$\text{(C12):} \quad I_{it} \in R, \quad i \in I_s, t \in I_p, \tag{15}$$

$$\text{(C13):} \quad a_{it} \geqslant 0, \quad i \in I_s, t \in I_p, \tag{16}$$

$$\text{(C14):} \quad x_{ijt} \in \{0,1\}, \quad i,j \in I_s, t \in I_p, \tag{17}$$

$$\text{(C15):} \quad x_{0jt} \in \mathbb{N}, \quad j \in I_s, t \in I_p, \tag{18}$$

$$\text{(C16):} \quad x_{i,N+1,t} \in \mathbb{N}, \quad i \in I_s, t \in I_p, \tag{19}$$

$$\text{(C17):} \quad x_{0,N+1,t} = 0, \quad t \in I_p, \tag{20}$$

$$\text{(C18):} \quad x_{i0t} = 0, \quad i \in I_s, t \in I_p, \tag{21}$$

$$\text{(C19):} \quad x_{N+1,j,t} = 0, \quad j \in I_s, t \in I_p, \tag{22}$$

$$\text{(C20):} \quad q_{ijt} \geqslant 0, \quad i \in I_s, j \in I'_s, t \in I_p, \tag{23}$$

$$\text{(C21):} \quad q_{0jt} = 0, \quad j \in I_s, t \in I_p. \tag{24}$$

The objective function defined by Eqs. (2) and (3) comprises both inventory costs (holding and backlogging) and transportation costs (variable travel costs and vehicle fixed cost). We note that the fixed transportation cost consists of the fixed cost incurred per trip and the constant cost of vehicles returning to the station from the retailer.

Constraint (C1) is the inventory balance equation for each product, while (C2) is the product flow conservation equations, assuring the flow balance at each supplier and eliminating all subtours. Constraint (C3) assures the accumulative picked up quantities at the retailer and (C4) and (C5) ensure that the number of vehicles leaving a supplier, the retailer or the station is equal to the number of its arrival vehicles. We note that constraint (C5) is introduced because each vehicle has to visit the retailer before returning to the station. Constraint (C6) guarantees that the vehicle capacity is respected and gives the logical relationship between q_{ijt} and x_{ijt}, which allows for split pick ups.

Constraint (C7) is introduced due to the limited fleet size. Constraint (C8) ensures that the cumulative demand for every product will be satisfied, while (C9) is imposed to ensure that either $q_{ijt} = 0$ with $x_{ijt} = 0$ or $q_{ijt} \geqslant 1$ with $x_{ijt} \geqslant 1$. Moreover, (C10) ensures that the pick up quantities are limited by the number of vehicles and their capacities. Constraint (C11) ensures that if there is a vehicle to travel from one supplier to another, then this vehicle should previously arrive to the first supplier from another one (or the station). This is necessary to avoid fake closed loops of vehicles that may appear in the VR formulation of the problem. Finally, (C12) implies that the demand can be backlogged. The rest are non-negativity constraints imposed on the variables. We note that (C17), (C18), and (C19) ensure the absence of direct links from the station to the retailer, from a supplier to the station, and from the retailer to a supplier, respectively.

3   Proposed Approach

3.1   Ant Colony Optimization

Ant Colony Optimization (ACO) constitutes a general metaheuristic methodology for solving combinatorial optimization problems [12]. It draws inspiration from the collective behavior of termites during social activities such as foraging. The emergent behavior of such primitive entities is based on a mechanism called stigmergy, which allows them to coordinate their actions based on stimulation from the environment. The stimulation is their response to pheromone's chemical traces left in the environment by their mates and them.

ACO can be elegantly introduced in the framework of the Traveling Salesman Problem (TSP) as a group (swarm) of agents that individually construct a route from a start city to an end city visiting all other cities just once. At each stage of the route construction, the agent makes a decision of its next move based on a probabilistic scheme that gives higher probability to the alternatives that have more frequently been visited by the rest of the swarm and, thus, they possess higher pheromone levels. The general procedure flow of ACO approaches can be summarized in the following actions:

```
// Procedure ACO
WHILE (termination condition is false)
    Construct_Solutions()
    Further_Actions()
    Update_Pheromones()
END WHILE
```

Due to space limitation the reader is referred to devoted texts such as [5, 12] for a thorough introduction of the general framework of ACO and its most popular variants.

3.2 Solution Representation

We will now try to put the considered IRP problem in a form that can be handled from the considered ant-based approaches that will be later described. Let the problem consist of N suppliers that employ M vehicles to transport their products over a time horizon of T periods. Following the notation presented in Section 2, we consider the sets of indices, I_s, I_v, and I_p, defined in Eq. (1). Our approach is vehicle-centric, i.e., each vehicle constructs its order for visiting the suppliers at each time period. The decision of not visiting a specific supplier implies that the supplier is absent in the constructed visiting order. This formulation is adequate to transform the IRP into an equivalent VR problem as described below.

Putting it formally, let $p_{it}^{[j]}$ denote the position of supplier i in the visiting order of vehicle j at time period t. Then, it holds that:

$$p_{it}^{[j]} \in \{0, 1, 2, \ldots, N\}, \quad \text{for all } i \in I_s, j \in I_v, t \in I_p, \tag{25}$$

where $p_{it}^{[j]} = 0$ simply denotes that supplier i is not visited by vehicle j at time period t. The quantities $p_{it}^{[j]}$ for all i, j, and t, are adequate to provide the visiting frequencies x_{ijt} in our IRP model described in Section 2. Indeed, the set of indices of the vehicles that contain the directed arc (i, j) with $i, j \in I_s$, $i \neq j$, in their routes at time period t, is defined as:

$$K_{(i,j,t)} = \left\{ k \in I_v \text{ such that } p_{it}^{[k]} = l - 1, \ p_{jt}^{[k]} = l, \ l \in I_s \setminus \{1\} \right\}.$$

Moreover, let the sets:

$$K_{(0,i,t)} = \left\{ k \in I_v \text{ such that } p_{it}^{[k]} = 1 \right\},$$

$$K_{(i,N+1,t)} = \left\{ k \in I_v \text{ such that } p_{it}^{[k]} > p_{jt}^{[k]}, \ \forall j \in I_s \setminus \{i\} \right\},$$

which define the vehicles that visit supplier i first, and the vehicles that visit supplier i last, just before completing their route at the retailer, respectively. Then, if $|K_{(i,j,t)}|$ denotes the cardinality of $K_{(i,j,t)}$, we can easily infer that:

$$x_{ijt} = \begin{cases} |K_{(i,j,t)}|, & \text{if } K_{(i,j,t)} \neq \emptyset, \\ 0, & \text{otherwise,} \end{cases} \quad \text{for all } i, j, t. \tag{26}$$

This equation determines all the visiting frequencies in the IRP, solely using the constructed visiting orders of the vehicles.

Apart from the visiting order, each vehicle also follows a policy regarding the quantities that are picked up from each supplier. The policy is based on the reasonable assumption that a vehicle shall satisfy all demand and backlogging requirements as long as it is permitted by its capacity. In other words, if $L_{it}^{[k]}$ denotes the k-th vehicle's load when visiting supplier i at time t, and $a_{it}^{[k]}$ is the quantity of products that it will pick up from the supplier, it shall hold that $a_{it}^{[k]} = \min \left\{ C - L_{it}^{[k]},\ d_{it} - I_{i,t-1} \right\}$, where C is the vehicle's capacity, while d_{it} and $I_{i,t-1}$ stand for the demand and the current inventory, respectively. Obviously, if $p_{it}^{[k]} = 0$ (i.e., the vehicle does not visit supplier i) then the picked-up quantity will be $a_{it}^{[k]} = 0$. Then, the total quantity picked up by supplier i at time t by all vehicles, is given by:

$$a_{it} = \sum_{k=1}^{M} a_{it}^{[k]}.$$

We can easily notice that, according to this equation, it is possible that the total amount picked up by supplier i becomes larger than the quantity dictated by the system's demands. This may be observed in the case where a number of vehicles with adequate free capacity visit the same supplier, each one picking an amount equal to $d_{it} - I_{i,t-1}$. However, such a solution will be infeasible due to the constraint of Eq. (11) and, eventually, it will be rejected by the algorithm.

The rest of the model's parameters, i.e., the inventory levels I_{it} and the transported quantities q_{ijt}, can be straightforwardly determined by taking into consideration the constraints of the IRP model. Specifically, I_{it} is given directly from Eq. (4), while the transported quantities are given as:

$$q_{ijt} = \sum_{k \in K_{(i,j,t)}} \left(L_{it}^{[k]} + a_{it}^{[k]} \right), \qquad i, j \in I_s,\ i \neq j.$$

Thus, the vehicles' visiting orders can offer all the necessary information to describe the whole system's operation, rendering the VR problem an equivalent formulation of the original IRP.

Based on this formulation, we considered a solution representation scheme that consists of the visiting orders of all vehicles for all time periods, as follows:

$$\left(\quad \cdots \quad \underbrace{p_{1t}^{[1]}, \ldots, p_{Nt}^{[1]}}_{\text{vehicle 1}}, \quad \cdots, \quad \underbrace{p_{1t}^{[M]}, \ldots, p_{Nt}^{[M]}}_{\text{vehicle M}}, \quad \cdots \quad \right), \qquad (27)$$

$$\underbrace{\phantom{p_{1t}^{[1]}, \ldots, p_{Nt}^{[1]}, \quad \cdots, \quad p_{1t}^{[M]}, \ldots, p_{Nt}^{[M]}}}_{\text{time period } t}$$

where $p_{it}^{[j]}$ is defined as in Eq. (25). Obviously, for a problem with N suppliers, M vehicles, and T time periods, this scheme requires a fixed-size vector of $N \times M \times T$ components to represent a candidate solution.

For example in the case of a problem with 2 suppliers, 2 vehicles, and 2 time periods ($N = M = T = 2$), a candidate solution would be an 8-dimensional vector:

$$\left(\underbrace{\underbrace{p_{11}^{[1]},\ p_{21}^{[1]}}_{\text{vehicle 1}},\ \underbrace{p_{11}^{[2]},\ p_{21}^{[2]}}_{\text{vehicle 2}}}_{\text{time period 1}},\ \underbrace{\underbrace{p_{12}^{[1]},\ p_{22}^{[1]}}_{\text{vehicle 1}},\ \underbrace{p_{12}^{[2]},\ p_{22}^{[2]}}_{\text{vehicle 2}}}_{\text{time period 2}} \right),$$

where $p_{it}^{[j]}$ is defined as described above. Each ant in our approach has to construct such vectors, based on the procedures described in the following section.

3.3   Algorithm Operators and Procedures

The employed algorithm is based on the general framework and operation of established ACO variants. More specifically, it considers a group of agents, called *ants*, which iteratively construct a set of potential solutions of the form described in the previous section. The solution components are stochastically selected from a set of possible values (states), similarly to the stochastic selection of a route between several cities. Thus, each component value assumes a weight that is used for the computation of its selection probability. The weights constitute the *pheromone* values that guide the ants, and they are retained in a continuously updated pheromone table.

In our approach, the components' values that are selected more frequently in the best solutions during the run of the algorithm, increase their pheromone levels and, consequently, they are more frequently selected by the ants. Pheromone restarting is also applied after a number of iterations to alleviate search stagnation. All these operations are thoroughly described in the following paragraphs.

In our case, the algorithm assumes K ants, which iteratively construct candidate solutions while retaining in memory the best one from the beginning of the run. Each ant constructs a solution vector of the form of Eq. (27), in a componentwise manner. The solution construction process is based on the probabilistic selection of each component's value, based on the table of pheromones.

More specifically, there is a pheromone value, $\tau_{it}^{[j]}(l)$, $l = 0, 1, \ldots, N$, for each one of the l possible values (states) of the variables $p_{it}^{[j]}$ defined in Eq. (25), i.e.:

$$\text{States of } p_{it}^{[j]} : \{\quad 0,\qquad 1,\quad \ldots,\quad N\quad \}$$
$$\qquad\qquad\qquad\qquad \uparrow \qquad\quad \uparrow \qquad\qquad \uparrow$$
$$\text{Pheromones:}\quad \tau_{it}^{[j]}(0)\quad \tau_{it}^{[j]}(1)\ \cdots\ \tau_{it}^{[j]}(N)$$

The corresponding probability of taking $p_{it}^{[j]} = l$ is computed as:

$$\rho_{it}^{[j]}(l) = \frac{\tau_{it}^{[j]}(l)}{\sum\limits_{k=0}^{N} \tau_{it}^{[j]}(k)}, \qquad \forall i, j, t.$$

It is trivial to verify the necessary condition:

$$\sum_{l=0}^{N} \rho_{it}^{[j]}(l) = 1, \qquad \forall\, i, j, t.$$

Each ant uses these probabilities to decide for the assigned component value (similarly as deciding among cities in the TSP) through the well-known *fitness proportionate selection* (also known as *roulette-wheel selection*) procedure [4].

In practice, there are some limitations in this procedure. For example, $p_{it}^{[j]}$ cannot take the same value with a previously determined $p_{kt}^{[j]}$, with $k \neq i$, since this would imply that suppliers i and k are concurrently visited by vehicle j at time t. These constraints can be handled either by penalizing the corresponding solutions in the objective function or by allowing only the proper states to participate in the selection procedure above. We followed the latter approach since such restrictions can be straightforwardly incorporated in our algorithm, while it does not add further constraints to the (already over-constrained) problem.

As soon as a candidate solution is constructed, it is evaluated with the objective function and all constraints are evaluated. If the solution is infeasible, then its objective value is penalized on the basis of the number and magnitude of constraints violations. We postpone the description of the penalty function until the next section.

Immediately after the construction and evaluation of all K candidate solutions, there is an update procedure for the best solution detected from the beginning of the run. Specifically, each constructed solution is compared to the best solution and, if superior, it replaces it. In order to avoid strict feasibility restrictions that could lead to reduced search capability of the algorithm, we allow infeasible solutions to be constructed, although adopting the following common rules for updating the best solution:

(a) Between feasible solutions, the one with the smallest objective value is selected.
(b) Between infeasible solutions, the one with the smallest total penalty is selected.
(c) Between a feasible best solution and an infeasible new candidate, the feasible best solution is always selected.
(d) Between an infeasible best solution and a feasible new candidate, the feasible one is selected.

These rules allow infeasible solutions to be accepted (which is the case mostly in the first iterations of the algorithm), while favoring solutions that lie closer or inside the feasible region. Obviously, no feasible initial solutions are required in this case.

After updating the best solution, the pheromone update takes place. This procedure is one of the features that distinguish between different ACO variants. In our approach, we combined features from different variants that were found to fit the studied problem better. Specifically, the first step in pheromone update is the evaporation, where each pheromone value is reduced as follows:

$$\tau_{it}^{[j]}(l) \leftarrow (1 - R_{\text{ev}})\tau_{it}^{[j]}(l), \qquad \text{for all } i, j, t, l,$$

where $R_{\text{ev}} \in (0, 1)$ is the *pheromone evaporation rate* and determines the algorithm's rate of "forgetting" previous states in most ACO variants [12]. After the evaporation, the pheromones are updated again, as follows:

$$\tau_{it}^{[j]}(l) \leftarrow \tau_{it}^{[j]}(l) + \Delta_{it}^{[j]}(l), \qquad \text{for all } i, j, t, l,$$

where:

$$\Delta_{it}^{[j]}(l) = \begin{cases} \frac{\Delta\tau}{K}, & \text{if the best solution contains } p_{it}^{[j]} = l, \\ 0, & \text{otherwise,} \end{cases} \qquad (28)$$

and $\Delta\tau$ is a fixed quantity that, in combination with R_{ev}, determines how strongly the algorithm promotes the best detected solution. Similarly to the $\mathcal{MMAS}$ approach [19], we considered only the best solution to add pheromone instead of all ants. In the same spirit, we considered a lower bound for the pheromone values, $\tau_{it}^{[j]}(l) \geqslant 10^{-3}$, in order to avoid the complete exclusion of specific values of $p_{it}^{[j]}$ during the search. Although, an upper bound was not found to benefit the algorithm.

In addition, preliminary experiments revealed that the algorithm could stagnate if a very good solution was prematurely detected. This weakness was addressed by adopting a restarting mechanism as in $\mathcal{MMAS}$ [19]. Thus, every r_{re} iterations all pheromones are randomly re-initialized, offering the necessary perturbation to unstuck the algorithm from possible local minimizers. The algorithm is terminated as soon as a user-defined stopping criterion is satisfied.

The initialization of the pheromones follows a special yet reasonable scheme. In the initial step of the algorithm, decisions shall be unbiased with respect to the selected components values. Hence, we shall assign equal probability of visiting or not a supplier. Also, if the decision is to visit a supplier, then its position in the vehicle's visiting order shall be selected with equal probability among the different states. For this reason, we assign the following initial selection probabilities of the components values:

$$\rho_{it}^{[j]}(l) = \begin{cases} 0.5, & \text{for } l = 0, \\ 0.5/N, & \text{for } l = 1, 2, \dots, N. \end{cases}$$

Thus, the state $l = 0$ (i.e., supplier i is not visited by vehicle j at time t) is selected with probability 0.5, while the rest are selected with equal probability among them.

3.4   Penalty Function

Let $f(P)$ be the objective function under minimization, which is defined in Eq. (2), with the candidate solution vector P being defined as in Eq. (27).

Table 1. The considered problem instances and the dimensions of the corresponding optimization problems

Suppliers (N)	Vehicles (M)	Time periods (T)	Problem dimension
4	5	5	100
4	8	10	320
6	6	5	180
6	8	10	480
6	12	10	720
8	6	5	240
8	8	5	320
8	10	10	800
10	6	5	300
10	10	10	1000

All parameters of the model are determined as described in Section 3.2. As mentioned in previous sections, infeasible solutions are penalized by using a penalty function that takes into account the number and the degree of violations.

Thus, if $VC(P)$ denotes the set of violated constraints for the candidate solution P, then the penalized objective function becomes:

$$PF(P) = f(P) + \sum_{i \in VC(P)} |MV(i)|, \qquad (29)$$

where $MV(i)$ is the magnitude of violation for constraint i. A constraint is considered as violated if the magnitude of violation exceeds a small, fixed tolerance value, $\varepsilon_{tol} > 0$.

Moreover, we shall notice that the constraints (C12)-(C21), defined in Eqs. (15)-(24), are *de facto* satisfied by the solution representation that is used in our approach. Thus, there is no need to include them in the penalty function. All other constraints were equally considered and penalized since a solution shall satisfy all of them in order to be useful for the considered IRP model.

4   Experimental Results

The proposed algorithm was applied on a set of test problems with various numbers of suppliers, vehicles, and time periods. More specifically, the problem instances reported in Table 1 were derived from the data set[1] provided in [15]. Each problem instance was considered along with the parameter setting reported in Table 2. Notice that, in contrast to the model studied in [17] using the same

---

[1] Available at http://www.mie.utoronto.ca/labs/ilr/IRP/

Table 2. Problem and algorithm parameters

Parameter	Description	Value(s)
C	Vehicle capacity	10
F	Fixed vehicle cost per trip	20
V	Travel cost per distance unit	1
s_i	Backlogging cost (with respect to h_i)	$3 \times h_i$
K	Number of ants	5, 10, 15
$E_{\max}$	Maximum function evaluations	60×10^6
ε_{tol}	Constraints violation tolerance	10^{-8}
R_{ev}	Pheromone evaporation rate	10^{-3}
$\Delta\tau$	Pheromone increment	$10^{-2}/N$
r_{re}	Evaluations for pheromone restart	5×10^6

test problems, we considered a limited flee size instead of unlimited. This makes the problem significantly harder even in its small instances.

We followed the experimental setting in [17] whenever possible, except the number of experiments, which was set to 20 per problem instance, instead of 10 for the GA-based approach in [17]. The best solution detected by the proposed algorithm was recorded for each problem and compared to the solution provided by the CPLEX software (Ver. 12.2) for our model within 4 hours of execution.

Following the basic analysis in similar papers [17], the algorithm's performance was assessed in basis of the gap between the best obtained solution and the CPLEX solution. The gap is computed as follows:

$$\text{gap} = \frac{\text{solution value} - \text{CPLEX solution value}}{\text{CPLEX solution value}} \times 100\%.$$

Also, the required percentage fraction of the maximum computational budget $E_{\max}$ for the detection of each solution, was computed as follows:

$$E_{\text{FR}} = \frac{\text{required number of function evaluations}}{E_{\max}} \times 100\%,$$

and it was used as a measure of the algorithm's running-time requirements. Finally, each experiment was replicated for different numbers of ants, namely $K = 5$, 10, and 15. The experiments were conducted on a desktop computer with Intel® i7 processor and 8GB RAM. The running time for the obtained solutions was also reported.

The most successful setting proved to be that of $K = 5$, since it was able to achieve the same solution gaps with the rest, but with smaller demand in resources. The corresponding results are reported in Table 3. The results are promising, since the obtained gaps from optimality were kept in lower values, exhibiting a reasonable increasing trend with the problem's dimension. This is a well-known effect in approximation algorithms and it is tightly related to the *curse of dimensionality* [18].

Table 3. The obtained results for $K = 5$ ants.

Problem			Gap	E_{FR}	Time
N	M	T	(%)	(%)	(sec)
4	5	5	0.00	14.9	67.8
4	8	10	0.96	27.1	1218.3
6	6	5	1.84	3.1	1329.4
6	8	10	4.34	14.8	3684.6
6	12	10	5.62	17.9	3688.3
8	6	5	4.87	7.4	1842.3
8	8	5	6.64	20.6	3317.8
8	10	10	8.63	60.1	5035.3
10	6	5	4.00	45.9	2778.5
10	10	10	10.05	54.9	7346.6

Moreover, the required computational budget was slightly higher than half of the available one for the hardest cases, as we can see in the last column of the table. However, it was not proportional to the corresponding problem's dimension. This can be primarily attributed to the stochastic nature of the algorithm, as well as to the unique structure of each problem instance.

The scaling properties of the algorithm with respect to K were assessed on the basis of the *gap growing factor* (ggf), which is defined as follows:

$$\text{ggf}_{K_1 \to K_2} = \frac{\text{Solution gap for } K = K_2}{\text{Solution gap for } K = K_1}.$$

The obtained values of $\text{ggf}_{5 \to 10}$ and $\text{ggf}_{5 \to 15}$, are illustrated in Fig. 1. Notice that there is no data for the problem instance 4-5-5 due to the zero values of the gaps reported in Table 3. In almost all cases, we observe an increasing tendency of the gap growing factor.

In a first thought, the later may seem to be a counterintuitive evidence, since the addition of ants would be expected to boost the algorithm's search capability and performance. However, it is a straightforward consequence of the pheromone update scheme. More specifically, as described in Section 3.3, only the overall best ant updates the pheromones at each iteration. Also, the amount of update is inversely proportional to the number of ants as defined in Eq. (28). Hence, larger number of ants implies smaller pheromone increments and, consequently, higher number of iterations for the domination of the best ant's components against the rest.

Therefore, under the same computational budget, smaller values of K shall be expected to converge faster to the optimal solution than the higher ones. Nevertheless, we can observe that the rate of growth of the gap is sub-linearly associated with the corresponding growth in K, since it exceeds it only in 2 cases.

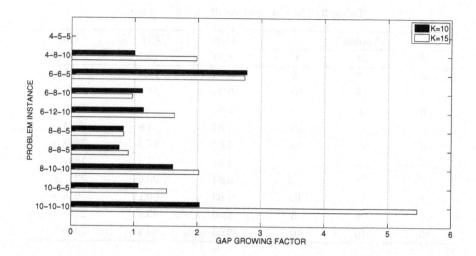

Fig. 1. Scaling properties of the algorithm with respect to the gap growing factor for $K = 10$ ($ggf_{5 \to 10}$) and $K = 15$ ($ggf_{5 \to 15}$) ants

5   Conclusions

We proposed a unified framework for solving IRPs as VR problems through an ant-based algorithm. We considered a model where a fleet of capacitated homogeneous vehicles transport different products from multiple suppliers to a retailer to meet the demand in each period over a finite planning horizon, while shortages are allowed and unsatisfied demand is backlogged.

The corresponding VR problem was solved with an ant-based optimization approach. Experiments on different test problems offered preliminary insight regarding the algorithm's potential. The solution gaps between the algorithm and CPLEX solutions were kept in reasonably low values, while offering perspective for further improvement by proper parameter tuning.

Also, the stochastic nature of the algorithm as well as its tolerance to new operators and representations, allows the inclusion of problem-based specialized operations. Finally, the algorithm tackles the problem without the necessity for breaking it into sub-problems and solving each one separately. Naturally, additional work is required to fully reveal the strengths and weaknesses of the algorithm. Parallel cooperative schemes could be beneficial for reducing the time complexity, and this is the primary goal of our ongoing efforts.

Acknowledgement. This research has been co-financed by the European Union (European Social Fund - ESF) and Greek national funds through the Operational Program "Education and Lifelong Learning" of the National Strategic Reference Framework (NSRF) - Research Funding Program: ARCHIMEDES III. Investing in knowledge society through the European Social Fund.

References

1. Abdelmaguid, T.F., Dessouky, M.M.: A genetic algorithm approach to the integrated inventory–distribution problem. International Journal of Production Research 44, 4445–4464 (2006)
2. Abdelmaguid, T.F., Dessouky, M.M., Ordonez, F.: Heuristic approaches for the inventory–routing problem with backlogging. Computers and Industrial Engineering 56, 1519–1534 (2009)
3. Andersson, H., Hoff, A., Christiansen, M., Hasle, G., Lokketangen, A.: Industrial aspects and literature survey: Combined inventory management and routing. Computers & Operations Research 37, 1515–1536 (2010)
4. Bäck, T.: Evolutionary Algorithms in Theory and Practice. Oxford University Press, New York (1996)
5. Bonabeau, E., Dorigo, M., Théraulaz, G.: Swarm Intelligence: From Natural to Artificial Systems. Oxford University Press, New York (1999)
6. Campbell, A.M., Clarke, L., Kleywegt, A.J., Savelsbergh, M.W.P.: The inventory routing problem. In: Crainic, T., Laporte, G. (eds.) Fleet Management and Logistics, pp. 95–113. Kluwer Academic Publishers, Boston (1998)
7. Campbell, A.M., Savelsbergh, M.W.P.: A decomposition approach for the inventory–routing problem. Transportation Science 38(4), 488–502 (2004)
8. Chan, L., Federgruen, A., Simchi-Levi, D.: Probabilistic analyses and practical algorithms for inventory–routing models. Operations Research 46(1), 96–106 (1998)
9. Chien, T.W., Balakrishnan, A., Wong, R.T.: An integrated inventory allocation and vehicle routing problem. Transport Science 23, 67–76 (1989)
10. Dorigo, M., Di Caro, G.: The ant colony optimization meta–heuristic. In: Corne, D., Dorigo, M., Glover, F. (eds.) New Ideas in Optimization, pp. 11–32. McGraw-Hill (1999)
11. Dorigo, M., Gambardella, L.M.: Ant colony system: A cooperative learning approach to the traveling salesman problem. IEEE Transactions on Evolutionary Computation 1(1), 53–66 (1997)
12. Dorigo, M., Stützle, T.: Ant Colony Optimization. MIT Press (2004)
13. Federgruen, A., Zipkin, P.: A combined vehicle routing and inventory allocation problem. Operations Research 32(5), 1019–1036 (1984)
14. Hvattum, L.M., Lokketangen, A.: Using scenario trees and progressive hedging for stochastic inventory routing problems. Journal of Heuristics 15, 527–557 (2009)
15. Lee, C.-H., Bozer, Y.A., White III, C.C.: A heuristic approach and properties of optimal solutions to the dynamic inventory routing problem. Working Paper (2003)
16. Moin, N.H., Salhi, S.: Inventory routing problems: a logistical overview. Journal of the Operational Research Society 58, 1185–1194 (2007)
17. Moin, N.H., Salhi, S., Aziz, N.A.B.: An efficient hybrid genetic algorithm for the multi–product multi–period inventory routing problem. International Journal of Production Economics 133, 334–343 (2011)
18. Powell, W.B.: Approximate Dynamic Programming: Solving the Curses of Dimensionality. Wiley (2007)
19. Stützle, T., Hoos, H.H.: Max min ant system. Future Generation Computer Systems 16, 889–914 (2000)

Measuring Multimodal Optimization Solution Sets with a View to Multiobjective Techniques

Mike Preuss and Simon Wessing

Computational Intelligence Group, Chair of Algorithm Engineering,
Department of Computer Science, TU Dortmund, Germany
{mike.preuss,simon.wessing}@tu-dortmund.de

Abstract. As in multiobjective optimization, multimodal optimization generates solution sets that must be measured in order to compare different optimization algorithms. We discuss similarities and differences in the requirements for measures in both domains and suggest a property-based taxonomy. The process of measuring actually consists of two subsequent steps, a subset selection that only considers 'suitable' points (or just takes all available points of a solution set) and the actual measuring. Known quality indicators often rely on problem knowledge (objective values and/or locations of optima and basins) which makes them unsuitable for real-world applications. Hence, we propose a new subset selection heuristic without such demands, which thereby enables measuring solution sets of single-objective problems, provided a distance metric exists.

Keywords: multimodal optimization, multiobjective optimization, performance measuring, solution sets, subset selection, archive, indicator.

1 Introduction

What is multimodal optimization? While the works on applications of *evolutionary algorithms* (EA) to single-objective multimodal problems go back at least to the 1960s, most EA based research done on these problems in the last decades is scattered over many different subfields. Relevant algorithmic developments have been flagged as belonging to (real-valued) *genetic algorithms* (GA), *evolution strategies* (ES), *differential evolution* (DE), *particle swarm optimization* (PSO) and others. As a consequence, this part of the EA research lacks clear definitions concerning what is searched and how it is measured. For global optimization (single global optimum to be found in shortest time), the set of benchmarking problems, the experimental setup and the performance measuring actually employed for new publications are more and more converging to the *black box optimization benchmarking* (BBOB) definitions as quasi-standard [1].

However, for multimodal optimization, we are currently far from that encouraging development, and one of the main reasons for that is probably that it is actually not very clear what kinds of problems shall be tackled and what the performance of a specific solution set should be. Currently, the organizers of the CEC 2013 niching competition [2] undertake a valiant effort to consolidate

M. Emmerich et al. (eds.), *EVOLVE - A Bridge between Probability, Set Oriented Numerics,* 123
and Evolutionary Computation IV, Advances in Intelligent Systems and Computing 227,
DOI: 10.1007/978-3-319-01128-8_9, © Springer International Publishing Switzerland 2013

the area and establish a common set of test problems. While we explicitly support this endeavor, it may be questioned if focusing on global optima only and assessing algorithms by means of the peak ratio measure only can pave the way for evolving better optimization algorithms for (usually multimodal) real-world applications. Speaking of real-world optimization requirements is of course not unambiguous because the details depend on the actual application. However, we presume that in many settings, we find similar requirements:

1. Limited time
2. Delivery of a small set of very good solutions
3. Diversity of provided solutions
4. Robustness of provided solutions

Within this work, we ignore the first and the last item. This is not because they are not considered important but because dealing with the other two is challenging enough for the time being. Especially when the set of very good solutions is large and the objective values within this set differ, measuring gets very difficult as one has to deal with tradeoffs: shall a more distant but slightly worse solution be valued higher than one that is near to another but possesses a better obective value? It is rather difficult to take such decisions for benchmark functions where the locations of all optima are known. However, one may encounter situations where only very little knowledge about the optimized problem is available, and most of the standard measures cannot be computed.

Interestingly, in a related subfield of evolutionary optimization, very similar problems have to be treated, as one also deals with sets of solutions instead of the single best solution. This is the area of evolutionary multiobjective optimization (EMO), which now has seen more than a decade of extensive research. The positive results of these prolonged efforts are clearly visible. Besides a solid theoretical foundation, there exist established test problems and widely accepted guidelines for the experimental evaluation of multiobjective optimization algorithms [3]. As stated before, this cannot be said about multimodal optimization, where the user is also not only interested in a single globally optimal solution, but a set of locally optimal solutions that is as diverse as possible. Diversity and optimality are conflicting and there is no obvious way to achieve a tradeoff.

We therefore propose the consequent adoption of EMO principles, language, and tools in multimodal optimization. Even today, the similarities in the applied methods in multimodal and multiobjective optimization are already striking. Common properties are:

– Incorporation of archives
– Population-based algorithms
– Multiobjective selection approaches
– Necessity for quality indicators to assess performance

Within this work, we survey currently employed measures from multimodal and multiobjective optimization and compare their effectiveness experimentally for varying solution sets on two typical multimodal test problems. Our task is to

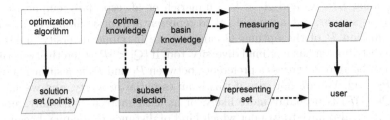

Fig. 1. Data flow from the solution set obtained by an optimization algorithm up to the final scalar measure value. Problem knowledge is an optional input for the subset selection and/or measuring part.

obtain a taxonomy of measures according to their most important properties, and then derive rough guidelines for choosing the measures to use. We also explicitly deal with the case where no problem knowledge is available and thus most methods fail, by suggesting a heuristic that resolves the tradeoff in a well-defined way.

2   What Do We Want?

We interpret measuring the quality of a solution set for a multimodal problem as consisting of several modules as depicted in Fig. 1. At first, an optimization algorithm provides a solution set. This may e. g. be the final population or an archive of recorded good solutions. The cardinality of this set should depend on the tackled problem but is at current usually determined by the employed algorithm alone. This is inappropriate, because it may firstly bias the quality assessment and secondly make a human inspection of the set too costly. Thus, we explicitly include a subset selection step in the evaluation process. We may term the result of the subset selection a *representing set*, and it is important to note that all the other solutions contained in the solution set have no influence on the result of the measurement. In Sec. 4, we will suggest a method to choose a subset without explicit problem knowledge, which could then be combined with an already existing quality assessment method.

Emmerich et al. [4] analyze many quality indicators for level set approximations. Again, there are many similarities to multimodal optimization, so most of the indicators are also relevant here. They define two properties that are important criteria for good level set approximations:

- *Representativeness:* Distances of elements of the level set L to the approximation set A should be as small as possible.
- *Feasibility:* Symmetrically, distances of elements of the approximation set A to the level set L should be as small as possible.

In the following we will denote the set of optima by Q and the representing or approximation set by P. The two requirements can principally be carried over by assuming $L = Q$ and $A = P$. Emmerich et al. especially emphasize that diversity is similar, but not identical to representativeness. Similarly, a perfect approximation P of the optima Q cannot have maximal diversity, unless the optima distribution has maximal diversity, too. If $|Q| < |P|$, then the two criteria would not be able to express a preference between P_1 and P_2 in a situation where $P_1 \supset P_2 \supset Q$. We are not sure if this is an essential property, and we suppose that $|Q| \geq |P|$ is probably more likely, or at least more user-friendly, anyway. However, we also have to wonder which kind of distance to assume, because not only the search space distance is relevant, but also the objective space. In the following, we will denote the distance to the nearest neighbor of $x \in P$ with

$$d_{\mathrm{nn}}(x, P) = \min\{\mathrm{dist}(x, y) \mid y \in P \setminus \{x\}\}$$

and the distance to the nearest better neighbor as

$$d_{\mathrm{nb}}(x, P) = \min\{\mathrm{dist}(x, y) \mid f(y) < f(x) \wedge y \in P\}.$$

The nearest- and nearest better neighbor will be denoted with $\mathrm{nn}(x, P)$ and $\mathrm{nbn}(x, P)$, respectively, by using the above definitions with arg min instead of min. Minimization is assumed for the optimization problem $f : S \subseteq \mathbb{R}^n \to \mathbb{R}$.

3   Quality Indicators

We classify the presented quality indicators according to the amount of information that is necessary for their application. Throughout the section, $P = \{x_1, \ldots, x_\mu\}$ ($\mu < \infty$) denotes the approximation set that is to be assessed.

3.1   Indicators without Problem Knowledge

Solow-Polasky Diversity (SPD). Solow and Polasky [5] developed an indicator to measure a population's biological diversity and showed that it has superior theoretical properties compared to the sum of distances and other indicators. Ulrich et al. [6] discovered its applicability to multiobjective optimization. They also verified the inferiority of the sum of distances experimentally by directly optimizing the indicator values. To compute this indicator for P, it is necessary to build a $\mu \times \mu$ correlation matrix $\mathbf{C}$ with entries $c_{ij} = \exp(-\theta \, \mathrm{dist}(x_i, x_j))$. The indicator value is then the scalar resulting from $\mathrm{SPD}(P) := e^\top \mathbf{C}^{-1} e$, where $e = (1, \ldots, 1)^\top$. As the matrix inversion requires time $O(\mu^3)$, the indicator is only applicable to relatively small sets. It also requires a user-defined parameter θ, which depends on the size of the search space. The choice of this parameter is unfortunately critical, as we will show in Sec. 3.4.

Sum of Distances (SD). As already mentioned, the sum of distances

$$\mathrm{SD}(P) := \sqrt{\sum_{x \in P} \sum_{y \in P, y \neq x} \mathrm{dist}(x, y)}$$

appears in [5,6] as a possible diversity measure. To obtain indicator values of reasonable magnitude, we suggest to take the square root of the sum.

Sum of Distances to Nearest Neighbor (SDNN). As Ulrich et al. [6] showed that SD has some severe deficiencies, we also consider the sum of distances to the nearest neighbor

$$\text{SDNN}(\mathcal{P}) := \sum_{i=1}^{\mu} d_{\text{nn}}(\boldsymbol{x}_i, \mathcal{P}).$$

In contrast to SD, SDNN penalizes the clustering of solutions, because only the nearest neighbor is considered. Emmerich et al. [4] mention the arithmetic mean gap $\frac{1}{\mu} \text{SDNN}(\mathcal{P})$ and two other similar variants. We avoid the averaging here to reward larger sets. However, it is still possible to construct situations where adding a new point to the set decreases the indicator value.

Statistics of the Distribution of Objective Values. Regarding the assessment of the population's raw performance, few true alternatives seem to exist. To us, the only things that come to mind are statistics of the objective value distribution, with the mean or median as the most obvious measures. Values from the tail of the distribution, as the best or worst objective value, do not seem robust enough to outliers. Thus we include the average objective value in our indicator collection:

$$\text{AOV}(\mathcal{P}) := \frac{1}{\mu} \sum_{i=1}^{\mu} f(\boldsymbol{x}_i).$$

3.2 Indicators Requiring Knowledge of the Optima

All indicators in this section require a given set of locally optimal positions $\mathcal{Q} = \{\boldsymbol{z}_1, \ldots, \boldsymbol{z}_k\}$ ($k < \infty$) to assess $\mathcal{P}$. This means they can only be employed in a benchmarking scenario on test problems that were specifically designed so that $\mathcal{Q}$ is known. Note, however, that $\mathcal{Q}$ does not necessarily have to contain all existing optima, but can also represent a subset (e. g. only the global ones).

Peak Ratio (PR). Ursem [7] defined the number of found optima $\ell = |\{\boldsymbol{z} \in \mathcal{Q} \mid d_{\text{nn}}(\boldsymbol{z}, \mathcal{P}) \le \epsilon\}|$ divided by the total number of optima as peak ratio $\text{PR}(\mathcal{P}) := \ell/k$. The indicator requires some constant ϵ to be defined by the user, to decide if an optimum has been approximated appropriately.

Peak Distance (PD). This indicator simply calculates the average distance of a member of the reference set $\mathcal{Q}$ to the nearest individual in $\mathcal{P}$:

$$\text{PD}(\mathcal{P}) := \frac{1}{k} \sum_{i=1}^{k} d_{\text{nn}}(\boldsymbol{z}_i, \mathcal{P})$$

A first version of this indicator (without the averaging) was presented by Stoean et al. [8] as "distance accuracy". With the $1/k$ part, peak distance is analogous to the indicator inverted generational distance [9], which is computed in the objective space of multiobjective problems.

Peak Inaccuracy (PI). Thomsen [10] proposed the basic variant of the indicator

$$\text{PI}(\mathcal{P}) := \frac{1}{k} \sum_{i=1}^{k} |f(\boldsymbol{z}_i) - f(\text{nn}(\boldsymbol{z}_i, \mathcal{P}))|$$

under the name "peak accuracy". To be consistent with PR and PD, we also add the $1/k$ term here. We allow ourselves to relabel it to peak inaccuracy, because speaking of accuracy is a bit misleading as the indicator must be minimized. PI has the disadvantage that the representativeness of $\mathcal{P}$ is not directly rewarded, because it is possible for one solution to satisfy several optima at once. On the other hand, comparing the indicator value to a baseline performance, e. g. calculated as the peak inaccuracy for the global optimum alone, might relativize seemingly good performances. Note that PI is somewhat related to the maximum peak ratio (MPR) by Miller and Shaw [11]. MPR is also extensively used by Shir [12].

Averaged Hausdorff Distance (AHD). This indicator can be seen as an extension of peak distance due to its relation to the inverted generational distance. It was defined by Schütze et al. [13] as

$$\text{AHD}(\mathcal{P}) := \Delta_p(\mathcal{P}, \mathcal{Q})$$

$$= \max \left\{ \left(\frac{1}{k} \sum_{i=1}^{k} d_{\text{nn}}(\boldsymbol{z}_i, \mathcal{P})^p \right)^{1/p}, \left(\frac{1}{\mu} \sum_{i=1}^{\mu} d_{\text{nn}}(\boldsymbol{x}_i, \mathcal{Q})^p \right)^{1/p} \right\}.$$

The definition contains a parameter p that controls the influence of outliers on the indicator value (the more influence the higher p is). For $1 \leq p < \infty$, AHD has the property of being a semi-metric [13]. We choose $p = 1$ throughout this paper, analogously to [4]. The practical effect of the indicator is that it rewards the approximation of the optima (as PD does), but as well penalizes any unnecessary points in remote locations. This makes it an adequate indicator for the comparison of approximation sets of different sizes.

3.3   Indicators Requiring Knowledge of the Basins

Even more challenging to implement than the indicators in Sec. 3.2 are indicators that require information about which basin each point of the search space belongs to. This information can either be provided by a careful construction of the test problem, or by running a hill climber for each $\boldsymbol{x} \in \mathcal{P}$ as a start point during the assessment and then matching the obtained local optima with the points of the known $\mathcal{Q}$. Regardless of how it is achieved, we assume the existence of a function

$$b(\boldsymbol{x}, \boldsymbol{z}) = \begin{cases} 1 & \text{if } \boldsymbol{x} \in \text{basin}(\boldsymbol{z}), \\ 0 & \text{else.} \end{cases}$$

The rationale of indicators for covered basins instead of distances to local optima is that the former also enables measuring in early phases of an optimization, when the peaks have not been approximated well yet. If the basin shapes are not very regular, the latter indicator type may be misleading in this phase.

Basin Ratio (BR). The number of covered basins is calculated as

$$\ell = \sum_{i=1}^{k} \min \left\{ 1, \sum_{j=1}^{\mu} b(\boldsymbol{x}_j, \boldsymbol{z}_i) \right\}.$$

The basin ratio is then $\mathrm{BR}(\mathcal{P}) := \ell/k$, analogous to PR. This indicator can only assume $k+1$ distinct values, and in lower dimensions it should be quite easy to obtain a perfect score by a simple random sampling of the search space. It makes sense especially when not all of the existing optima are relevant. Then, its use can be justified by the reasoning that the actual optima can be found relatively easily with a hill climber, once there is a start point in each respective basin.

Basin Inaccuracy (BI). This is a novel combination of basin ratio and peak inaccuracy. We define it as

$$\mathrm{BI}(\mathcal{P}) := \frac{1}{k} \sum_{i=1}^{k} \begin{cases} \min \{|f(\boldsymbol{z}_i) - f(\boldsymbol{x})| \mid \boldsymbol{x} \in \mathcal{P} \wedge b(\boldsymbol{x}, \boldsymbol{z}_i)\} & \exists \boldsymbol{x} \in \mathrm{basin}(\boldsymbol{z}_i), \\ f_{\max} & \text{else,} \end{cases}$$

where $f_{\max}$ denotes the difference between the global optimum and the worst possible objective value. For each optimum, the indicator calculates the minimal difference in objective values between the optimum and all solutions that are located in it's basin. If no solution is present in the basin, a penalty value is assumed for it. Finally, all the values are averaged. The rationale behind this indicator is to enforce a good basin coverage, while simultaneously measuring the deviation of objective values $f(\boldsymbol{x})$ from $f(\boldsymbol{z}_i)$.

3.4   Extending and Comparing Indicators

In the following, we will try to give assessment criteria for the indicators' utility in practical situations. Table 1 shows an overview of the indicators with some important properties. We included a column that indicates if an indicator utilizes the actual objective values, because they of course contain important information. Consider for example the situation in Fig. 2. The optimum of this function is located at $x = 2$. Two points a and b have the same distance to the optimum, but clearly varying objective values. If an indicator disregards $f(a)$ and $f(b)$, both points seem equally valuable. Even worse, if a was closer to the optimum than b, we would have a conflict between objective value and distance. One possible workaround could be to augment the search points with their objective value. Instead of $\mathcal{P}$, we would then assess the points $\mathcal{P}' = \{(\boldsymbol{x}_1, f(\boldsymbol{x}_1)), \ldots, (\boldsymbol{x}_\mu, f(\boldsymbol{x}_\mu))\}$. This approach may be especially reasonable with PD and AHD, although in practice different scales in search- and objective space can greatly bias the method.

Table 1. Overview of the quality indicators. Where no reference is given, we are not aware of any previous publication.

Indicator	Ref.	Best	Worst	Regards $f(x)$	Use with var. μ	Without optima	Without basins	Without params
SPD	[5]	μ	1	✗	✓	✓	✓	✗
SD	[6]	> 0	0	✗	✗	✓	✓	✓
SDNN	[4]	> 0	0	✗	✗	✓	✓	✓
AOV		$f(x^*)$	$> f(x^*)$	✓	✓	✓	✓	✓
PR	[7]	1	0	✗	✗	✗	✓	✗
QAPR		1	0	✗	✓	✗	✓	✗
PD	[8]	0	> 0	✗	✗	✗	✓	✓
APD		0	> 0	✓	✗	✗	✓	✓
PI	[10]	0	> 0	✓	✗	✗	✓	✓
AHD	[13]	0	> 0	✗	✓	✗	✓	✗
AAHD		0	> 0	✓	✓	✗	✓	✗
BR		1	0	✗	✗	✗	✗	✓
QABR		1	0	✗	✓	✗	✗	✓
BI		0	> 0	✓	✗	✗	✗	✓

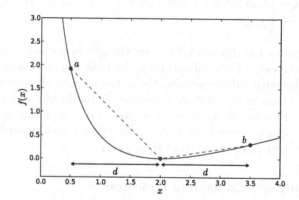

Fig. 2. A problematic situation for indicators relying solely on distance values

Note, however, that also differently scaled decision variables and ill-conditioned problems are completely disregarded in this paper and in most of the literature regarding multimodal optimization. For indicators in Sec. 3.2 and 3.3, $\mathcal{Q}$ may be used for normalization. The two new approaches shall be denoted as augmented peak distance $\text{APD}(\mathcal{P}) := \text{PD}(\mathcal{P}')$ and augmented averaged Hausdorff distance $\text{AAHD}(\mathcal{P}) := \text{AHD}(\mathcal{P}')$.

With the notable exception of AOV and (A)AHD, all presented indicators are inclined to favor approximation sets with unduly large sizes μ. While this makes sense for diversity indicators, it should be avoided for those in Sec. 3.2 and 3.3. We therefore suggest to incorporate a quantity-adjustment by replacing the $1/k$

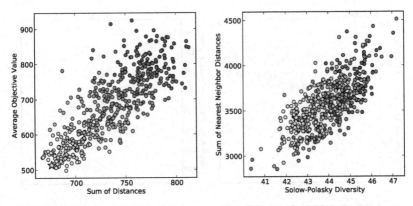

Fig. 3. Test data on problem f_1, $n = 2$. Left: $\mathcal{Q}$ can be dominated. Right: Diversity indicators show relatively low correlation. Marker color indicates high (yellow) or low (red) number of contained optima.

terms with $1/\sqrt{k \cdot \mu}$ for the two indicators that are to be maximized, PR and BR. Consider for example the quantity-adjusted peak ratio $\mathrm{QAPR}(\mathcal{P}) := \ell/\sqrt{k \cdot \mu}$. For any fixed k, finding $\ell = 1$ optimum with $\mu = 2$ points is evaluated as good as finding two optima with eight points. Even better would be to find two optima with four points, which is just as rewarding as finding one with one point. The formula is also convenient, because the best indicator value is still 1 and for $k = \mu$ the results are identical to the basic version.

We now conduct an experimental comparison of the indicators. To do this, we use Schwefel's problem $f_1(\boldsymbol{x}) = \sum_{i=1}^{n} -x_i \cdot \sin(\sqrt{|x_i|})$ and Vincent's problem $f_2(\boldsymbol{x}) = -\frac{1}{n} \sum_{i=1}^{n} \sin(10 \cdot \log x_i)$ as test instances. The search spaces are $[-500, 500]^n$ and $[0.25, 10]^n$, respectively. f_1 has $k = 7^n$ local optima, of which only one is global, while f_2 has 6^n global optima. To generate some data, we take for each $i \in \{0, \ldots, k\}$ a random sub-sample of the optima of size i. If necessary, the set is filled up with random points until $\mu = k$. This procedure is repeated a number of times to obtain enough data.

Figure 3 shows a scatter plot of indicator values for data generated on f_1 with $n = 2$ (ten repeats). In its left panel, we see SD versus AOV. The set of optimal solutions $\mathcal{Q}$ is marked with a star. We see that it is dominated if we interpret optimizing AOV and SD as a bi-objective problem. This result also holds for the two other diversity indicators (not shown here). It is such an important observation, because the only currently known assessment methods for multimodal real-world problems are based on AOV and diversity indicators. Another observation is that the correlation of diversity indicators is surprisingly weak (see e.g. SPD vs. SDNN in the right panel of Fig. 3). Figure 4 compares all diversity indicators, still using the same data as before. Here we see the adjustment problem of θ for SPD on f_1 ($n = 3$, 100 repeats). Depending on θ, SPD can give oppositional assessments. This result agrees with observations of Emmerich et al. [4]. Overall, SDNN seems to be the most reliable candidate for assessing diversity, at least as long as sets with equal sizes are compared.

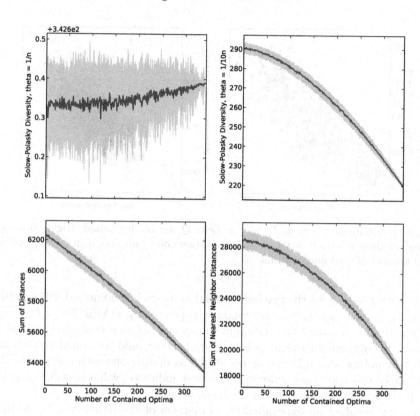

Fig. 4. Mean indicator values versus the number of optima contained in the approximation sets (f_1, $n = 3$). The shaded areas indicate standard deviations.

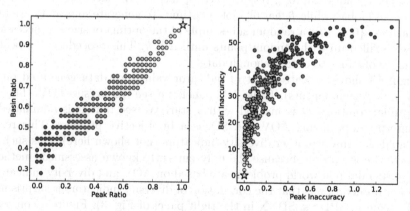

Fig. 5. Test data on problem f_2, $n = 2$. Left: BR is naturally easier to satisfy than PR. A random sample of 36 points covers > 30% of the basins. Right: BI is more challenging than PI.

Algorithm 1. Nearest-better clustering (NBC)

---

1 compute all search points mutual distances;
2 create an empty graph with $|search\ points|$ nodes;
3 **forall the** *search points* **do**   `// make spanning tree`
4 ⌐ find nearest search point that is better; create edge to it;

   `// cut spanning tree into clusters:`
5 delete edges of length $> \phi \cdot$ mean(*lengths of all edges*) ;
6 find connected components;   `// find clusters`

Figure 5 analyzes data on f_2 ($n = 2$). It shows the differences between the related pairs PR/BR and PI/BI. While BR is naturally much easier to optimize than PR, BI seems to be much more challenging than PI. This is just what we intended, but further experiments with optimization algorithms will be necessary to verify the benefit of BI.

4 Subset Selection Heuristics

In the previous section, we have enlisted 14 indicators (see Tab. 1), but except the group of diversity measures and AOV, all these employ information about optima and/or basins that may not be available for a real-world application. If neither the number of 'good enough' local optima nor the best attainable objective value is known, heuristics naturally come into play. Depending on the task of the optimization, one can either try to identify the basins within a population or find a small but diverse and good subset that may be handed to the user in order to take a final decision. The former can be established, at least to a certain extent, by means of the nearest better clustering method that has been suggested in [14]. For the latter, we propose the *Representative 5 Selection* (R5S) that is described in the following.

4.1 Nearest Better Clustering (NBC)

The NBC method was originally conceived for basin identification within a population in order to establish subpopulations or local searches. However, it may also be used as subset selection method for quality assessment. In search spaces with a small number of dimensions and low numbers of basins $|basins| \ll |\mathcal{P}|$, it has shown to be quite accurate, but for $n > 10$, we cannot recommend the method. Algorithm 1 shows how the basic method works, an improvement has been made in [15] but is not discussed here. The parameter ϕ is usually set to 2 as a robust default.

4.2 R5S: Representative 5 Selection

The R5S method is inspired from a demand that regularly occurs in real-world applications: to present a small set of solutions (on the order of 5) to the customer, who can then make the final decision on which one to implement. There

---

Algorithm 2. Representative 5 selection (R5S)

---

1 compute all search points' mutual distances;
2 **forall the** *search points* **do**
3   └ find nearest search point and nearest point that is better, record distances;
4 sort points after objective values from best to worst;
5 **forall the** *search points* **do**
6   │ $ds(point) = $ sum of distances to all worse points with weight $1/2^i$,
7   └ for $i = 1 : |worse\ points|$;
8 **if** $|search\ points| > 2$ **then**   `// remove probably uninteresting points`
9   └ delete points where nearest neighbor distance equals nearest better distance
   `// remove dominated points`
10 **forall the** *search points in reverse order up to 3rd best* **do**
11   └ remove point if better point with higher $ds(point)$ exists;
12 return remaining points;

---

are several reasons why it is usually not possible to rule out alternative solutions automatically:

- The optimization problem cannot be formulated as exact as it would be desirable; this is partly because not all criteria are 'hard' and can be given numerical values.
- Sensitivity of the solution against small changes is important but usually not taken into account by the optimization.
- Multiple solutions enable an impression of what can be expected if the optimization is continued, possibly with another method.

The number 5 is often given as orientation for the approximate number of expected good solutions. Much larger solution sets are usually undesired because they would make the post-optimization decision process tiresome for humans. Our method does not always return 5 solutions, for smaller solution sets to choose from this is obviously not possible anyway. However, it mostly returns 3 to 6 solutions, rarely more or less, although this is not explicitly coded into it.

Algorithm 2 describes the method in pseudo-code. Its main approach is multiobjective: we compute an accumulated distance from all points in the original set to all worse points with an exponentially decreasing weight and then detect the non-dominated set over the objective values and the accumulated distance values. As the maximum distance found in any given set is bounded and the sum of all weights can approach but never reach 1, the maximum for the accumulated distance is also known. Points for which the nearest neighbor distance equals the nearest better neighbor distance are removed from the set because in this case, the nearest better neighbor is usually the better representative for this search space area. The whole method is a heuristic, of course, but it is explicitly designed for cases in which almost nothing about the optimized function is known. After its application, the set of remaining points can be assigned a scalar value,

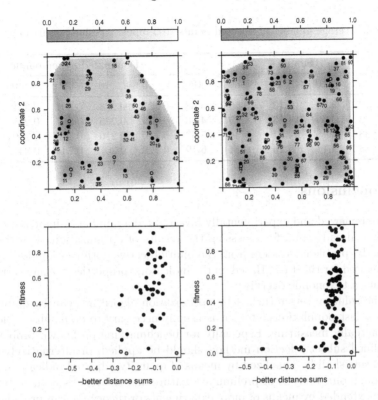

Fig. 6. Top: Example of R5S chosen solutions for 50 and 100 randomly placed points on the cosine problem, with an interpolation of the fitness landscape in the back. Bottom: The chosen solutions in the objective value/accumulated distance sum space.

e.g. by computing the hypervolume they dominate, as usually done in assessing populations in multiobjective optimization [3].

As an application example, Fig. 6 shows the selected points of 50 and 100 random samples on a separable cosine function $f_3(\boldsymbol{x}) = \sum_{i=1}^{n} \cos(6\pi x_i)$ for $n = 2$ (note that the function has 3^n global optima). The imprinted numbers give the objective value rank, the green dot represents the best point that is always chosen, the orange points the other selected points. Naturally, the method has a tendency to select the best points, but avoids them if they are located too near to an already chosen point. Tests on other simple multimodal functions obtained very similar results. Table 2 shows the average number of selected points and the averaged dominated hypervolume of the selected points on the same function over different sample sizes and number of dimensions. Note that geometry-based measuring increasingly fails if the number of dimensions raises towards 20, which does not only render this method, but all distance-based methods (e. g. diversity measures) practically useless for large dimensional spaces.

Table 2. Average number of selected points, 100 repeats, cosine test function $f_3(x)$

	Samples					Dominated hypervolume				
n	5	10	20	50	100	5	10	20	50	100
2	2.12	3.29	3.72	3.85	3.88	0.137	0.238	0.307	0.331	0.342
5	2.26	3.69	4.63	4.79	4.86	0.139	0.225	0.257	0.288	0.309
10	2.38	4.13	4.81	5.34	5.53	0.130	0.216	0.249	0.269	0.275
20	2.38	4.11	5.26	6.24	6.19	0.128	0.207	0.234	0.247	0.257

5   Conclusions

We have reviewed and experimentally investigated a number of indicators which
are either already used for assessing the results of optimization algorithms on
multimodal problems or stem from the multiobjective optimization domain and
could be used for this task. Based on the indicators' properties, we have arranged
them into a taxonomic overview.

As the solution sets returned by optimization algorithms can be quite large
(archives, last populations, etc.), it is usually necessary to do a subset selection
before actually measuring. Especially for benchmarking on known problems, a
well defined subset selection method should be applied, or alternatively, large
solution sets shall be penalized by means of quantity-adjusted indicators.

Although our study is of preliminary nature and the results need to be verified and extended by means of more experiments or theory, we can provide some
rough guidelines concerning the use of indicators for multimodal optimization.
Benchmarking: If contestants are allowed to submit arbitrary large solution sets
in benchmarking competitions, we can only recommend the indicators (A)AHD
and QAPR (given that its parameter is chosen reasonably well). If only a subset
of all optima are requested to be found as in [2], maybe also QABR is an acceptable choice. If benchmarks contain a requirement to return a fixed and probably
small number of solutions (or maybe even their own subset selection as part
of the assessment), we can additionally recommend BI. *Real-world applications:*
As no problem knowledge is available, only AOV, SDNN, SPD, or related measures are available for the assessment, possibly prepended with a subset selection
method as R5S.

References

1. Hansen, N., Auger, A., Finck, S., Ros, R.: Real-Parameter Black-Box Optimization
   Benchmarking: Experimental Setup (2013),
   `http://coco.lri.fr/downloads/download13.05/bbobdocexperiment.pdf`
   (accessed March 22, 2013)
2. Li, X., Engelbrecht, A., Epitropakis, M.: Benchmark functions for CEC 2013 special session and competition on niching methods for multimodal function optimization. Technical report, RMIT University, Evolutionary Computation and Machine
   Learning Group, Australia (2013)

3. Zitzler, E., Knowles, J., Thiele, L.: Quality assessment of pareto set approximations. In: Branke, J., Deb, K., Miettinen, K., Słowiński, R. (eds.) Multiobjective Optimization. LNCS, vol. 5252, pp. 373–404. Springer, Heidelberg (2008)
4. Emmerich, M.T.M., Deutz, A.H., Kruisselbrink, J.W.: On quality indicators for black-box level set approximation. In: Tantar, E., Tantar, A.-A., Bouvry, P., Del Moral, P., Legrand, P., Coello Coello, C.A., Schütze, O. (eds.) EVOLVE- A Bridge between Probability, Set Oriented Numerics and Evolutionary Computation. SCI, vol. 447, pp. 153–184. Springer, Heidelberg (2013)
5. Solow, A.R., Polasky, S.: Measuring biological diversity. Environmental and Ecological Statistics 1(2), 95–103 (1994)
6. Ulrich, T., Bader, J., Thiele, L.: Defining and optimizing indicator-based diversity measures in multiobjective search. In: Schaefer, R., Cotta, C., Kołodziej, J., Rudolph, G. (eds.) PPSN XI. LNCS, vol. 6238, pp. 707–717. Springer, Heidelberg (2010)
7. Ursem, R.K.: Multinational evolutionary algorithms. In: Angeline, P.J. (ed.) Proceedings of the Congress of Evolutionary Computation (CEC 1999), vol. 3, pp. 1633–1640. IEEE Press, Piscataway (1999)
8. Stoean, C., Preuss, M., Stoean, R., Dumitrescu, D.: Multimodal optimization by means of a topological species conservation algorithm. IEEE Transactions on Evolutionary Computation 14(6), 842–864 (2010)
9. Coello Coello, C.A., Cruz Cortés, N.: Solving multiobjective optimization problems using an artificial immune system. Genetic Programming and Evolvable Machines 6(2), 163–190 (2005)
10. Thomsen, R.: Multimodal optimization using crowding-based differential evolution. In: IEEE Congress on Evolutionary Computation, vol. 2, pp. 1382–1389 (2004)
11. Miller, B.L., Shaw, M.J.: Genetic algorithms with dynamic niche sharing for multimodal function optimization. In: International Conference on Evolutionary Computation, pp. 786–791 (1996)
12. Shir, O.M.: Niching in evolutionary algorithms. In: Rozenberg, G., Bäck, T., Kok, J.N. (eds.) Handbook of Natural Computing, pp. 1035–1069. Springer, Heidelberg (2012)
13. Schütze, O., Esquivel, X., Lara, A., Coello Coello, C.A.: Using the averaged hausdorff distance as a performance measure in evolutionary multiobjective optimization. IEEE Transactions on Evolutionary Computation 16(4), 504–522 (2012)
14. Preuss, M., Schönemann, L., Emmerich, M.: Counteracting genetic drift and disruptive recombination in $(\mu +/, \lambda)$-EA on multimodal fitness landscapes. In: Proceedings of the 2005 Conference on Genetic and Evolutionary Computation, GECCO 2005, pp. 865–872. ACM (2005)
15. Preuss, M.: Improved topological niching for real-valued global optimization. In: Di Chio, C., et al. (eds.) EvoApplications 2012. LNCS, vol. 7248, pp. 386–395. Springer, Heidelberg (2012)

A Benchmark on the Interaction of Basic Variation Operators in Multi-objective Peptide Design Evaluated by a Three Dimensional Diversity Metric and a Minimized Hypervolume

Susanne Rosenthal and Markus Borschbach

University of Applied Sciences, FHDW
Faculty of Computer Science, Chair of Optimized Systems,
Hauptstr. 2, D-51465 Bergisch Gladbach, Germany
{Susanne.Rosenthal,Markus.Borschbach}@fhdw.de

Abstract. Peptides play a key role in the development of drug candidates and diagnostic interventions. The design of peptides is cost-intensive and difficult in general for several well-known reasons. Multi-objective evolutionary algorithms (MOEAs) introduce adequate in silico methods for finding optimal peptide sequences which optimize several molecular properties. A mutation-specific fast non-dominated sorting GA (termed MSNSGA-II) is especially designed for this purpose.

In addition, an advanced study is conducted in this paper on the performance of MSNSGA-II when driven by further mutation and recombination operators. These operators are application-specific developments or adaptions. The fundamental idea is to gain an insight in the interaction of these components with regard to an improvement of the convergence behavior and diversity within the solution set according to the main challenge of a low number of generations. The underlying application problem is a three-dimensional minimization problem and satisfies the requirements in biochemical optimization: the objective functions have to determine clues for molecular features which - as a whole- have to be as generic as possible. The fitness functions are provided by the BioJava library. Further, the molecular search space is examined by a landscape analysis. An overview of the fitness functions in the molecular landscapes is given and correlation analysis is performed.

Keywords: multi-objective peptide design, variation operators, dynamic mutation schemes, dynamic recombination operators, landscape analysis.

1 Introduction

Multi-objective optimization (MOO) is an important issue in the fields of biochemistry, medicine and biology, especially for molecular modeling and drug-design methodologies. The aim of MOO in the chemoinformatic is to discover the optimal solution in a complex search space for several and often competing objective functions representing biological or chemical properties. In this field,

M. Emmerich et al. (eds.), *EVOLVE - A Bridge between Probability, Set Oriented Numerics,* 139
and Evolutionary Computation IV, Advances in Intelligent Systems and Computing 227,
DOI: 10.1007/978-3-319-01128-8_10, © Springer International Publishing Switzerland 2013

multi-objective genetic algorithms are established tools [1] [2] [3] as they proved themselves as effective and robust methods. In related work [31], the state-of-the-art evolutionary algorithm (10+1)-SMS-EMOA [32], [33] is applied for 1189 iterations to explore the bi-objective space consisting of 23^{22} (23 amino acids and 22-mer peptides) to design an isoform-selective ligand of the 14-3-3 protein. Like SMS-EMOA, the most popular GA's are Pareto-based: NSGA[4], NSGA-II [5], SPEA [6], SPEA2 [27]. Generally, the GA procedure is based on optimization of populations of individuals which are improved in a multi-objective sense by the genetic components: recombination (or crossover), mutation and selection of a variation of the fittest individuals for the next population. Variants of these components improve the convergence behavior and diversity significantly within the restriction and main challenge of a low number of generations. Especially the configuration and type of recombination and interaction with the mutation operator determines the search characteristic of a GA as they directly influence the 'broadness' of solution space exploration. Therefore, various works have been published, proposing guidelines and recommendations for a number of multi-objective optimization problems [7], [8]. In [9] a multi-objective genetic algorithm (MSNSGA-II) evolved for biochemical optimization was introduced. It comprises three variants of mutations: The deterministic dynamic mutation operator of Bäck and Schütz [10], the self-adaptive mutation scheme proposed by Bäck and Schütz [10] and a method which varies the number of mutations via a Gaussian distribution [35].

The suitable choice of components is an ongoing challenge [35] and usually bases on empirical analysis. Intensive research have been done according to recombination methods. A detailed study of various recombination methods can be found in [11]. Recombination operators are usually adapted for the encoding of a GA. For real-coded GAs the recombination operators are categorized in mean-centric and parent-centric recombination approaches [12]. Here, unimodal normal distribution crossover (UNDX) [13], simplex (SPX) [14] and blend crossover (BLX) [15] are mean-centric operators, whereas the well-known simulated binary crossover (SBX) [16] used in NSGA-II and the parent-centric recombination operator (PCX) [11] are parent-centric approaches. The following recombination operators are used within a character-encoded GA [17]: single point, double point, distance bisector, multi point, uniform and shuffle crossover. In each recombination, two parents are chosen randomly for recombination at randomly chosen points except for distance bisector crossover with a fixed central crossover point.

In an ensuing empirical study [23], MSNSGA-II are extended by two different recombination operators. One operator varies the number of recombination points over the generations via a linearly decreasing function and is termed 'LiDeRP'. The other operator is a dynamic 2-point recombination where the recombination points move linearly to the edges of the sequence and is termed '2-point-edges'. The evaluation of this study reveals that the convergence behavior as well as the diversity within the solutions is mainly governed by the mutation operators. The recombination operators are supposed to take the role to support the convergence behavior and diversity by its interaction with the mutation operator.

In this work, MSNSGA-II is extended by two further mutation methods and another recombination operator which are evolved application-specifically. The mutation methods vary the number of mutations over the generations via a linear and a quadratic decreasing function. The recombination operator determines the number of recombination points via an exponential function. A three-dimensional minimization problem that is as generic as possible is used to benchmark and the objective functions are established reference functions: the three objective functions are taken from the BioJava library [18]. For a closer understanding of the optimization difficulty, landscape analysis of the molecular search spaces is performed. This analysis is based on a random walk on the molecular search space.

2   Procedure and Components

The fundamental idea of MSNSGA-II is to improve upon the convergence behavior of the traditional NSGA-II and to customize NSGA-II for biochemical applications by exchanging the default component encoding, mutation, recombination and selection mechanisms with the ones specifically tailored for the proposed application. The genetic workflow of MSNSGA-II still corresponds to NSGA-II. The procedure of MSNSGA-II proceeds as follows:

Input: N Population size, T total number of generations

1. Random initialization of $P(t)$ with N individuals
2. Evaluation of objective functions
3. Pareto ranking and determination of crowding for each individual
4. Selection of parents for recombination and mutation
5. Evaluation of objective functions
6. Pareto ranking and determination of crowding distance
7. Selection of $P(t+1)$
8. Repeat steps 4.-7. if $t+1 < T$

The encoding applied is chosen to be intuitive for biochemical applications, in that the individuals are implemented by a fixed length string of 20 characters, each representing one of the natural canonical amino acids at each position. The remaining components differing to those of the standard NSGA-II are described in the following.

2.1   Recombination Operators

Every henceforth proposed recombination operator uses three randomly selected individuals as parents. The use of three parents for recombination proved to be suitable for a higher diversity within the solutions compared to the typical choice of two parents. The three parent recombination is based on the investigations of multi-parent recombination in evolutionary strategies of Eiben and Bäck [19].

Our empirical experiences reveal an overall low convergence if more predecessors (parents) are used, due to an enforced high exploration realized by high diversity. The basic idea of the following recombination operator is that a higher number of recombination points in early generations leads to a broader search in the solution space, whereas a lower number in later generations supports the local search. It is, however, questionable which functional approach provides at most adequate average rates of change.

The recombination operator 'LiDeRP' varies the number of recombination points over the generations via a linearly decreasing function:

$$x_R(t) = \frac{l}{2} - \frac{l/2}{T} \cdot t \tag{1}$$

which depends on the length of the individual l, the total number of the GA generations T and the index of the actual generation t.

A further recombination operator varies the number of recombination points over the generations via a exponetially decreasing function:

$$x_R(t) = 2 + (0,2 \cdot l - 1) \cdot 2^{\frac{-l/2}{T} \cdot t} \tag{2}$$

This operator is further termed as 'ExpoDeRP'. In general, the recombination points are chosen randomly.

2.2 Mutation Operators

In [9], it was shown that the dynamic deterministic mutation is much more successful in MSNSGA-II than the self-adaptive mutation. Commonly, self-adaptive mutation does not work well in MOO [28]. The motivation for the following newly introduced mutation schemes is due to the basic idea of dynamic mutations: high mutation rates in early generations lead to a good exploration and lower mutation rates in later generations lead to a good exploitation. Fine-tuning of these mutation rates over the generations is challenging and therefore different dynamic schemes are examined.

The first newly introduced mutation operater is the the mutation 'LiDeMut' which varies the number of mutations via a linearly decreasing function according to the recombination operator:

$$x_M(t) = \frac{l}{5} - \frac{l/5}{T} \cdot t. \tag{3}$$

The second newly introduced mutation operator is the mutation 'QuadDeMut' which determines the number of mutations via a quadratic decreasing function:

$$x_M(t) = \frac{1 - l/4}{T^2} \cdot t^2 + \frac{l}{4}. \tag{4}$$

The third mutation operator is the deterministic dynamic operator of Bäck and Schütz [10] which determines the mutation probabilities by the function

$$p_{BS} = \left(a + \frac{l-2}{T-1} t\right)^{-1}, \tag{5}$$

with $a = 2$ and the corresponding variables mentioned in the last section. The mutation rate is bounded by $(0; \frac{1}{2}]$. In combination with the proposed recombination operators, the function by Bäck and Schütz has been adapted to a lower initial mutation rate: $a = 4$ in (3).

In general, the mutation operators substitute certain characters with other ones chosen from a mutation pool and comprised of 20 different characters representing amino acids.

2.3   Selection Operator

The basic aggregation principles have been developed to combine different multi-objective goals solving the Rubics' cube [24], [25] - a problem of even higher solution space complexity. This selection process is not based on the critical component crowding distance. It is commonly known that the crowding distance does not provide good results if the number of objectives increases [29]. The crowding distance is critical even in the bi-objective case: Simplified, the crowding distance of an individual is the cumulative sum of the individual distances according to each objective function. The goal is to remove the individuals with the smallest values to ensure a high diversity. In the case of two individuals close to each other but far from the others, these two reveal quite close values though one of these should be removed for better spacing [30]. In other words, the general disadvantage of the crowding distance is that the crowding distance of individuals are based on the positions of its neighbors and not on the position of the individuals themselves.

The following pseudo-code demonstrates the workings of the standard aggregate-selection-operator in MSNSGA-II. This operator was first introduced in [9].

Algorithm 1. Procedure of AggregateSelection

1. **select x** (= tournament size) **random individuals** from population
2. **Pareto-rank** the tournament individuals

(a) **0.5 probability**
(b) preselect **front 0**
(c) **randomly chose one** to put into the selection pool

or opposite

(a) **0.5 probability**
(b) **SUS by front size** to preselect individuals from some front
(c) **randomly chose one** to put into selection pool

3. **if** (selection pool size $= \mu$) then done **else** back to 1.

In the first step, a pre-defined number of individuals is randomly selected from the population. The second step starts with the Pareto-ranking of these individuals. The following main part of the selection process is based on the biological discovery which shows that peptides with inferior fitness qualities may still have high quality subsequences of amino acids. These may become important for producing high quality solutions in later generations by recombination. Therefore, the selection procedure is two-part: With a probability of 50%, an individual of the first front is randomly selected to take part in the selection pool (left column). With a probability of 50% an individual of the other fronts is selected by stochastic universal sampling (right column). The selection process stops, when the pool size is reached, otherwise the process continues with step 1.

The specific design and parameter values are the product of an extensive empirical analysis and benchmarking process, using the proposed fitness functions. A more thorough and selection-focused examination of selection behavior under different GA-configurations such as different population and tournament sizes and optimization problems poses a highly interesting prospect and subject to ongoing work.

2.4   Fitness Functions

A three-dimensional minimization problem serves as benchmark test. To keep the runtime complexity at a low level, the designed fitness functions are chosen within limited computational complexity. Therefore, the three objective functions were selected of the BioJava library: As the individuals are encoded as character strings, this enables the use of Needleman Wunsch algorithm (NMW) [20], Molecular Weight (MW) and *hydrophilicity* (hydro). These fitness functions act comparatively: individuals are compared to a predefined reference-solution. Therefore, these three objective functions have to be minimized. A brief description of these functions is given in the following:

Needleman Wunsch Algorithm. The NMW algorithm performs a global sequence alignment of two sequences by using an iterative matrix method of calculation. More precisely, NMW is an objective function in MSNSGA-II that measures similarity of an individual to a pre-defined reference individual.

Molecular Weight. The fitness values for MW of an individual of the length l are determined from the individual amino acids (a_i) for $i = 1, ..., l$ [18]: Molecular weight is computed as the sum of mass of each amino acid plus a water molecule: $\sum_{i=1}^{l} mass(a_i) + 17.0073(OH) + 1.0079(H)$. (According to the periodic system of elements: Oxygen (O), hydrogen (H))

Hydrophilicity. The fitness value for hydrophilicity of an individual is calculated analogous to MW: $\frac{1}{l} \cdot (\sum_{i=1}^{l} hydro(a_i))$.

For a closer understanding and the source-code of these three objective functions, please refer to [18], [20].

2.5   Landscape Analysis

Landscape analysis is important to gain an insight into the difficulty of optimization in the molecular search space. The molecular search space of the proposed three-dimensional minimization problem has a dimension of 20^{20} possible solutions (20-mer peptides consisting of 20 canonical amino acids). The analysis methods are based on a random walk by a variation operator on the molecular landscape as applied in [34]. 20 random walks consisting of 100 steps are applied

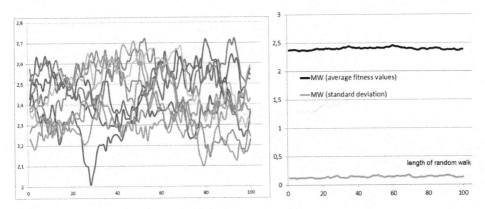

Fig. 1. MW fitness values in the molecuar landscape

Fig. 2. Average fitness values and standard deviation of MW

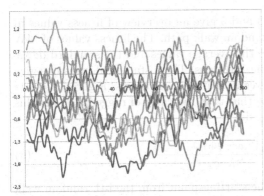

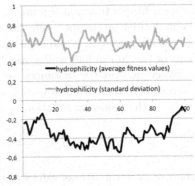

Fig. 3. Hydrophilicity fitness values in the molecuar landscape

Fig. 4. Average fitness values and standard deviation of hydrophilicity

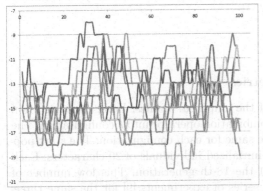

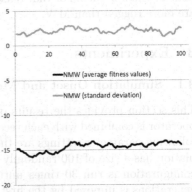

Fig. 5. NMW fitness values in the molecuar landscape

Fig. 6. Average fitness values and standard deviation of NMW

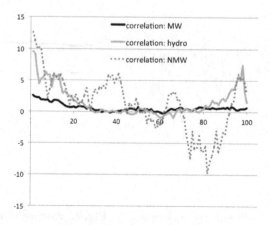

Fig. 7. Correlation Analysis

for each fitness function. Figure 1, 3 and 5 give an overview of fitness values in the molecular landscape along the random walk path. The fitness values of MW are scaled. Figure 2, 4 and 6 picture the average values and the standard deviations of the 20 random walks over the 100 steps. These statistical measures give an overview over the distances between different generations. Figure 7 depicts the results of the correlation analysis that is an indicator of ruggedness. For this analysis, the following function is used:

$$p_s = \frac{\frac{1}{n}\sum_{i=1}^{n}(x_{i0} - \mu_0)(x_{is} - \mu_s)}{\sigma_0 \cdot \sigma_s} \tag{6}$$

where x_{i0} are the fitness values of the starting point from the i-th random walk and x_{is} the fitness value after s steps of the i-th random walk. μ_0 and μ_s are the average fitness values and σ_0 and σ_s are the standard deviations of the starting point and after s steps. The correlation values of NMW and hydrophilicity are scaled by factor 1000. For that reason, the landscape of NMW and hydrophilicity are more rugged than MW.

3   Experiments

3.1   Simulation Onset and Metrics

This section provides the results of the empirical studies. Here, each mutation operator is combined with each recombination operator into a configuration. The following parameter settings are the same for each configuration: The start population has a size of 100 randomly initialized individuals of 20 characters. Each configuration is run 30 times until the 18-th generation. This low number of generations is imposed by the first series of experiments [9] for the same three-dimension optimization problem - minimization of the objective functions NMW,

Recombination: LiDeRP

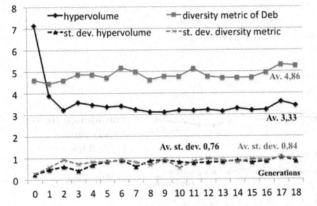

av.
confidence limits
hypervolume:
[2,95; 3,71]
diversity:
[4,44 ; 5,28]

Fig. 8. Mutation: deterministic dynamic operator by Bäck and Schütz

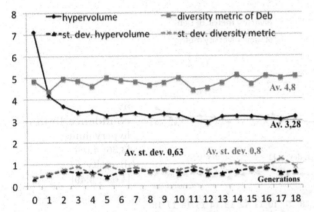

av.
confidence limits
hypervolume:
[2,97; 3,59]
diversity:
[4,4 ; 5,2]

Fig. 9. Mutation: LiDeMut

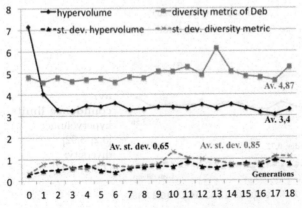

av.
confidence limits
hypervolume:
[3,08; 3,72]
diversity:
[4,45 ; 5,29]

Fig. 10. Mutation: QuadDeMut

Recombination: ExpoDeRP

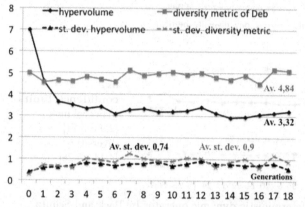

av.
confidence limits
hypervolume:
[2,95; 3,69]
diversity:
[4,39 ; 5,29]

Fig. 11. Mutation: deterministic dynamic operator by Bäck and Schütz

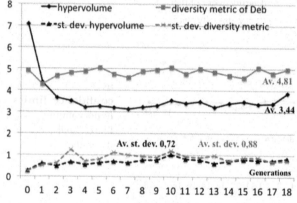

av.
confidence limits
hypervolume:
[3,36; 4,08]
diversity:
[4,65 ; 5,49]

Fig. 12. Mutation: LinDeMut

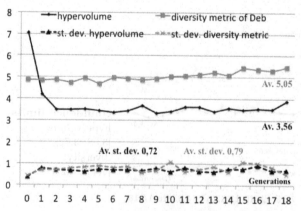

av.
confidence limits
hypervolume:
[3,2; 3,92]
diversity:
[4,66 ; 5,44]

Fig. 13. Mutation: QuadDeMut

MW and hydro. In these experiments, all MSNSGA-II configurations result in early convergence, meaning that an unusual low number (mainly under ten) is already sufficient to point out convergence behavior. Furthermore, MSNSGA-II performs better when compared to the traditional NSGA-II for this optimization problem on the chosen metrics.

Two metrics are consulted for evaluation: The S-metric or hypervolume of Zitzler [22] is used as a measure for convergence of the Pareto front. The true Pareto front for this benchmark test is unknown. As this test problem is to be minimized, the element $(0/0/0)$ is a theoretical minimal limit of the Pareto front. The hypervolume measures the space spanned by a set of non-dominated solutions to a pre-defined reference point - in this case **0**. We focus on early convergence, meaning a high number of at most optimal solutions well spread over the objective space is desired in a really low number of solutions. The restriction of the low number of generations is that only a low number of solutions is generated over the generations in general. In biochemical applications a high number of high quality solutions is required. Thus, we have to face the convergence behavior of the whole generation to the Pareto front and the hypervolume is determined for the whole generation after each iteration.

MOO has two aims: to receive optimal convergence behavior and to produce well-spread non-dominated solution sets at the same time. Regarding the latter, the diversity metric or 'Spacing metric' of Deb [21] is used which characterizes the homogeneity and the evenness of the solutions' distribution over the solution space. The diversity metric of Deb is defined as

$$\Delta = \sum_{i=1}^{|F|} \frac{|d_i - \bar{d}|}{|F|} \tag{7}$$

whereas $|F|$ is the number of non-dominated solutions and d_i is the Euclidean distance of two consecutive vectors. $\bar{d}$ is the average Euclidean distance over all distances. This metric is only suitable for two-dimensional objective spaces, as consecutive vectors cannot be defined in higher dimensions [26]. Hence, this metric has been adapted for the application on higher dimensions to

$$\Delta = \sum_{i,j=1, i<j, i\neq j} \frac{|d_{ij} - \bar{d}|}{N} \quad \text{with } N = \binom{n}{2} = \frac{n(n-1)}{2}. \tag{8}$$

In words, the Euclidean distance of each possible combination of solutions is determined (without repetition and without taking account of the order), symbolized by d_{ij}. n is the number of solutions, here $n = 100$ is the number of solutions in the generation. As in the case of the hypervolume, the diversity of the whole generation is determined. N is the number of calculated distances. $\bar{d}$ is the average distance over all determined distances.

The average metric values as well as the standard deviations over 30 runs for each configuration are depicted in Figs. 1-5. The metric values are scaled. The objective function values of individuals are normalized w.r.t. a fixed reference point (in the case of MW and hydrophilicity) and a non-varying reference individual (in the case of NMW). The average metric values are tested according to statistical significance by one-tailed t-tests at a significant level of $0,05$: For the hypervolume an upper tailed t-test and for the diversity metric a lower-tailed t-test are performed. All average metric values receive a clear statistical significance. Furthermore, below Figs. 1-5 confidence limits are listed at the same significant level of $0,05$ each for the hypervolume and for the diversity metric.

3.2 Evaluation

The best performance is achieved if the hypervolume is as small as possible - this is due to the choice of $(0/0/0)$ as reference point for this minimization problem, as mentioned in the last section - and a diversity value as large as possible. Considering this, the best performance is achieved in the configuration with the mutation LiDeMut and the recombination LiDeRP. In [23] the configuration LiDeRP and the deterministic dynamic mutation operator results in the best performance and is therefore the benchmark for further evaluations. The thesis [23]: *'the convergence behavior as well as the diversity within the solutions is mainly governed by the mutation operators. Therefore, the recombination operators are supposed to take the role to support the convergence behavior and diversity by its interaction with the mutation operator.'* holds true for the deterministic dynamic mutation (see Fig. 1 and 2). The mutation LiDeMut in combination with LiDeRP results in improved performance with regard to the hypervolume and the lower standard deviations (Fig. 3) compared to the configuration LiDeMut with ExpoDeRP (Fig. 4). The mutation QuadDeMut in combination with ExpoDeRP results in the best performance with regard to the diversity at the cost of convergence (Fig. 6). The combination of QuadDeMut with these two recombinations results in different convergence and diversity behavior (Fig. 5 and 6): The configuration of QuadDeMut and LinDeRP outperforms the configuration with ExpoDeRP. The above thesis is not true for the LiDeMut and QuadDeMut mutations: the configuration LiDeRP and LiDeMut (Fig. 3) outperforms ExpoDeRP and LiDeMut (Fig. 4) in view of convergence whereas the configuration QuadDeMut with LiDeRP (Fig. 5) outperforms the configuration QuadDeMut and ExpoDeRP (Fig. 6) in view of convergence and diversity.

In general, the deterministic dynamic mutation may be considered as the most solid mutation operator independent of the chosen recombination operator as the metric values for convergence and diversity are in the mid-range.

4 Conclusions

This paper presents an advanced empirical study about the performance of the character-encoded mutation-specific GA MSNSGA-II extended by two new mutations and a recombination operator. The main interest of this study is to gain an insight in the interaction of these components with regard to a possible improvement of the convergence behavior and the diversity within the solutions. The performance was tested on a synthetic three-dimensional minimization problem. The objective functions were drawn from the BioJava library to simultaneously optimize molecular features. Every possible combination of mutation- and recombination operators was inspected as a distinct GA configuration. It can be noted that the thesis [23]: *'the convergence behavior and the diversity within the solutions is mainly governed by the mutation operators. The recombination operators are supposed to take the role to support the convergence behavior and the diversity by its interaction with the mutation operators.'* holds true for the deterministic dynamic mutation. The best performance with regard to the convergence yields the mutation LiDeMut in combination with LiDeRP at a slightly decreasing of the diversity. The best performance with regard to the diversity within the solutions is achieved by the configuration ExpoDeRP and QuadDeMut at the cost of convergence. Concluding, the above thesis is not true for the mutations LiDeMut and QuadDeMut. Both mutations in combination with LiDeRP outperform the combination with ExpoDeRP. The most solid performance yields the deterministic dynamic mutation as the results of convergence and diversity are in the mid-range.

The difficulty of optimization for this benchmark problem is revealed by a landscape analysis of the molecular search spaces. The landscapes of NMW and hydrophilicity are very rugged, whereas the landscape of MW is relatively smooth.

As this approach seems promising for biochemical research, a closer look is necessary to gain a deeper insight into the convergence velocity rate. Future research focus on the theoretical analysis of the interaction between mutation and recombination operators and the related performance. Feasible theoretical results will be utilized in the design and improvement of mutation and recombination variants. Here, customized mutation operators with direct feedback in view of the solution quality are in focus. Furthermore, we will evolve and benchmark further selection procedures against the critical component crowding distance.

References

1. Vainio, M.J., Johnson, M.S.: Generating conformer ensembles using a multiobjective genetic algorithm. J. Chem. Inf. Model. 47(6), 2462–2474 (2007)
2. Nicolaou, C.A., Brown, N., Pattichis, C.S.: Molecular optimization using computational multi-objective methods. Drug Discovery & Development 10(3), 316–324 (2007)
3. Knapp, B., Gicziv, V., Ribarics, R.: PeptX: Using genetic algorithms to optimize peptides for MHC binding. BMC Bioinformatics 12, 241 (2011)
4. Srinivas, N., Deb, K.: Multiobjective optimization using nondominated sorting in genetic algorithms. J. Evol. Comput. 2(3), 221–248 (1994)
5. Deb, K., Pratap, A., Agarwal, S., Meyarivan, T.: A fast and elitist multiobjective genetic algorithm: NSGA-II. IEEE Trans. Evol. Comput. 6(2), 182–197 (2002)
6. Zitzler, E., Thiele, L.: An evolutionary algorithm for multiobjective optimization: The strength Pareto approach. Technical report 43, Computer engineering and Networks Laboratory (TIK), Swiss Federal Institute of Technology, ETH (1999)
7. Kötzinger, T., Sudholt, D.: How crossover helps in Pseudo-Boolean Optimization. In: Proceedings of the Genetic and Evolutionary Computation Conference GECCO, pp. 989–996 (2011)

8. Sato, H., Aquire, H.: Improved S-CDAS using Crossover Controlling the Number of Crossed Genes for Many-Objective Optimization. In: Proceedings of the Genetic and Evolutionary Computation Conference GECCO, pp. 753–760 (2011)
9. Rosenthal, S., El-Sourani, N., Borschbach, M.: Introduction of a Mutation Specific Fast Non-dominated Sorting GA Evolved for Biochemical Optimization. In: Bui, L.T., Ong, Y.S., Hoai, N.X., Ishibuchi, H., Suganthan, P.N. (eds.) SEAL 2012. LNCS, vol. 7673, pp. 158–167. Springer, Heidelberg (2012)
10. Bäck, T., Schütz, M.: Intelligent mutation rate control in canonical genetic algorithm. In: Proc. of the International Symposium on Methodology for Intelligent Systems, pp. 158–167 (1996)
11. Deb, K., Anand, A., Joshi, D.: A computationally Efficient Evolutionary Algorithm for Real Parameter Optimization, KanGAL report: 2002003
12. Deb, K., Joshi, D., Anand, A.: Real-coded evolutionary algorithms with parent-centric recombination. KanGAL Report No. 2001003 (2001)
13. Ono, I., Kobayashi, S.: A real-coded genetic algorithm for functional optimization using unimodal normal distribution crossover. In: Proceedings of th 7th International Conference on Genetic Algorithms (ICGA-7), pp. 246–253 (1997)
14. Tsusui, S., Yamamura, M., Higuchi, T.: Multi-parent recombination with simplex crossover in real-coded genetic algorithms. In: Proceedings of the Genetic and Evolutionary Computing Conference (GECCO 1999), pp. 657–664 (1999)
15. Eshelman, L.J., Schaffer, J.D.: Real-coded genetic algorithms and interval schemata. In: Whitley, D. (ed.) Foundation of Genetic Algorithm II, pp. 187–202 (1993)
16. Deb, K., Agrawal, R.B.: Simulated binary crossover for continuous search space. Complex System 9, 115–148 (1995)
17. Knapp, B., Gicziv, V., Ribarics, R.: PeptX: Using genetic algorithms to optimize peptides for MHC binding. BMC Bioinformatics 12, 241 (2011)
18. BioJava: CookBook, release 3.0, http://www.biojava.org/wiki/BioJava
19. Eiben, A.E., Bäck, T.: Empirical investigation of multiparent recombination operators in evolutionary strategies. Evolutionary Computation 5(3), 347–365 (1997)
20. Needleman, S.B., Wunsch, C.D.: A general method applicable to the search for similarities in the amino acid sequence of two proteins. Journal of Molecular Biology 48(3), 443–453 (1970)
21. Deb, K., Pratap, A., Agarwal, S., Meyarivan, T.: A fast and elitist multiobjective genetic algorithm: NSGA-II. IEEE Trans. Evol. Comput. 6(2), 182–197 (2002)
22. Zitzler, E., Thiele, L.: Multiobjective optimization using evolutionary algorithms- a comparative case study. In: Eiben, A.E., Bäck, T., Schoenauer, M., Schwefel, H.-P. (eds.) PPSN 1998. LNCS, vol. 1498, pp. 292–301. Springer, Heidelberg (1998)
23. Rosenthal, S., El-Sourani, N., Borschbach, M.: Impact of Different Recombination Methods in a Mutation-Specific MOEA for a Biochemical Application. In: Vanneschi, L., Bush, W.S., Giacobini, M. (eds.) EvoBIO 2013. LNCS, vol. 7833, pp. 188–199. Springer, Heidelberg (2013)
24. El-Sourani, N., Borschbach, M.: Design and comparison of two evolutionary approaches for solving the rubik's cube. In: Schaefer, R., Cotta, C., Kołodziej, J., Rudolph, G. (eds.) PPSN XI. LNCS, vol. 6239, pp. 442–451. Springer, Heidelberg (2010)
25. Borschbach, M., Grelle, C., Hauke, S.: Divide and evolve driven by human strategies. In: Deb, K., et al. (eds.) SEAL 2010. LNCS, vol. 6457, pp. 369–373. Springer, Heidelberg (2010)

26. Grosan, C., Oltean, M., Dumitrescu, P.: Performance Metrics for Multiobjective Evolutionary Algorithms. In: Proceedings of Convergence on Applied and Industrial Mathematics, Romania (2003)
27. Zitzler, E., Laumanns, M., Thiele, L.: SPEA2: Improving the Strength Pareto Evolutionary Algorithm. TIK-Report 103 (2001)
28. Laumanns, M.: Analysis and Applications of Evolutionary Multiobjective Optimization Algorithms. PhD thesis no. 15251, Swiss Federal Institute of Technology Zurich (2003)
29. Coello, C.A.C.: 20 Years of Evolutionary Multi-Objective Optimization: What has been done and What remains to be done. In: Yen, G.Y., Fogel, D.B. (eds.) Computational Intelligence: Principles and Practise, pp. 73–88. IEEE Computer Society (2006)
30. Li, M., Zheng, J., Wu, J.: Improving NSGA-II Algorithm based in Minimum Spanning Tree. In: Li, X., et al. (eds.) SEAL 2008. LNCS, vol. 5361, pp. 170–179. Springer, Heidelberg (2008)
31. Sanchez-Faddeev, H., Emmerich, M.T.M., Verbeek, F.J., Henry, A.H., Grimshaw, S., Spaink, H.P., van Vlijmen, H.W., Bender, A.: Using Multiobjective Optimization and Energy Minimization to Design an Isoform-Selective Ligand of the 14-3-3 Protein. In: Margaria, T., Steffen, B. (eds.) ISoLA 2012, Part II. LNCS, vol. 7610, pp. 12–24. Springer, Heidelberg (2012)
32. Emmerich, M., Beume, N., Naujoks, B.: An EMO Algorithm using the Hypervolume Measure as Selection Criterion. In: Coello Coello, C.A., Hernández Aguirre, A., Zitzler, E. (eds.) EMO 2005. LNCS, vol. 3410, pp. 62–76. Springer, Heidelberg (2005)
33. Beume, N., Naujoks, B., Emmerich, M.: SMS-EMOA: Multiobjective Selection based on Dominated Hypervolume. European Journal of Operation Research 181(3), 1653–1669 (2007)
34. Lee, B.V.Y.: Analysing Molecular Landscape Using Random Walks and Information Theory. Master Thesis, Leiden Institute for Advanced Computer Science, Leiden University (2009)
35. Röckendorf, N., Borschbach, M., Frey, A.: Molecular evolution of peptide ligands with custom-tailored characteristics. PLOS Computational Biology, Open Access Journal (December 2012)

Logarithmic-Time Updates in SMS-EMOA and Hypervolume-Based Archiving

Iris Hupkens and Michael Emmerich

Leiden Institute of Advanced Computer Science, Faculty of Science, Leiden
University, 2333-CA Leiden, The Netherlands
{ihupkens,emmerich}@liacs.nl

Abstract. The hypervolume indicator is frequently used in selection
procedures of evolutionary multi-criterion optimization algorithms
(EMOA) and in bounded size archivers for Pareto non-dominated points.
We propose and study an algorithm that updates all hypervolume contributions and identifies a minimal hypervolume contributor after the
removal or insertion of a single point in $\mathbb{R}^2$ in amortized time complexity $O(\log n)$. This algorithm will be tested for the efficient update of
bounded-size archives and for a fast implementation of the steady state
selection in the bi-criterion SMS-EMOA. To achieve an amortized time
complexity of $O(\log n)$ for SMS-EMOA iterations a constant-time update method for establishing a ranking among dominated solutions is
suggested as an alternative to non-dominated sorting. Besides the asymptotical analysis, we discuss empirical results on several test problems and
discuss the impact of the overhead caused by maintaining additional
AVL tree data structures, including scalability studies with very large
population size that will yield high resolution approximations.

1 Introduction

In evolutionary multiobjective optimization algorithms *(EMOA)* [6] and
algorithms that keep bounded-size archives for Pareto optimal points [14]
(bounded-size archiving), fast ranking schemes that take into account diversity
and proximity to Pareto fronts are required. The hypervolume indicator (also:
hypervolume, S-Metric, Lebesgue measure) is a commonly applied measure for
the quality of a Pareto front approximation. As it can be computed without
knowledge of the true Pareto front, it can be used not only in performance assessment, but also in selection schemes of EMOA and in bounded-size archiving.
The maximization of the hypervolume indicator over bounded-size approximation sets, yields approximation sets that are distributed across the Pareto front
and cover a diverse set of attainable objective function vectors (see Fonseca and
Fonseca [5])[1].

The focus of this work will be on EMOA and bounded-size archivers that use
single point replacement schemes to keep the size of a population bounded. After

---

[1] An objective function vector is attainable, if it is either possible to find a preimage
of this vector in the search space or a preimage of a vector that dominates it.

M. Emmerich et al. (eds.), *EVOLVE - A Bridge between Probability, Set Oriented Numerics,* 155
and Evolutionary Computation IV, Advances in Intelligent Systems and Computing 227,
DOI: 10.1007/978-3-319-01128-8_11, © Springer International Publishing Switzerland 2013

a point is added to a population or archive of size n, the size of this population is kept constant by selecting a subset from the now $n + 1$ points that maximizes the hypervolume indicator. This is equivalent to removing a point with minimal exclusive contribution to the hypervolume indicator (hypervolume contribution). Single point replacement is used in bounded size archivers [14] and also in the SMS-EMOA [8] which is one of the first and most commonly applied EMOA using hypervolume-based selection.

A disadvantage of using hypervolume indicators in selection schemes has so far been, that as compared to simpler selection schemes the required computational overhead caused by repeated computations of hypervolume indicators is high. This is in particular the case for many-objective optimization problems, but also in the very common case of bicriteria optimization hypervolume indicators where hypervolume contributions can be computed with time complexity in $O(n \log n)$. In particular when working with fast-computable objective functions and large population sizes and number of iterations, hypervolume computations can dominate the computational effort and strongly limit the algorithms' performance.

The goal of this paper is to provide provably fast incremental algorithms for implementing the hypervolume-based selection in SMS-EMOA and online updates of hypervolume-based bounded archiving. The focus is on bi-criterion problems, and we will study both asymptotical time complexity and empirical performance.

In Section 2 some of the concepts used throughout this paper are explained. Section 3 describes SMS-EMOA, and some other work related to this paper. In Section 4 we prove that updating hypervolume contributions is possible with amortized time complexity of $O(\log n)$ by presenting an algorithm that proves that theorem. In Section 5 we discuss how this theorem can be used to improve the time complexity of SMS-EMOA. In Section 6 the performance of this proposed method is investigated by measuring the time needed to update hypervolume contributions in an implementation of a 2-D archiver which uses the proposed update scheme, and the empirical performance of a fast SMS-EMOA implementation is tested. Finally, Section 7 contains some concluding remarks.

2   Technical Preliminaries

Here we consider minimization of k objective functions $f_i : X \to \mathbb{R}$, $i = 1, \ldots, k$ over some decision space X. In particular we will focus on the case $k = 2$ (bi-criterion optimization), but some techniques can be generalized to more dimensions. Most of our discussion will be on finite point sets in the objective space $\mathbb{R}^k$ that contains the image vectors (or: objective vectors) of the mapping provided by $(f_1, \ldots, f_k)^T$. As selection and archiving procedures operate on sets of objective vectors, it is of no concern for the discussion of these algorithms where the preimages of these vectors are located. We will use the notation $p.x$ to denote the f_1 coordinate of an objective function vector p and $p.y$ to denote the f_2 coordinate of an objective vector p. In this case Pareto dominance is defined as follows: An objective vector p dominates an objective vector p', if and only if $p.x \le p'.x$ and $p.y \le p'.y$ and $p \ne p'$.

Each objective vector in a bi-criterion optimization problem can be visualized as a 2-dimensional point in the space spanned by the values of f_1 (horizontal axis) and the values of f_2 (vertical axis).

See Figure 1. Green points (stars) depict non dominated objective vectors, blue points denote dominated solutions, and the red curve separates the attained subspace from the non-attained subspace.

In Pareto optimization with bounded-size archives/populations we are interested in finding a set of n non-dominated solutions that approximate a Pareto front in the objective space. A set of n objective vectors will be termed an approximation set of size n. Here the Pareto front is the set of all non-dominated objective vectors that can be obtained from solutions in X.

It is usually infeasible to calculate the entire Pareto front. The amount of solutions in it might be infinitely large, or simply too large to be practical. Therefore it becomes important to find an approximation set without knowing the true Pareto front. In this case a method for measuring the quality of a finite approximation set is needed. The hypervolume indicator is one such measure.

2.1 Hypervolume and Hypervolume Contributions

The dominated hypervolume of a set of points representing solutions is the area (or in arbitary dimensions the Lebesgue measure) of the set of all points in the objective space that is dominated by that set and bounded from above by a reference point. In Figure 1, the area above the attainment curve is the dominated hypervolume of the Pareto optimal set. The higher the dominated hypervolume of a set of points, the better an approximation it is considered to be. The dominated hypervolume contribution of a point in this set represents the volume of the search space that is dominated by that point, and not by any other points in the set. If a point needs to be removed from the set to make it smaller, the point with the lowest hypervolume contribution is a good candidate, as this removal is equivalent to maintaining the subset of size $n - 1$ which dominates the largest hypervolume.

2.2 AVL Trees

An AVL tree [11] is a binary search tree that keeps an internal balance factor for every node, representing the difference in height between its left and right subtrees. The AVL tree is considered balanced if this difference is not larger than 1 for any node in the tree, in which case its height is bounded by approximately $1.44 * \log n$. Nodes on the path from a newly inserted/deleted node to the root are rotated if the imbalance exceeds 1. Because the height of the AVL tree is bounded, and because all overhead required to keep it that way has a complexity of $O(\log n)$, the lookup, insertion and deletion operations on an AVL tree all have a worst-case complexity of $O(\log n)$. Red-black trees have been suggested as an alternative to AVL trees, but essentially offer the same complexity for the operations considered in this work.

3   SMS-EMOA and Related Work

SMS-EMOA [8] is an evolutionary multiobjective algorithm that uses the hypervolume measure as selection criterion. In each iteration of the algorithm one new individual is created by using random variation operations. After adding the new individual, an individual with minimal contribution to the hypervolume indicator is removed. All dominated solutions have a contribution of zero. Therefore, in order to decide among multiple dominated solutions which one to delete, SMS-EMOA partitions the approximation set into layers of mutually non-dominated points (fronts) of decreasing rank and deletes an individual on the worst ranked layer. Calculating the differently ranked fronts is done using the fast-nondominated-sort algorithm from NSGA-II [7], and has a worst-case complexity of $O(mn^2)$ with m being the number of dimensions. When there is only a single front and the problem is two-dimensional, calculating the hypervolume contributions of the points in that front is dominated by the sorting algorithm and has a complexity of $\Theta(n \log n)$. Calculating the hypervolume contribution of any one point can be done in constant time plus the time it takes to find the points that come right before and right after it in the sorting order, and therefore takes only $O(n)$ time if the points are already sorted.

Bringmann and Friedrichs [3] show that not just calculating, but also approximating the least hypervolume contributor is #P-Hard in the number of dimensions d. If d is constant, the problem has polynomial complexity. They have also devised an algorithm for calculating the hypervolume contributions of sets in higher dimensions[4] which runs in $O(n^{d/2} \log n)$. For two and three dimensions calculating the minimal hypervolume contributor and all contributions has time complexity $\Theta(n \log n)$ [9].

Other work that has been done in the area of updating hypervolume contributions quickly includes [2], in which a method is given for not having to fully recalculate the hypervolume contributions of points in three dimensions by keeping them the same if they can be shown to not have changed, and by recalculating hypervolume contributions after the removal of a point based on the contributions of that point (which is usually smaller and simpler to calculate since it is the worst point). No complexity analysis is given for this method, though a large empirical speedup is shown during experiments.

4   Logarithmic Time Updates

In this section, we will prove the following theorem by providing a complexity analysis of an algorithm that implements it:

Theorem 1. *Let P denote a set of n mutually non-dominated points on a 2-D plane. When a new point is inserted into P, all points dominated by that point*

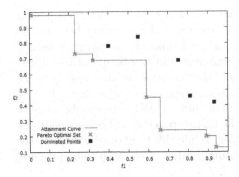
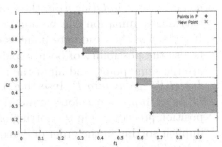

Fig. 1. A visual representation of a set of objective vectors. Blue squares: dominated points. Green stars: Non-dominated points.

Fig. 2. A visualization of the hypervolume contributions of a set of mutually non-dominated points, and how the insertion of a new point affects them

are removed. Updating P and the hypervolume contributions of the points in P after adding a single point can be done with a time complexity of amortized $O(\log n)$.

Proof. Points in P are kept in an AVL tree, sorted in order of ascending x value. Because the points in P are mutually non-dominated, each point that comes after a point in the sorting order will need to have a smaller y value than that point to not be dominated by it. Therefore, the set of points is implicitly also sorted in order of descending y value.

Since the points are on a 2-D plane, the hypervolume contribution of a point p in P is equal to the rectangle which has p at its lower left corner, with the height being equal to the difference in y value between p and the point in P with the next lowest y value and the width being equal to the difference in x value between p and the point in P with the next lowest x value. See Figure 2 for a visualization. The size of the hypervolume is also bounded by a reference point representing the 'worst possible' point in the search space. This only affects the hypervolume contributions of the two outermost points, which lack a neighboring point with a lower x or y value that would otherwise bound their hypervolume contribution. In SMS-EMOA we always want to keep those points, and therefore we can sidestep the problem of choosing a reference point by setting the hypervolume contribution of these points to be infinitely large. Whenever we have a point p which is not part of P, but which we are considering adding to it, we have to perform the following operations:

1. *Check if p is non-dominated by the points already in P by finding a neighbor in the x value*: If it is dominated, then we do not want to include it, so we can skip the following steps. Only the point q with equal or next lower x value to $p.x$ needs to be considered, because points s with $s.x > p.x$ can never dominate p, and if q does not dominate p, no other point can dominate it

since all the points s with $s.y < q.y$ have $s.x > p.x$. In an AVL tree, finding q can be done in $O(\log n)$ in the worst case. Once the point has been found, checking domination can be done in constant time by comparing y values. If $q.y \leq p.y$, the candidate point is dominated and can be discarded. There is also the possibility of both x and y being the same: in that case, p and q are the same point, and all of the following steps can also be skipped.

2. *Insert the point into P.* Inserting a point into an AVL tree can be done in $O(\log n)$. If $p.x = q.x$, for q found in the first step, p should be inserted in the position preceding q in the sorting order. While this temporarily violates the implicit sorting in the order of descending y value, it causes q to be properly deleted in the next step.

3. *Delete points that are dominated by p.* Points with a lower x value will never become dominated, so only the points that occur later in the sorting order have to be considered, and only their y value has to be compared to $p.y$. Since the points are sorted in descending order of y value, we can start at the point which occurs after p in the sorting order, process points in descending order until we have either reached the last point in P or found a point that is not dominated by p, and delete the interval of points in between p and that point.

   Deleting a single point can be done in $O(\log n)$, because that is the worst-case complexity of both deletion and lookup in an AVL tree. In the worst case, up to n points will be deleted, so there is an upper bound on the worst case complexity for this step of the algorithm of $O(n \log n)$.

4. *Update the hypervolume contributions of points in P.* When a point is added to or removed from P, up to three points need to have their hypervolume contribution changed: the point that was inserted, and the points that come right before and after it in the sorting order. Note that it is not necessary to update the hypervolume contribution after deleting a point that is dominated by p, as done in step 3, because, since the points that are deleted are always in a row, the neighbors of a point that was just deleted are either the same neighbors that will need to be updated due to the insertion of p, or points that are themselves dominated by p.

   The calculation of the new hypervolume contribution of a point can be done in constant time, but in the worst case $O(\log n)$ time can be required to find a neighbor point in the AVL tree. Therefore the worst-case complexity of this step is also $O(\log n)$.

The worst-case complexity of adding a point to P stems from the complexity of the second operation described above, in which points are deleted from P. It is the only step which has a complexity greater than $O(\log n)$. However, each point can only be deleted from P once, and the number of points in P can only increase (by one) if a point is added without deleting any points. If this update scheme is used to add every point to the set, then for every k points that are deleted there have to have previously been at least k updates of the set which did not require the deletion of any points. Therefore, the amortized complexity of this update scheme is on average $O(\log n)$ per update after any n steps of the algorithm.

A few shortcuts naturally present themselves when implementing the proposed update scheme. It is usually not necessary to find the neighbors that are used in step 4, because they are already found in previous steps: if the right neighbor exists, it is always the point that was found in step 3 as the first point that is not dominated by p. The upper neighbor is point q as found in step 1, unless the x value of q and p was equal. In that specific case, the true upper neighbor does need to be found in the AVL tree before being able to update the hypervolume contribution of p, but its own hypervolume contribution will not need to be updated since the only part of the right neighbor that is used to calculate the hypervolume is its x value, which stayed the same.

The update scheme described can be extended to allow for removing points with the lowest hypervolume contribution from P:

Corollary 1. *A replacement is a transaction consisting of an insertion of a new point to a set P in $\mathbb{R}^2$ and subsequent removal of the point in P with lowest hypervolume contribution. The amortized time complexity for n replacements is $O(\log n)$ per replacement after any n replacements.*

Proof. A second AVL tree can be maintained that is sorted by hypervolume contributions. This tree will be called the *HC-tree*, while the tree that is sorted by x values will be called the *X-tree*. The HC-tree needs to be updated (with complexity $O(\log n)$) whenever a hypervolume contribution is changed, a point is deleted, or a point is inserted, each of which already requires an operation on an AVL tree with a complexity of $O(\log n)$. When this second AVL tree is maintained, finding the point with the worst hypervolume contribution can be done with a single lookup of complexity $O(\log n)$. Deleting this point can then be done in $O(\log n)$, including the necessary updates of the hypervolume contribution of the point that comes right before it in the sorting order and the point that comes right after it.

Remark 1. By extending the update scheme in this way, it is possible to implement a *bounded 2-D archiver* which works with amortized time complexity $O(\log n)$, where n is the maximum size of the archive.

It is often desirable to maintain the value of the hypervolume indicator for an archive, in order to measure progress:

Corollary 2. *Starting from the empty set and a stream of points to be inserted one by one to the archive, online updates of the hypervolume indicator can be computed in amortized $O(\log n)$ time per point.*

Proof. One possibility to achieve updates of the hypervolume indicator in 2-D in logarithmic time is to keep track of the cumulated hypervolume contributions in the above algorithm. Another possibility, is to use a modified version of the dimension sweep algorithm of Fonseca et al. [1,10] for computing the hypervolume indicator in 3-D. This proceeds in increasing order of the third (or z) value of the objective function vector and successively updates the area dominated by a set of points in their projection to the xy-plane. For any n such updates the algorithm requires time in $O(n \log n)$. The modification is as follows: One can simply use the update scheme of the hypervolume indicator in the projection to the xy-plane. (This way, the z value can be interpreted as the time when a point enters the archive to be minimized.)

Remark 2. Using the second idea in the proof of corollary 2, the dimension sweep algorithm for computing the hypervolume indicator in 4-D that was proposed by Guerreiro et al. [12] can be used for an amortized $O(n)$-time incremental update scheme of the hypervolume indicator for tri-criterion optimization. In this case points are inserted in the order of the fourth dimension into a 3-D hypervolume-based archive of point projections to the xyz-hyperplane. After n updates the amortized time complexity is $O(n^2)$. A more detailed analysis of the higher dimensional cases and unbounded archiving is beyond the scope of this paper and left to future work.

5   Improving SMS-EMOA

SMS-EMOA keeps a population which includes not just mutually non-dominated points, but dominated points as well. While storing all dominated points in an additional data structure after they are removed from P is a trivial addition, SMS-EMOA requires the hypervolume of the worst-ranked front to be calculated rather than the best-ranked front. It is possible to use the update scheme described above to update all the fronts, however this comes with an increase in complexity. The situation shown in Figure 3 illustrates that there can be a large number of changing ranks as a consequence of the insertion of a single point. This can reoccur an arbitrary number of times through the runtime of the algorithm.

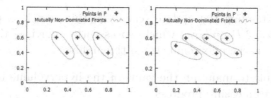

Fig. 3. If dominated fronts are kept in sets of mutually non-dominated points, then inserting a single point can necessitate recalculating all sets - and all hypervolume contributions

As a fast alternative method it is proposed to rank the dominated points based on how long ago they were removed from the set of non-dominated points. If only small numbers of points are dominated, then the difference in ranking caused by not calculating hypervolume contributions will be marginal. In situations where many points are dominated, though, the modification of the scheme might have a bigger impact. It will thus be tested in the next section.

6   Empirical Performance

All empirical tests were performed on the same computer and using the same compiler: A GNU C++ compiler, O3 optimized, using MINGW/Windows 7 on an Intel i7 quadcore CPU with 2.1 GHz clockspeed and 8Gb RAM. For time measurements a stopwatch method based on the processor's high resolution performance counter was used. Source codes are available on request to the authors.

In order to test the performance of the proposed update scheme, a bounded-size 2-D archiver was implemented. This archiver only calculates the hypervolume contributions of points that are not dominated by any other points in P. Dominated points are stored in a double-ended queue, from which the oldest point is removed whenever room needs to be made for a new point.

6.1   Performance of Incremental versus Non-incremental Hypervolume Updates

The performance of the incremental and non-incremental update schemes using AVL trees was tested using the following method: Pseudorandom numbers used in the tests below were generated using a 64-bit linear congruential generator of which only the 32 highest-order bits were used. In a first test, each n, a set P of n different points was prepared for which the x value was an even pseudorandom number between 0 and 2^{31}, and the y value was equal to $2^{31} - x$. This choice of values causes all possible points to be mutually non-dominated or the same point. Hypervolume values were computed in 64-bit. Then, 100000 insertions were performed by repeatedly generating a point p for which the x value was an odd pseudorandom number between 0 and 2^{31} and the y value was again equal to $2^{31} - x$, then inserting p into P and removing the point which had the worst hypervolume contribution afterwards. The insertion and the removal that resulted from it were included in the time measurements, the preparation of P and the generation of the random numbers were not.

Figure 4 shows the results averaged over 20 runs as compared to a non-incremental update scheme. The non-incremental scheme recomputes all hypervolume contributions in each iteration, but already keeps the population sorted using an AVL tree.

The linear time complexity of the non-incremental hypervolume update scheme is especially noticeable, as it almost immediately dominates the logarithmic time complexity of the AVL tree operations necessary to insert and remove points from the set. For the implementation used in this test, an incremental update scheme starts outperforming a non-incremental update scheme when P is bounded to values higher than about 70.

To better visualize the time complexity of the incremental hypervolume update scheme, the test was also performed for larger values of n, where n started at 4 and was repeatedly multiplied by 1.5. Figure 5 shows the results of this test averaged over 20 runs. With logarithmic time complexity, the expected result was a diagonal line when plotted logarithmically, however this was not the case.

The time taken increases slowly for a while, then increases faster, then starts to increase more slowly again.

Examining the operations performed during the tests shows the likely cause. Figure 6 shows the number of rotations performed in each of the AVL trees during the test; again, averaged over 20 runs, and there was little variance. For every point that is removed from the X-tree, up to three points need to be removed from the HC-tree and up to two points need to be inserted, and for every point that is inserted into the X-tree, up to two points need to be removed from the HC-tree and up to three points need to be inserted. Therefore, the number of rotations can be five times as large.

When n is very small, the limited amount of possible structures that an AVL tree with n points can have causes the values to jump around a bit at first before settling into a pattern. The number of rotations stays much the same both when inserting and deleting points in the X-tree. The HC-tree behaves differently. The rotations required after deleting a point in the tree shrinks slowly for the first part, then starts rising and reaches its peak after n has reached six digits. In the timing graph, the slope of the HC-tree deletion rotations is noticeable in the slope of the timing of the test as a whole. It is flatter in the part where the HC-tree deletion rotations are slowly decreasing, then becomes steeper when the HC-tree deletion rotations start to increase.

Once n is above ca. 10^6, the effect of the HC-tree deletion rotations is replaced by that of the HC-tree insertion rotations. These initially stay pretty much constant, just like the X-tree rotations, but once the HC-tree deletion rotations become more common, this type of rotation starts to become more numerous as well. The amount of HC-tree insertion rotations increases very quickly for a while after the HC-tree deletion rotations start going down again. Due to memory limitations, it was not feasible to test whether the HC-tree insertion rotations stay at this new level or if they will gradually decrease in number after this point.

The behavior of the HC-tree might be a consequence of how it is used. Nodes are not equally likely to be inserted or removed. The leftmost node is removed every time, as it represents the hypervolume contribution of the node that is removed to keep the size of P constant. The other nodes that are removed from the HC-tree represent neighbors of a point that was inserted or removed, meaning they will be replaced by nodes that are smaller or larger respectively, and thus more likely to end up in the sides of the tree than in the middle. Either way, the value of n for which the amount of rotations starts increasing is sufficiently large for this to be of no concern for practical purposes, and even after it increases the algorithm is still fast.

Next, initialization time or time for incrementally updating an unbounded archive was empirically evaluated. See Figure 7 for a plot of the time taken to either fill an AVL tree with n points (for the non-incremental update scheme), or fill an AVL tree with n points and incrementally update the hypervolume contributions of each of these n points after each insertion of a point (for the incremental update scheme). As can be seen, the incremental update scheme has an almost linear increase in a linearly scaled diagram. This shows that the logarithmic factor in the $O(n \log n)$ time complexity for n insertions is almost negligible. Used in the initialization phase of an unbounded archiver, the incremental scheme would take several times longer than a scheme using only a X-tree. For both methods, the cost of initializing, even for large archives is still only in the range of milliseconds. Note, that the cost of not using any tree would be prohibitively high for the population sizes reported in the diagram.

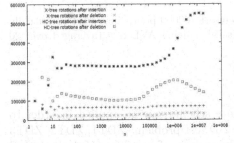

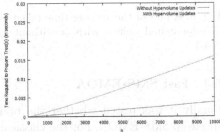

Fig. 4. Performance of the update scheme described above, versus an non-incremental hypervolume update scheme

Fig. 5. Logarithmic plot of the performance of the incremental and non-incremental hypervolume update schemes

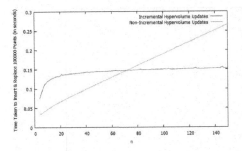

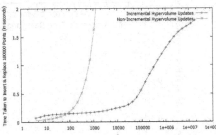

Fig. 6. Average number of rotations required for AVL tree updates during each test

Fig. 7. A plot of the time taken to prepare an initial population of n mutually non-dominated points (averaged over 20 runs)

6.2   AVL Trees versus Non-self-balancing Binary Search Trees

Although the complexity of a regular binary search tree is $O(n)$ in the worst case, this cannot be expected to occur in the average case, and there is some overhead associated with the self-balancing of the AVL tree. To measure whether this overhead is significant, the first test described in Section 6.1 was repeated, except the incremental update scheme was compared to a version of itself without the self-balancing parts of the AVL tree code (turning the trees into regular binary search trees). As seen in Figure 8, the average performance of the AVL tree is very similar to that of the regular binary search tree. On average, it appears to get slightly slower with increasing n. Not pictured: the non-self-balancing binary search tree starts rising much more sharply after n becomes larger than about a million. That result is consistent with the results found in Section 6.1, where the number of rotations required to keep the HC-tree balanced after insertions increases after that point. However, it needs to be investigated still whether this comparison provides the same results for less symmetrical test problems.

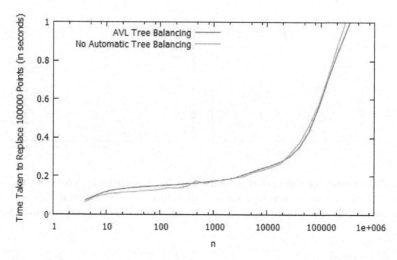

Fig. 8. A plot of the average time taken to insert 100000 points into a set of mutually non-dominated points, with or without a self-balancing AVL tree (averaged over 20 runs)

6.3   Fast SMS-EMOA

Finally it was tested whether the update scheme can be applied in order to speed-up SMS-EMOA. For the classical 2-D test problems ZDT1, ZDT2, ZDT3 and ZDT6, a number of 10 runs per problem were performed for (1) the canonical SMS-EMOA described in Emmerich et al. [8] in its original implementation and (2) for an SMS-EMOA using fast incremental updates and a queue to maintain dominated solutions. Settings were as in [8], e.g. population size was 30 and 20000 iterations were performed. The reference point was $\{10000000\}^2$. The resulting histories of the hypervolume indicator are depicted in Figure 10 to Figure 17. As can be seen from the graphs, the changed selection scheme does not lead to worse results. Also by visual inspection of a final result, it was observed that both methods achieve very accurate results (cf. Figure 9) However, instead of taking a total time of 1.585 seconds for all selection steps, a single run with fast SMS-EMOA now takes only 0.02936 seconds for all selection steps. This means an improvement by a factor of no less than 50.

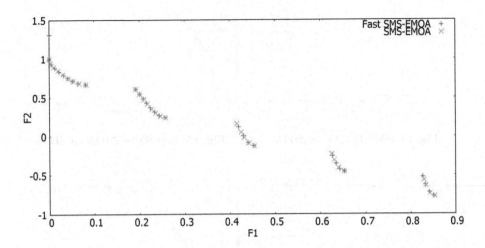

Fig. 9. Result on Zdt3 with SMS-EMOA and Fast SMS-EMOA

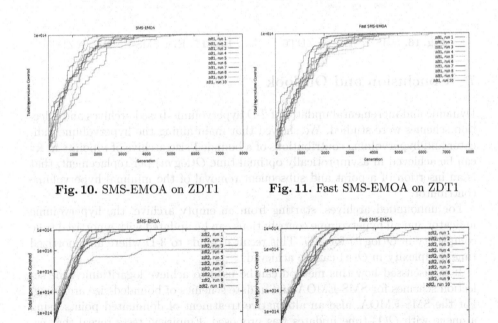

Fig. 10. SMS-EMOA on ZDT1   **Fig. 11.** Fast SMS-EMOA on ZDT1

Fig. 12. SMS-EMOA on ZDT2   **Fig. 13.** Fast SMS-EMOA on ZDT2

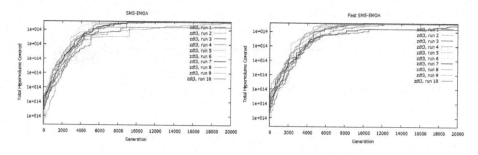

Fig. 14. SMS-EMOA on ZDT3   **Fig. 15.** Fast SMS-EMOA on ZDT3

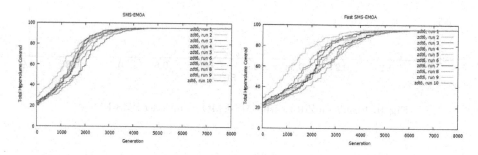

Fig. 16. SMS-EMOA on ZDT6   **Fig. 17.** Fast SMS-EMOA on ZDT6

7   Conclusion and Outlook

Dynamic and incremental updates of 2-D hypervolume-based archives and selection schemes were studied. We showed that maintaining the hypervolume indicator and hypervolume contributions of a bounded-size archive of n points in $\mathbb{R}^2$ can be achieved in asymptotically optimal time $O(\log m)$ per replacement, that is an insertion of a point and subsequent removal of the minimal hypervolume contributor.

For unbounded archives, starting from an empty archive, the hypervolume indicator and all hypervolume contributions can be updated with amortized time complexity in $O(\log n)$ for 2-D. This result extends to 3-D where an amortized time complexity in $O(n)$ can be achieved.

It is discussed how this method can be used to achieve logarithmic time selection schemes for SMS-EMOA and update schemes of bounded-size archiving. For the SMS-EMOA, also an alternative treatment of dominated points using a queue with $O(1)$-time updates was proposed. Empirical tests reveal that especially for large population sizes ($\gg 100$), that would be needed to achieve high-resolution Pareto front approximations, using a fast scheme will be of crucial importance in practice. But even in SMS-EMOA with a population size of only 30 individuals a remarkable acceleration of the selection procedure by a factor of 50 was achieved on standard benchmarks.

Future work will have to deal with a full generalization of results to the higher dimensional case, and include additional benchmarks. Moreover, algorithms similar to SMS-EMOA, e.g. MOO-CMA [13], could benefit from fast update schemes in indicator-based selection, and other alternative selection schemes to non-dominated sorting [15] could be compared.

References

1. Beume, N., Fonseca, C.M., López-Ibáñez, M., Paquete, L., Vahrenhold, J.: On the complexity of computing the hypervolume indicator. IEEE Trans. Evolutionary Computation 13(5), 1075–1082 (2009)
2. Bradstreet, L., Barone, L., While, L.: Updating exclusive hypervolume contributions cheaply. In: Proceedings of the Eleventh Conference on Congress on Evolutionary Computation, CEC 2009, pp. 538–544. IEEE Press, Piscataway (2009)
3. Bringmann, K., Friedrich, T.: Approximating the least hypervolume contributor: NP-hard in general, but fast in practice. In: Ehrgott, M., Fonseca, C.M., Gandibleux, X., Hao, J.-K., Sevaux, M. (eds.) EMO 2009. LNCS, vol. 5467, pp. 6–20. Springer, Heidelberg (2009)
4. Bringmann, K., Friedrich, T.: An efficient algorithm for computing hypervolume contributions. Evol. Comput. 18(3), 383–402 (2010)
5. da Fonseca, V.G., Fonseca, C.M.: The relationship between the covered fraction, completeness and hypervolume indicators. In: Hao, J.-K., Legrand, P., Collet, P., Monmarché, N., Lutton, E., Schoenauer, M. (eds.) EA 2011. LNCS, vol. 7401, pp. 25–36. Springer, Heidelberg (2012)
6. Deb, K.: Multi-objective optimization using evolutionary algorithms. John Wiley & Sons, Hoboken (2001)
7. Deb, K., Pratap, A., Agarwal, S., Meyarivan, T.: A fast and elitist multi-objective genetic algorithm, Nsga-ii (2000)
8. Emmerich, M., Beume, N., Naujoks, B.: An EMO algorithm using the hypervolume measure as selection criterion. In: 2005 Intl. Conference, pp. 62–76. Springer (March 2005)
9. Emmerich, M.T.M., Fonseca, C.M.: Computing hypervolume contributions in low dimensions: asymptotically optimal algorithm and complexity results. In: Takahashi, R.H.C., Deb, K., Wanner, E.F., Greco, S. (eds.) EMO 2011. LNCS, vol. 6576, pp. 121–135. Springer, Heidelberg (2011)
10. Fonseca, C.M., Paquete, L., Lopez-Ibanez, M.: An improved dimension-sweep algorithm for the hypervolume indicator, pp. 1157–1163 (July 2006)
11. Landis, E.M., Adelson-Velskii, G.: An algorithm for the organization of information. Proceedings of the USSR Academy of Sciences 146, 263–266
12. Guerreiro, A.P., Fonseca, C.M., Emmerich, M.T.M.: A fast dimension-sweep algorithm for the hypervolume indicator in four dimensions. In: CCCG, pp. 77–82 (2012)
13. Igel, C., Hansen, N., Roth, S.: Covariance matrix adaptation for multi-objective optimization. Evol. Comput. 15(1), 1–28 (2007)
14. Knowles, J.D., Corne, D.W., Fleischer, M.: Bounded archiving using the Lebesgue measure. In: Proceedings of the IEEE Congress on Evolutionary Computation, pp. 2490–2497. IEEE Press (2003)
15. Naujoks, B., Beume, N., Emmerich, M.T.M.: Multi-objective optimisation using S-metric selection: application to three-dimensional solution spaces, vol. 2, pp. 1282–1289 (September 2005)

Computing the Set of Approximate Solutions of a Multi-objective Optimization Problem by Means of Cell Mapping Techniques

Carlos Hernández[1], Jian-Qiao Sun[2], and Oliver Schütze[1]

[1] CINVESTAV-IPN,
Computer Science Department
Av. IPN 2508, Col. San Pedro Zacatenco
07360 Mexico City, Mexico
chernandez@computacion.cs.cinvestav.mx,
schuetze@cs.cinvestav.mx
[2] University of California Merced
School of Engineering
Merced, CA 95344, USA
jsun3@ucmerced.edu

Abstract. Here we address the problem of computing the set of approximate solutions of a given multi-objective optimization problem (MOP). This set is of potential interest for the decision maker since it might give him/her additional solutions to the optimal ones for the realization of the project related to the MOP. In this study, we make a first attempt to adapt well-known cell mapping techniques for the global analysis of dynamical systems to the problem at hand. Due to their global approach, these methods are well-suited for the thorough investigation of small problems, including the computation of the set of approximate solutions. We conclude this work with the presentation of three academic bi-objective optimization problems including a comparison to a related evolutionary approach.

Keywords: multi-objective optimization, approximate solutions, cell mapping techniques, global optimization.

1 Introduction

In many real-world engineering problems one is faced with the problem that several objectives have to be optimized concurrently leading to a multi-objective optimization problem (MOP). Such problems can be stated as

$$\min_{x} F : Q \subset \mathbb{R}^n \to \mathbb{R}^k. \tag{1}$$

The solution set of a MOP, the Pareto set P_Q, typically forms a $(k-1)$-dimensional entity, where k is the number of objectives involved in the problem. Hence, the approximation of P_Q already represents a challenge for many search procedures, in particular if the objectives are nonlinear. However, in some applications it might be interesting to know in addition to the optimal solutions

M. Emmerich et al. (eds.), *EVOLVE - A Bridge between Probability, Set Oriented Numerics,* 171
and Evolutionary Computation IV, Advances in Intelligent Systems and Computing 227,
DOI: 10.1007/978-3-319-01128-8_12, © Springer International Publishing Switzerland 2013

also nearly optimal ones. The reason is that if for two vectors $x, y \in Q$ the objective values $F(x)$ and $F(y)$ are close, this does not have to hold for x and y. Hence, if the decision maker (DM) is willing to accept a certain small deterioration ϵ in the performance (which is measured by F), he/she might be given several backup alternatives to realize the project. The approximation of the set of nearly optimal solutions $P_{Q,\epsilon}$ has caught little attention in literature so far. The main reason for this lack might be that this set is n-dimensional, where n is the dimension of the decision space.

In this work, we propose to use cell mapping techniques ([20]) for the approximation of $P_{Q,\epsilon}$. Methods of that kind divide the domain Q into a set of small n-dimensional cells and perform a cell-to-cell mapping of the given dynamical system g (here we will use a particular system derived from a Pareto descent method). In this way, a global view on the dynamics of g as well as the fitness landscape of F on Q are obtained, and is thus well-suited for the problem at hand. The result of the algorithm is ideally a tight covering of $P_{Q,\epsilon}$, but also the discretization of this set is straightforward (e.g., one representative of each cell can be chosen) which is required for the presentation of the result to the DM. Due to its approach, the method is restricted to small dimensions of the parameter space. However, there exist small dimensional problems where a thorough investigation is desirable. Such models e.g. arise in preliminary space mission design (e.g., [12,36,35]) or in the design of electrical circuits (e.g., [3,2]).

In literature, there exists a huge variety of methods for the approximation of the Pareto set P_Q (respectively its image, the Pareto front). There exist, for instance, point-wise iterative mathematical programming techniques that lead to single solutions of a MOP. An approximation of P_Q can be obtained by choosing a clever sequence of these problems (e.g., [8,23,15,14]). Further, there exist set oriented methods such as multi-objective evolutionary algorithms (MOEAs, see e.g., [9,5]) or subdivision techniques ([13,22,37]) that aim for the approximation of P_Q in one run of the algorithm. None of them, however, are designed for the computation of $P_{Q,\epsilon}$. The only studies in that direction seem to be the works [32,35], where stochastic search techniques such as MOEAs are investigated.

The remainder of this paper is organized as follows: In Section 2, we present the background required for the understanding of the sequel. In Section 3, we propose a cell mapping technique for the computation of the set of approximate solutions. In Section 4, we present some numerical results, and the final Section states the conclusions and future work.

2   Background

Here we will shortly recall the required background: The concept of multi-objective optimization including our definition of nearly optimality and a brief review of cell mapping techniques.

A multi-objective optimization problem (MOP) can be expressed as follows:

$$\min_{x \in Q}\{F(x)\}, \qquad\qquad \text{(MOP)}$$

where F is the map that consists of the objective functions $f_i : Q \to \mathbb{R}$ under consideration, i.e.,

$$F : Q \to \mathbb{R}^k, \qquad F(x) = (f_1(x), \dots, f_k(x)).$$

The domain $Q \subset \mathbb{R}^n$ of F can in general be expressed by inequality and equality constraints:

$$Q = \{x \in \mathbb{R}^n \mid g_i(x) \leq 0,\ i = 1, \dots, l,\ \text{and}\ h_j(x) = 0,\ j = 1, \dots, m\}.$$

In this work we will merely consider inequality constraints.

Next, we have to define optimal solutions of a given MOP. This can e.g. be done using the concept of *dominance* ([27]).

Definition 1. *(a) Let $v, w \in \mathbb{R}^k$. Then the vector v is less than w (in short: $v <_p w$), if $v_i < w_i$ for all $i \in \{1, \dots, k\}$. The relation $\leq_p$ is defined analogously.*
(b) A vector $y \in Q$ is called dominated by a vector $x \in Q$ ($x \prec y$) with respect to (MOP) if $F(x) \leq_p F(y)$ and $F(x) \neq F(y)$, else y is called non-dominated by x.

If a vector x dominates a vector y, then x can be considered to be 'better' according to the given MOP. The definition of optimality (i.e., of a 'best' solution) of a given MOP is now straightforward.

Definition 2. *(a) A point $x \in Q$ is called (Pareto) optimal or a Pareto point of (MOP) if there is no $y \in Q$ that dominates x.*
(b) The set of all Pareto optimal solutions is called the Pareto set, i.e.,

$$\mathcal{P} := \{x \in Q\ :\ x \text{ is a Pareto point of } (MOP)\}. \tag{2}$$

(c) The image $F(\mathcal{P})$ of $\mathcal{P}$ is called the Pareto front.

Pareto set and Pareto front typically form $(k - 1)$-dimensional objects under certain mild assumptions on the MOP, see [17] for a thorough discussion.

We now define another notion of dominance which we use to define approximate solutions.

Definition 3 ([31]). *Let $\epsilon = (\epsilon_1, \dots, \epsilon_k) \in \mathbb{R}^k_+$ and $x, y \in Q$.*

(a) x is said to ϵ-dominate y ($x \prec_\epsilon y$) with respect to (MOP) if $F(x) - \epsilon \leq_p F(y)$ and $F(x) - \epsilon \neq F(y)$.
(b) x is said to $-\epsilon$-dominate y ($x \prec_{-\epsilon} y$) with respect to (MOP) if $F(x) + \epsilon \leq_p F(y)$ and $F(x) + \epsilon \neq F(y)$.

The notion of $-\epsilon$-dominance is of course analogous to the 'classical' ϵ-dominance relation [24] but with a value $\tilde{\epsilon} \in \mathbb{R}^k_-$. However, we highlight it here since we use it to define our set of interest:

Definition 4. *Denote by $P_{Q,\epsilon}$ the set of points in $Q \subset \mathbb{R}^n$ that are not $-\epsilon$-dominated by any other point in Q, i.e.,*

$$P_{Q,\epsilon} := \{x \in Q \mid \nexists y \in Q : y \prec_{-\epsilon} x\}. \tag{3}$$

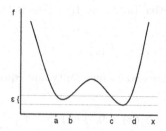
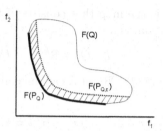

Fig. 1. Two different examples for sets $P_{Q,\epsilon}$. At the left, we show the case for $k = 1$ and in parameter space with $P_{Q,\epsilon} = [a, b] \cup [c, d]$. Note that the image solutions $f([a, b])$ are nearly optimal (measured in objective space), but that the entire interval $[a, b]$ is not 'near' to the optimal solution which is located within $[c, d]$. At the right, we show an example for $k = 2$ in image space, $F(P_{Q,\epsilon})$ is the approximate Pareto front (taken from [31]).

The set $P_{Q,\epsilon}$ contains all ϵ-efficient solutions, i.e., solutions which are optimal up to a given (small) value of ϵ. Figure 1 gives two examples.

To compare our results we will measure the distance of the outcome sets to the set of interest. A natural choice would be the Hausdorff distance d_H (e.g., [16]). Since d_H punishes single outliers that can occur when using stochastic search algorithms (as we will do in our comparison) we will use the averaged Hausdorff distance[1] instead.

Definition 5 ([34]). *Let $p \in \mathbb{N}$, $A = \{a_1, \ldots, a_r\}$ and $B = \{a_1, \ldots, a_m\}$ be two finite sets. Then it is*

$$\Delta_p(A, B) = \max\left(\left(\frac{1}{r} \sum_{i=1}^{r} dist(a_i, B)^p \right)^{1/p}, \left(\frac{1}{m} \sum_{i=1}^{m} dist(b_i, A)^p \right)^{1/p} \right), \quad (4)$$

where $dist(x, B) := \min_{b \in B} \|x - b\|$ denotes the distance between a point x and a set B.

Note that for $p = \infty$, we have $\Delta_\infty = d_H$, and for finite values of p the distances in Δ_p are averaged. In [34], Δ_p is discussed as a performance indicator in the context of Pareto front approximations. In that case, the indicator can be viewed as a combination of slight variations of the well-known indicators *Generational Distance* (see [38]) and the *Inverted Generational Distance* (see [6]).

The cell mapping method was originally proposed by Hsu [18,20] for global analysis of nonlinear dynamical systems in the *state space*. Two cell mapping methods have been extensively studied, namely, the simple cell mapping and the generalized cell mapping. The cell mapping methods have been applied to optimal control problems of deterministic and stochastic dynamic systems [19,4,7].

---

[1] The averaged Hausdorff is in general not a distance.

The cell mapping methods transform the point-to-point dynamics into a cell-to-cell mapping by discretizing both phase space and the integration time. The simple cell mapping (SCM) offers an effective approach to investigate global response properties of the system. The cell mapping with a finite number of cells in the computational domain will eventually lead to closed groups of cells of the period equal to the number of cells in the group. The periodic cells represent approximate invariant sets, which can be periodic motion and stable attractors of the system. The rest of the cells form the domains of attraction of the invariant sets. For more discussions on the cell mapping methods, their properties and computational algorithms, the reader is referred to the book by Hsu [20].

3   A Cell Mapping Method for the Approximation of $P_{Q,\epsilon}$

In this section, we review the SCM method [20] together with our adaptions to the context of multi-objective optimization, and present the post-processing to get an approximation of $P_{Q,\epsilon}$.

In the following, we assume the problem is bounded by box constraints, which constitutes our domain Q

$$lb_i \leq x_i \leq ub_i, \ i = 1, \cdots, N.$$

Now, we can divide each interval in N_i sections of size

$$h_i = \frac{lb_i - ub_i}{N_i}.$$

By doing this, we get a finite subdivision of the domain, where each of these elements are called regular cells. The number of regular cells is noted by N_c and we label the set of regular cells with positive integers, $1, 2, \cdots, N_c$. Everything that is outside the domain is called the sink cell. With the introduction of the sink cell the total number of cells is $N_c + 1$. The sink cell is also labeled as the regular cells using the value 0.

At this point the evolution of the system is given cell-to-cell instead of point-to-point. The dynamics of a cell z is represented by its center and the cell-to-cell mapping is denoted by C. Now the mapping can be described by

$$z(n + 1) = C(z(n)), \quad z(n), z(n+1) \in \{0, \cdots, Nc\},$$
$$C(0) = 0.$$

A cell z^* that is mapped onto itself,

$$z^* = C(z^*),$$

is called an equilibrium cell.

Definition 6. *A periodic motion of period k for C is a sequence of k cells $z^*(l), l = 1, \cdots, k$, such that*

$$z^*(m + 1) = C^m(z^*(1)), \quad m = 1, \cdots, k - 1,$$
$$z^*(1) = C^k(z^*(1)).$$

Every element $z^*(l)$ is called a periodic cell of period k or $P - k$ group.

Definition 7. *We say a cell z has distance r from a periodic motion if r is the minimum positive integer such that*

$$C^r(z) = z^*(l),$$

where $z^*(l)$ is one of the $P - k$ cells of the motion. The set of such cells is called the r-step basin of attraction of the motion. If r goes to infinity we obtain the basin of attraction.

The evolution of the system starting with any regular cell z can lead only to one of the following three possible outcomes:

- The cell belongs to a P-Group: z is itself a periodic cell of a periodic motion. The evolution of the system simply leads to a periodic motion.
- The cell maps to the sink cell in r steps: Cell z is mapped into the sink cell in r steps. Then the cell belongs to the r-step domain of attraction of the sink cell.
- The cell maps to a P-group in r steps: Cell z is mapped into a periodic cell of a certain periodic motion in r steps. In this case the cell belongs to the r-step domain of attraction of that periodic motion.

To capture the global properties of a cell, the SCM algorithm uses the following sets:

- Group motion number (Gr): The group number uniquely identifies a periodic motion; it is assigned to every periodic cell of that periodic motion and also to every cell in the domain of attraction. The group numbers, positive integers, can be assigned sequentially.
- Period (Pe): Defines the period of each periodic motion.
- Number of steps to a P-group (St): Used to indicate how many steps it takes to map this cell into a periodic cell.

So far the SCM for general dynamical systems. In order to apply it to the context of multi-objective optimization, we have to define a suitable dynamical system. For this, we have chosen to take models that are derived from *descent directions*. A direction $v \in \mathbb{R}^n$ is called a descent direction at a given point $x_0 \in Q$ if a search in that direction leads to an improvement of all objectives. In other words, there exists a $\bar{t} \in \mathbb{R}_+$ such that $F(x_0 + tv) <_p F(x_0) \ \forall t \leq \bar{t}$. Descent directions are e.g. proposed in [21,29,25,1]. Since we consider bi-objective optimization problems in this work, we use descent direction from [1] which is given by

Theorem 1 ([1]). *Let $x \in \mathbb{R}^n$, and $f_1, f_2 : \mathbb{R}^n \to \mathbb{R}$ define a two-objective MOP. if $\nabla f_i(x) \neq 0$, for $i = 1, 2$, then the direction*

$$v(x_0) := -\left(\frac{\nabla f_1(x_0)}{||\nabla f_1(x_0)||} + \frac{\nabla f_2(x_0)}{||\nabla f_2(x_0)||} \right), \tag{5}$$

is a descent direction at x_0 of MOP. This descent direction yielded best results in our tests but is restricted to problems with two objectives. Using (5), the dynamical system

$$\dot{x}(t) = v(x(t)) \tag{6}$$

can now be used since it defines a pressure toward the Pareto set/front of the MOP at hand.

In order to be able to store those cells that are candidates to be Pareto optimal (to be more precise, cells that potentially contain a part of the Pareto set), we use a set called cPs. For these cells it holds $St(cell) = 0$ and $Gr(cell) \neq 1$. It is important to notice that because of the dynamical system periodic groups with size greater than 1 should not appear, however, due to discretization errors and too large step sizes periodic groups greater than 1 may be generated (i.e., an oscillation around the Pareto set can occur). Thus, cells that are involved in the current periodic group are also considered to be candidates.

Algorithm 1 shows the key elements of the SCM method for the treatment of MOPs. According to the previous discussion, the algorithm works as follows, until all cells are processed, the value of the group motion indicates the state of the current cell and it also points out the corresponding actions to the cell.

A value of $Gr(cell) = 0$ means, the cell has not been processed, hence the state of the cell changes to under process and then we follow the dynamical system to the next cell.

A value of $Gr(cell) = -1$ means, the cell is under processed, which means we have found a periodic group and we can compute the global properties of the current periodic motion.

A value $Gr(cell) > 0$ means, the cell has already been processed, hence we found a previous periodic motion along with its global properties, which can be used to complete the information of the cells under process.

Figure 2 shows an example of a group motion with $N_c = 10 \times 10$ and $Pe(z) = 12$.

After one run of the SCM algorithm, we have gathered the information on the global dynamics of the system and are hence able to approximate the set of interest in a post-processing step. For the problem at hand, the approximation of $P_{Q,\epsilon}$, we use the archiving technique $ArchiveUpdateTight2$ proposed in [26] as follows: Once the group number of the current periodic motion is discovered, we use Algorithm 2 to compute the set $P_{Q,\epsilon}$. Algorithm 2 updates the archive first with the periodic group of the current periodic motion and continues with the rest of the periodic motion. Once it finds a cell which is not in $P_{Q,\epsilon}$ it stops the procedure. The reason for this can be easily seen. Since each periodic group is a curve of dominated points, once a point $x_j \notin P_{Q,\epsilon}$ the other points would not be either, since by construction these points are dominated by x_j.

Algorithm 1. Simple Cell Mapping for MOPs.

Require: $DynamicalSystem, F, ub, lb, N, h, N_c$
Ensure: z, C, Gr, Pe, St, cPs

1: $current\_group \leftarrow 1$
2: $cPs = \{\}$
3: $Gr(i) \leftarrow 0, \forall i \in N_c$
4: **for all** $pcell \in N_c$ **do**
5:    $cell \leftarrow pcell$
6:    $i \leftarrow 0$
7:    **while** $newcell = $ **true do**
8:      $x_i \leftarrow$ center point of $cell$
9:      **if** $Gr(cell) = 0$ **then**
10:       $Gr(cell) \leftarrow -1$
11:       $p_{i+1} \leftarrow DynamicalSystem(x_i)$
12:       $ncell \leftarrow$ cell where p_{i+1} is located
13:       $C(cell) \leftarrow ncell$
14:       $cell \leftarrow ncell$
15:       $i \leftarrow i + 1$
16:      **end if**
17:      **if** $Gr(cell) > 0$ **then**
18:       $Gr(C^j(pcell)) \leftarrow Gr(cell), j \leftarrow 0, \cdots, i$
19:       $Pe(C^j(pcell)) \leftarrow Pe(cell), j \leftarrow 0, \cdots, i$
20:       $St(C^j(pcell)) \leftarrow St(cell) + i - j, j \leftarrow 0, \cdots, i$
21:       $cell \leftarrow C(cell)$
22:       $newcell \leftarrow$ **false**
23:      **end if**
24:      **if** $Gr(cell) = -1$ **then**
25:       $current\_group \leftarrow current\_group + 1$
26:       $Gr(C^k(pcell)) \leftarrow current\_group, k \leftarrow 0, \cdots, i$
27:       $j \leftarrow i^{th}$ value when period appears
28:       $Pe(C^k(pcell)) \leftarrow i - j, k \leftarrow 0, \cdots, i$
29:       $St(C^k(pcell)) \leftarrow j - k, k \leftarrow 0, \cdots, j - 1$
30:       $St(C^k(pcell)) \leftarrow 0, k \leftarrow j, \cdots, i$
31:       $cPs \leftarrow cPs \cup \{x_k\}, k \leftarrow j, \cdots, i$
32:       $cell \leftarrow C(cell)$
33:       $newcell \leftarrow$ **false**
34:      **end if**
35:    **end while**
36: **end for**

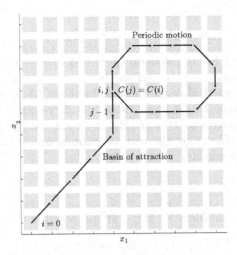

Fig. 2. Example of group motion

Algorithm 2. Post-processing to get $P_{Q,\epsilon}$

Require: $P_{Q,\epsilon}, C, pcell, i$
Ensure: $P_{Q,\epsilon}$
1: $j \leftarrow i - 1$
2: $isP_{Q,\epsilon} \leftarrow$ **true**
3: **while** $isP_{Q,\epsilon}$ **do**
4:      $x_j \leftarrow$ center point of $C^j(pcell)$
5:      $P_{Q,\epsilon} \leftarrow ArchiveUpdateTight2(P_{Q,\epsilon}, x_j, F(x_j), \epsilon, h)$
6:      **if** $x_j \notin P_{Q,\epsilon}$ **then**
7:         $isP_{Q,\epsilon} \leftarrow$ **false**
8:      **end if**
9:      $j \leftarrow j - 1$
10: **end while**

---

4 Numerical Results

Here we present some numerical results on three bi-objective benchmark models. In order to compare the results of the cell mapping technique we couple MOEAs with the archiving technique $ArchiveUpdateP_{Q,\epsilon}$ ([32]) as suggested in [35]. For the MOEAs we have chosen for the state-of-the-art algorithms NSGA-II ([10]) and MOEA/D ([39]).

4.1 Problem 1

First, we consider the MOP taken from [30] that is given by two objective functions $f_1, f_2 : \mathbb{R}^2 \to \mathbb{R}$,

$$f_1(x_1, x_2) = (x_1 - 1)^2 + (x_2 - 1)^4,$$
$$f_2(x_1, x_2) = (x_1 + 1)^2 + (x_2 + 1)^2. \tag{7}$$

Here, we have chosen for the domain $Q = [-3, 3] \times [-3, 3]$. The Pareto set P_Q forms a curve connecting the end points $x_1 = (-1, -1)^T$ and $x_2 = (1, 1)^T$. Figure 3 shows the numerical result obtained by the SCM algorithm and the two evolutionary algorithms. Tables 1 and 2 show the Δ_p values of the candidate sets (respectively their images) to $P_{Q,\epsilon}$ ($F(P_{Q,\epsilon})$) for this and all further examples. Apparently, SCM is able to get the best approximation of the set of interest, in particular in parameter space.

4.2   Problem 2

Next, we consider the problem $F : \mathbb{R}^2 \to \mathbb{R}^2$ proposed in [28]

$$F(x_1, x_2) = \begin{pmatrix} (x_1 - t_1(c + 2a) + a)^2 + (x_2 - t_2 b)^2 + \delta_t \\ (x_1 - t_1(c + 2a) - a)^2 + (x_2 - t_2 b)^2 + \delta_t \end{pmatrix}, \tag{8}$$

where

$$t_1 = \text{sgn}(x_1) \min \left(\left\lceil \frac{|x_1| - a - c/2}{2a + c} \right\rceil, 1 \right), t_2 = \text{sgn}(x_2) \min \left(\left\lceil \frac{|x_2| - b/2}{b} \right\rceil, 1 \right),$$

and

$$\delta_t = \begin{cases} 0 & \text{for } t_1 = 0 \text{ and } t_2 = 0 \\ 0.1 & \text{else} \end{cases}.$$

Using $a = 0.5$, $b = 5$, $c = 5$ the Pareto set of MOP (8) is connected and further there exist eight other connected components that are locally optimal. For $\epsilon > 0.1$, the set $P_{Q,\epsilon}$ consists of nine sets that contain these components. Figure 4 shows some numerical results. SCM computes a covering of the entire set of interest, while the evolutionary strategies do not always detect all components which is e.g. reflected by the averaged Δ_p values in Tables 1 and 2.

As a hypothetical decision making problem assume the DM is interested in the performance $Z = [0.2132, 0.2932]$ (measured in objective space) and is willing to accept a deterioration of $\epsilon = [0.1, 0.1]$. Then, e.g. the representatives of the cells those images are within the target regions can be presented to the DM leading here to the following 22 candidate solutions:

$(-6.04, -5.00)$, $(-0.04, -5.00)$, $(5.96, -5.00)$, $(-6.04, -0.04)$,
$(-0.12, -0.04)$, $(-0.04, -0.28)$, $(-0.04, -0.20)$, $(-0.04, -0.12)$,
$(-0.04, -0.04)$, $(-0.12, 0.04)$, $(-0.04, 0.04)$, $(-0.04, 0.12)$,
$(-0.04, 0.20)$, $(-0.04, 0.28)$, $(0.04, -0.12)$, $(0.04, -0.04)$,
$(0.04, 0.04)$, $(0.04, 0.12)$, $(5.96, -0.04)$, $(-6.04, 5.00)$,
$(-0.04, 5.00)$, $(5.96, 5.00)$.

The solutions are well-spread and come in this case from all nine components of $P_{Q,\epsilon}$. Since these components are located in different regions of the parameter space, the DM is hence given a large variety for the realization of his/her project.

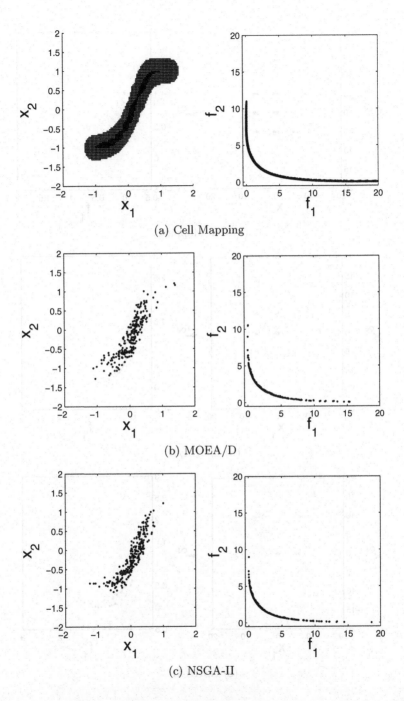

(a) Cell Mapping

(b) MOEA/D

(c) NSGA-II

Fig. 3. Numerical results for MOP (7). Black cells indicate Pareto optimal candidates and yellow cells regions in $P_{Q,\epsilon}$ that are not optimal.

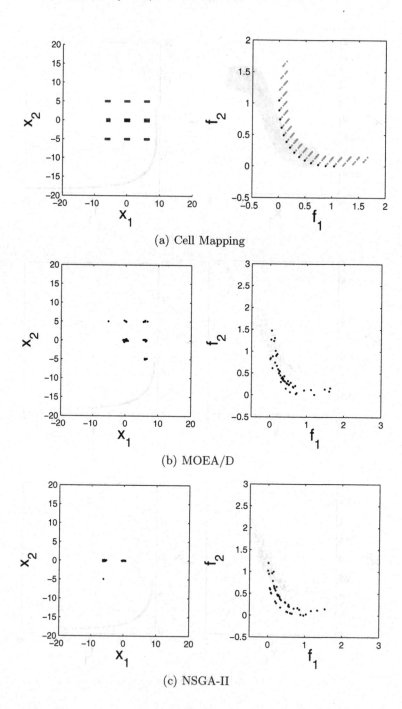

(a) Cell Mapping

(b) MOEA/D

(c) NSGA-II

Fig. 4. Numerical results for MOP (8). Black cells indicate Pareto optimal candidates and yellow cells regions in $P_{Q,\epsilon}$ that are not optimal.

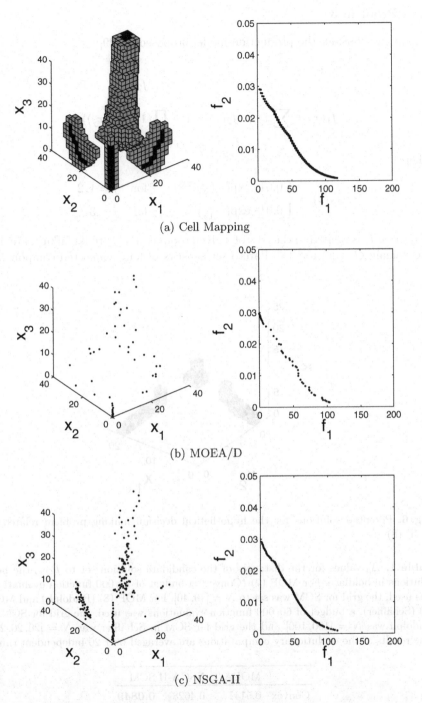

(a) Cell Mapping

(b) MOEA/D

(c) NSGA-II

Fig. 5. Numerical results for MOP (9). Black cells indicate Pareto optimal candidates and yellow cells regions in $P_{Q,\epsilon}$ that are not optimal.

4.3   Problem 3

Finally, we consider the production model proposed in [29]:

$$f_1, f_2 : \mathbb{R}^3 \to \mathbb{R}$$

$$f_1(x) = \sum_{j=1}^{3} x_j, \quad f_2(x) = 1 - \prod_{j=1}^{3}(1 - w_j(x_j)), \tag{9}$$

where

$$w_j(z) = \begin{cases} 0.01 \cdot \exp(-(\frac{z}{20})^{2.5}) & \text{for} \quad j = 1,2 \\ 0.01 \cdot \exp(-\frac{z}{15}) & \text{for} \quad j = 3 \end{cases}$$

Objective f_1 is related to the cost of a given product and f_2 to its failure rate. For the domain $Q = [0, 40]^3$ the Pareto set consists of four connected components.

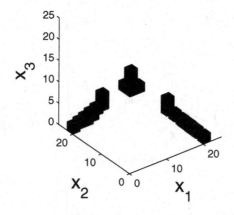

Fig. 6. Potential solutions for the hypothetical decision making problem related to MOP (8)

Table 1. Δ_p values for the distances of the candidate solution set to $P_{Q,\epsilon}$, the best solutions in boldface. For MOP (7) (Convex) a budget of $10,000$ function evaluations was used, the grid for SCM was set to $N = [40, 40]$. For MOP (8) (Rudolph) and MOP (9) (Schäffler), a budget of $60,000$ function evaluations was used, the grid for SCM in Rudolph was $N = [100, 100]$ and the grid for SCM in Schäffler was $N = [20, 20, 20]$. The results of the evolutionary computations are averaged over 20 independent runs.

	MOEA/D	NSGA-II	SCM
Convex	0.5141	0.4628	**0.0849**
Rudolph	5.0507	7.4737	**0.0632**
Schäffler	10.8365	10.9384	**0.8660**

Figure 5 shows a numerical result of the SCM and the evolutionary strategies. Also in this case, SCM obtains the best result in particular measured in parameter space (compare also to Tables 1 and 2). Figure 6 shows the resulting boxes of another hypothetical decision making problem where we have chosen $Z = [23, 0.02231]$ and $\epsilon = [2, 0.0004]$. Also here, the DM is offered an entire range of solutions with different parameter values.

Table 2. Δ_p values for the distances of the images of the candidate sets to $F(P_{Q,\epsilon})$, see Table 1 for details

	MOEA/D	NSGA-II	SCM
Convex	7.8902	8.0027	**2.4250**
Rudolph	0.4276	0.6317	**0.0524**
Schäffler	5.8152	2.6852	**1.5000**

5   Conclusion and Future Work

In this paper we have addressed the problem of computing the set of approximate solutions of a given multi-objective optimization problem. This set is of interest for the decision maker since it might enhance the set of options for him/her when compared to the set of optimal solutions, the Pareto set. To compute the set of approximate solutions we have adapted cell mapping techniques that were originally designed for the global analysis of dynamical systems. Since methods of that kind divide the search space into n-dimensional cells, where n is the dimension of the decision space of the MOP, they are well-suited for the problem at hand since they allow for an efficient approximation of the set of interest. We have tested the novel method on three academic functions and have compared it against two evolutionary methods. The results indicate that the cell mapping technique is able to reliably compute the set of approximate solutions, and is faster than the evolutionary approaches. The method, however, is restricted to small dimensions of the decision space.

For future work, there are several points to be addressed. First of all, it would be desirable to extend the applicability of the method to higher dimensional problems. For this, it seems promising to couple the cell mapping techniques with related set oriented methods such as subdivision techniques ([11,13]) or recovering algorithms ([13,33]). Next, the integration of constraint handling techniques has to be addresses which has been left out in this study. Finally, we plan to apply the new method to real-wold engineering problems.

Acknowledgement. The authors acknowledge support from the UC MEXUS-CONACyT project *"Cell-to-cell Mapping for Global Multi-objective Optimization."* The first author acknowledges support from CONACyT through a scholarship to pursue graduate studies at the Computer Science Department of CINVESTAV-IPN.

References

1. Lara, A.: Using Gradient Based Information to build Hybrid Multi-objective Evolutionary Algorithms. PhD thesis (2012)
2. Blesken, M., Chebil, A., Rückert, U., Esquivel, X., Schütze, O.: Integrated circuit optimization by means of evolutionary multi-objective optimization. In: Genetic and Evolutionary Computation Conference (GECCO 2011), pp. 807–812 (2011)
3. Blesken, M., Rückert, U., Steenken, D., Witting, K., Dellnitz, M.: Multiobjective optimization for transistor sizing of CMOS logic standard cells using set-oriented numerical techniques. In: 27th Norchip Conference (2009)
4. Bursal, F.H., Hsu, C.S.: Application of a cell-mapping method to optimal control problems. International Journal of Control 49(5), 1505–1522 (1989)
5. Coello Coello, C.A., Lamont, G., Van Veldhuizen, D.: Evolutionary Algorithms for Solving Multi-Objective Problems, 2nd edn. Springer (2007)
6. Coello Coello, C.A., Cruz Cortés, N.: Solving multiobjective optimization problems using an artificial immune system. Genetic Programming and Evolvable Machines 6(2), 163–190 (2005)
7. Crespo, L.G., Sun, J.Q.: Stochastic optimal control of nonlinear dynamic systems via bellman's principle and cell mapping. Automatica 39(12), 2109–2114 (2003)
8. Das, I., Dennis, J.: Normal-boundary intersection: A new method for generating the Pareto surface in nonlinear multicriteria optimization problems. SIAM Journal of Optimization 8, 631–657 (1998)
9. Deb, K.: Multi-Objective Optimization Using Evolutionary Algorithms. Wiley (2001)
10. Deb, K., Pratap, A., Agarwal, S., Meyarivan, T.: A fast and elitist multiobjective genetic algorithm: NSGA-II. IEEE Transactions on Evolutionary Computation 6(2), 182–197 (2002)
11. Dellnitz, M., Hohmann, A.: A subdivision algorithm for the computation of unstable manifolds and global attractors. Numerische Mathematik 75, 293–317 (1997)
12. Dellnitz, M., Ober-Blöbaum, S., Post, M., Schütze, O., Thiere, B.: A multiobjective approach to the design of low thrust space trajectories using optimal control. Celestial Mechanics and Dynamical Astronomy 105, 33–59 (2009)
13. Dellnitz, M., Schütze, O., Hestermeyer, T.: Covering Pareto sets by multilevel subdivision techniques. Journal of Optimization Theory and Applications 124, 113–155 (2005)
14. Eichfelder, G.: Adaptive Scalarization Methods in Multiobjective Optimization. Springer, Heidelberg (2008) ISBN 978-3-540-79157-7
15. Fliege, J.: Gap-free computation of Pareto-points by quadratic scalarizations. Mathematical Methods of Operations Research 59, 69–89 (2004)
16. Heinonen, J.: Lectures on Analysis on Metric Spaces. Springer, New York (2001)
17. Hillermeier, C.: Nonlinear Multiobjective Optimization - A Generalized Homotopy Approach. Birkhäuser (2001)
18. Hsu, C.S.: A theory of cell-to-cell mapping dynamical systems. Journal of Applied Mechanics 47, 931–939 (1980)
19. Hsu, C.S.: A discrete method of optimal control based upon the cell state space concept. Journal of Optimization Theory and Applications 46(4), 547–569 (1985)
20. Hsu, C.S.: Cell-to-cell mapping: a method of global analysis for nonlinear systems. In: Applied Mathematical Sciences. Springer (1987)
21. Fliege, J., Svaiter, B.F.: Steepest Descent Methods for Multicriteria Optimization. Mathematical Methods of Operations Research 51(3), 479–494 (2000)

22. Jahn, J.: Multiobjective search algorithm with subdivision technique. Computational Optimization and Applications 35(2), 161–175 (2006)

23. Klamroth, K., Tind, J., Wiecek, M.: Unbiased approximation in multicriteria optimization. Mathematical Methods of Operations Research 56, 413–437 (2002)

24. Loridan, P.: ϵ-solutions in vector minimization problems. Journal of Optimization, Theory and Application 42, 265–276 (1984)

25. Schütze, O., Lara, A., Coello Coello, C.A.: The Directed Search Method for Unconstrained Multi-Objective Optimization Problems. Technical Report COA-R1, CINVESTAV-IPN (2010)

26. Schütze, O., Laumanns, M., Tantar, E., Coello Coello, C.A., Talbi, E.-G.: Computing Gap Free Pareto Front Approximations with Stochastic Search Algorithms. Evolutionary Computation 18(1)

27. Pareto, V.: Manual of Political Economy. The MacMillan Press (1971); original edition in French (1927)

28. Rudolph, G., Naujoks, B., Preuß, M.: Capabilities of emoa to detect and preserve equivalent pareto subsets. In: Obayashi, S., Deb, K., Poloni, C., Hiroyasu, T., Murata, T. (eds.) EMO 2007. LNCS, vol. 4403, pp. 36–50. Springer, Heidelberg (2007)

29. Schäffler, S., Schultz, R., Weinzierl, K.: A stochastic method for the solution of unconstrained vector opimization problems. Journal of Optimization, Theory and Application 114(1), 209–222 (2002)

30. Schütze, O.: Set Oriented Methods for Global Optimization. PhD thesis, University of Paderborn (2004),
http://ubdata.uni-paderborn.de/ediss/17/2004/schuetze/

31. Schütze, O., Coello Coello, C.A., Talbi, E.-G.: Approximating the ϵ-efficient set of an MOP with stochastic search algorithms. In: Gelbukh, A., Kuri Morales, Á.F. (eds.) MICAI 2007. LNCS (LNAI), vol. 4827, pp. 128–138. Springer, Heidelberg (2007)

32. Schütze, O., Coello Coello, C.A., Tantar, E., Talbi, E.-G.: Computing finite size representations of the set of approximate solutions of an MOP with stochastic search algorithms. In: GECCO 2008: Proceedings of the 10th Annual Conference on Genetic and Evolutionary Computation, pp. 713–720. ACM, New York (2008)

33. Schütze, O., Dell'Aere, A., Dellnitz, M.: On continuation methods for the numerical treatment of multi-objective optimization problems. In: Branke, J., Deb, K., Miettinen, K., Steuer, R.E. (eds.) Practical Approaches to Multi-Objective Optimization. Dagstuhl Seminar Proceedings, vol. 04461. Internationales Begegnungs- und Forschungszentrum (IBFI), Schloss Dagstuhl, Germany (2005),
http://drops.dagstuhl.de/opus/volltexte/2005/349

34. Schütze, O., Esquivel, X., Lara, A., Coello Coello, C.A.: Using the averaged Hausdorff distance as a performance measure in evolutionary multi-objective optimization. IEEE Transactions on Evolutionary Computation 16(4), 504–522 (2012)

35. Schütze, O., Vasile, M., Coello Coello, C.A.: Computing the set of epsilon-efficient solutions in multiobjective space mission design. Journal of Aerospace Computing, Information, and Communication 8, 53–70 (2011)

36. Schütze, O., Vasile, M., Junge, O., Dellnitz, M., Izzo, D.: Designing optimal low thrust gravity assist trajectories using space pruning and a multi-objective approach. Engineering Optimization 41, 155–181 (2009)

37. Schütze, O., Vasile, M., Junge, O., Dellnitz, M., Izzo, D.: Designing optimal low thrust gravity assist trajectories using space pruning and a multi-objective approach. Engineering Optimization 41(2), 155–181 (2009)

38. Van Veldhuizen, D.A.: Multiobjective Evolutionary Algorithms: Classifications, Analyses, and New Innovations. PhD thesis, Department of Electrical and Computer Engineering. Graduate School of Engineering. Air Force Institute of Technology, Wright-Patterson AFB, Ohio (May 1999)
39. Zhang, Q., Li, H.: MOEA/D: A multi-objective evolutionary algorithm based on decomposition. IEEE Transactions on Evolutionary Computation 11(6), 712–731 (2007)

The Directed Search Method for Pareto Front Approximations with Maximum Dominated Hypervolume

Víctor Adrián Sosa Hernández[1], Oliver Schütze[1],
Günter Rudolph[2], and Heike Trautmann[3]

[1] Computer Science Department, CINVESTAV-IPN, Av. IPN 2508, Col. San Pedro
Zacatenco, 07360 Mexico City, Mexico
msosa@computacion.cs.cinvestav.mx, schuetze@cs.cinvestav.mx
[2] Fakultät für Informatik, Technische Universität Dortmund, 44221 Dortmund,
Germany
guenter.rudolph@tu-dortmund.de
[3] Statistics and Information Systems, University of Münster, Schlossplatz 2, 48149
Münster, Germany
trautmann@wi.uni-muenster.de

Abstract. In many applications one is faced with the problem that multiple objectives have to be optimized at the same time. Since typically the solution set of such multi-objective optimization problems forms a manifold which cannot be computed analytically, one is in many cases interested in a suitable finite size approximation of this set. One widely used approach is to find a representative set that maximizes the dominated hypervolume that is defined by the images in objective space of these solutions and a given reference point.

In this paper, we propose a new point-wise iterative search procedure, Hypervolume Directed Search (HVDS), that aims to increase the hypervolume of a given point in an archive for bi-objective unconstrained optimization problems. We present the HVDS both as a standalone algorithm and as a local searcher within a specialized evolutionary algorithm. Numerical results confirm the strength of the novel approach.

Keywords: multi-objective optimization, evolutionary computation, dominated hypervolume, local search, directed search.

1 Introduction

In many real-world applications the problem arises that several objectives have to be optimized concurrently leading to a *multi-objective optimization problem* (MOP). The solution set of a MOP, the so-called Pareto set, typically forms a $(k-1)$-dimensional manifold, where k is the number of objectives involved in the problem [7]. For the treatment of MOPs specialized evolutionary algorithms, multi-objective evolutionary algorithms (EMOAs), have caught the interest of many researchers (see, e.g., [4] and references therein). Reasons for this include

M. Emmerich et al. (eds.), *EVOLVE - A Bridge between Probability, Set Oriented Numerics,* 189
and Evolutionary Computation IV, Advances in Intelligent Systems and Computing 227,
DOI: 10.1007/978-3-319-01128-8_13, © Springer International Publishing Switzerland 2013

that EMOAs are applicable to a wide range of problems, are of global nature and hence in principle not dependent on the initial candidate set, and allow to compute a finite size representation of the Pareto set in a single run of the algorithm. On the other hand, it is known that EMOAs tend to converge slowly resulting in a relatively high number of function evaluations needed to obtain a suitable representation of the set of interest. As a possible remedy, researchers have proposed *memetic strategies* in the recent past (e.g., [8]). Algorithms of that type hybridize local search strategies mainly coming from mathematical programming with EMOAs in order to obtain fast and reliable global search procedures.

In this paper, we derive an algorithm that fits into the last category. To be more precise, we present a local search mechanism, HVDS, that aims to improve the dominated hypervolume [21] of a point or set for a given MOP. The new search procedure is based on the Directed Search Method [15,9] that is able to steer the search into any direction given in objective space $\mathcal{O}$ and which is hence well-suited for the problem at hand since the hypervolume is defined in $\mathcal{O}$. We present the HVDS both as standalone algorithm and as local search engine within SMS-EMOA [2] which is a state-of-the-art EMOA for approximations w.r.t. maximum dominated hypervolume. Numerical experiments show the benefit of the new approach.

The remainder of this paper is organized as follows: In Section 2, we state the background required for the understanding of the sequel. In Section 3, we present the algorithm HVDS which aims to improve the hypervolume as standalone algorithm and propose a possible integration of it into an EMOA in Section 4. In Section 5, we present some numerical results, and finally, we conclude in Section 6.

2   Background

A general multi-objective optimization problem (MOP) can be stated as follows:

$$\min_{x \in Q}\{F(x)\}, \tag{1}$$

where F is defined as the vector of the objective functions $F : Q \to \mathbb{R}^k$, $F(x) = (f_1(x), \ldots, f_k(x))$, and where each objective is given by $f_i : Q \to \mathbb{R}$. In this study we will focus on unconstrained bi-objective problems, i.e., problems of form (1) with $k = 2$ and $Q = \mathbb{R}^n$. The optimality of a MOP is defined by the concept of *dominance*.

Definition 1

(a) *Let $v, w \in \mathbb{R}^k$. Then the vector v is less than w ($v <_p w$), if $v_i < w_i$ for all $i \in \{1, \ldots, k\}$. The relation $\leq_p$ is defined analogously.*

(b) *A vector $y \in Q$ is dominated by a vector $x \in Q$ ($x \prec y$) with respect to (1) if $F(x) \leq_p F(y)$ and $F(x) \neq F(y)$, else y is called non-dominated by x.*

(c) *A point $x \in Q$ is called* (Pareto) optimal *or a* Pareto point *if there is no $y \in Q$ which dominates x.*

(d) *The set of all Pareto optimal solutions is called the* Pareto set *and its image the* Pareto front.

Recently, the Directed Search (DS) Method has been proposed that allows to steer the search from a given point into a desired direction $d \in \mathbb{R}^k$ in objective space [15]. To be more precise, given a point $x \in \mathbb{R}^n$, a search direction $\nu \in \mathbb{R}^n$ is sought such that

$$\lim_{t \searrow 0} \frac{f_i(x_0 + t\nu) - f_i(x_0)}{t} = d_i, \quad i = 1, \ldots, k. \tag{2}$$

Such a direction vector ν solves the following system of linear equations:

$$J(x_0)\nu = d, \tag{3}$$

where $J(x)$ denotes the Jacobian of F at x. Since typically $k \ll n$, we can assume that the system in Equation (3) is (highly) underdetermined. Among the solutions of Equation (3), the one with the least 2-norm can be viewed as the greedy direction for the given context. This solution is given by

$$\nu_+ := J(x)^+ d, \tag{4}$$

where $J(x)^+$ denotes the pseudo inverse of $J(x)$. Since there is no restriction on d the search can be steered in any direction, e.g., toward and along the Pareto set. See [15,14] for a Pareto descent method and a continuation method based on DS. In [9] a modification of the DS is presented that does not require gradient information.

A commonly accepted measure [20] for assessing the quality of an approximation is the so-called *dominated hypervolume* of a population.

Definition 2. *Let $v^{(1)}, v^{(2)}, \ldots v^{(\mu)} \in \mathbb{R}^k$ be a nondominated set and $R \in \mathbb{R}^k$ such that $v^{(i)} \prec R$ for all $i = 1, \ldots, \mu$. The value*

$$H(v^{(1)}, \ldots, v^{(\mu)}; R) = \Lambda_d \left(\bigcup_{i=1}^{\mu} [v^{(i)}, R] \right) \tag{5}$$

is termed the dominated hypervolume *with respect to reference point R, where $\Lambda_d(\cdot)$ denotes the Lebesgue measure in $\mathbb{R}^k$.*

This measure has a number of appealing properties but determining its value is getting the more tedious the larger the number of objectives is considered [1]. In case of two objectives ($k = 2$) and lexicographically ordered nondominated set $v^{(1)}, v^{(2)}, \ldots v^{(\mu)}$ the calculation of (5) reduces to

$$H(v^{(1)}, \ldots, v^{(\mu)}; R) = \left[r_1 - v_1^{(1)} \right] \cdot \left[r_2 - v_2^{(1)} \right] + \sum_{i=2}^{\mu} \left[r_1 - v_1^{(i)} \right] \cdot \left[v_2^{(i-1)} - v_2^{(i)} \right].$$

3   The Algorithm

Here we describe the adaption of the DS to the context of hypervolume approximations. For this, we will first consider the simplest case that the archive only consists of one element that has to be improved. In a next step we consider archives of general size. The reason for this is that we will reduce the general case to the one element problem.

3.1   One Element Archives

We assume that we are given the archive $A = \{x\}$, i.e., we are given one point $x \in Q$ that is assigned for local search. Further, we are given a reference point $R = (r_1, r_2)^T \in \mathbb{R}^2$ for the hypervolume calculations.

In the following, we divide the objective space into three different regions, and will propose a different movement in each of these regions (compare to Figure 1):

- **Region** *I.* The objective vector $F(x)$ is 'far away' from the Pareto front (denoted by '$F(x) \in I$'). In that case, a greedy search toward the rough location of the Pareto front is desired.
- **Region** *II.* $F(x)$ is 'in between', i.e., not far away nor near the Pareto front. In that case, a descent direction has to be selected such that a movement in that direction maximizes the hypervolume.
- **Region** *III.* $F(x)$ is 'near' to the Pareto front. In that case, a movement toward the Pareto front will lead to non-significant improvements of the dominated hypervolume. Instead, a search *along* the Pareto front will be performed.

To assign the objective vector $F(x)$ into one of these regions, we can utilize some properties of the descent cone of a MOP: If x is 'far away from the Pareto set, then the objectives gradients nearly point into the same direction, and if x is

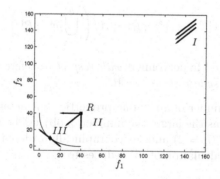

Fig. 1. Division of the objective space into distance regions

'close' then they point in opposite directions [3]. Hence, we can decide if $F(x)$ is in Region I, II or III by considering the angle between the gradients. Let

$$g_i := \nabla f_i(x), \quad i = 1, 2, \tag{6}$$

then the angle between g_1 and g_2 is defined by

$$\cos \alpha = \frac{g_1^T g_2}{\|g_1\| \, \|g_2\|} \in [-1, 1]. \tag{7}$$

If $\cos \alpha = 1$, both gradients point into the same direction ($\downarrow\downarrow$) which happens, roughly speaking, if x is infinitely far from the Pareto set. If $\cos \alpha = 0$, the gradients are orthogonal to each other ($\leftarrow\downarrow$). Finally, when $\cos \alpha = -1$, the grandients point into opposite directions ($\downarrow\uparrow$) which happens if x is on the Pareto set (i.e., zero distance). In order to divide the search space into three distance regions that can be numerically detected, we choose two values $a, b \in (-1, 1)$ with $b < a$ and define:

$$
\begin{array}{lll}
F(x) \in I & : \Leftrightarrow \cos \alpha \geq a, \\
F(x) \in II & : \Leftrightarrow \cos \alpha \in (b, a), \\
F(x) \in III & : \Leftrightarrow \cos \alpha \leq b,
\end{array}
$$

For the computations made in Section 5 we tested our approach using different values for a and b due to the problems behavior, finally the values taken to perform the experiments were $a = 0.8$ and $b = -0.8$, since they were the values that achieved better results. In a general case these values depend on how the cone (built by the gradient) behaves when is near the Pareto front. Now we describe the local search within each region.

Local Search in Region I. As shown in [14], large improvements in objective space can only be obtained when choosing

$$d_I = \begin{pmatrix} -1 \\ -|\lambda| \end{pmatrix}, \tag{8}$$

where $\|\nabla f_2(x)\|_2 = |\lambda| \|\nabla f_1(x)\|_2$, which defines a movement toward the rough location of the Pareto front. Hence, d_I can be chosen together with the DS approach. Alternatively, one can use Pareto descent methods since they define similar movements in Region I. For our computations we have used the method proposed in [10], namely the descent direction

$$\nu = \frac{1}{2} \left(\frac{g_1}{\|g_1\|} + \frac{g_2}{\|g_2\|} \right) \tag{9}$$

coupled with an Armijo-like step size control as used in [10].

Local Search in Region II. Given x such that $F(x) \in II$, the task is to find a search direction $d_{II} <_p 0$ such that a movement in that direction maximizes the hypervolume. Using DS, we can write the image of the new iteration x_{new} as

$$y_{new} = F(x) + td_{II}, \qquad (10)$$

where $t \in \mathbb{R}R$ is a given (fixed) step size and d_{II} is to be chosen such that it solves the two-dimensional problem

$$\max_{d \in \mathbb{R}^2} \nu(d) = (r_1 - f_1(x) - td_1) \times (r_2 - f_2(x) - td_2), \qquad (11)$$

$$\text{s.t.} \|d\|_2^2 = 1$$

If one replaces the 2-norm by the infinity norm in the constraint of Equation (11) (which drops the assumption that the movement is done with an equal step in objective space) a straightforward computation shows that

$$d_{II,\infty} = F(x) - R. \qquad (12)$$

solves the modified problem. We have used this direction for our implementations since it is easier to calculate and yields no difference in the performance of the algorithm.

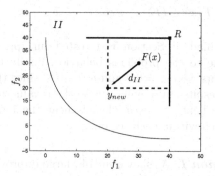

Fig. 2. Local search in Region II

Local Search in Region III. Finally, in the case $F(x)$ is in Region III, a movement along the Pareto front is desired to increase the dominated hypervolume. Here, we propose to linearize the Pareto front at $y = F(x)$ and to compute the optimal step size along direction d_{III} that describes the linearization (compare to Figure 3). Direction d_{III} can be computed as follows: Let x be a Karush-Kuhn Tucker (KKT) point. It is known that the corresponding weight vector α s.t. $\sum_{i=1}^{2} \alpha_i \nabla f_i(x) = 0$ is orthogonal to the linearized Pareto front at $F(x)$, and α solves the following quadratic optimization problem (see [12]):

$$\min_{\alpha \in \mathbb{R}^2} \left\{ \|\alpha_1 \nabla f_1(x) + \alpha_2 \nabla f_2(x)\|_2^2 : \alpha_i \geq 0, \ i = 1, 2, \ \alpha_1 + \alpha_2 = 1 \right\} \qquad (13)$$

Hence, one can compute a solution $\tilde{\alpha}$ of (13) and set

$$d_{III} = \begin{pmatrix} -\tilde{\alpha}_1 \\ \tilde{\alpha}_2 \end{pmatrix} \tag{14}$$

The maximization of the hypervolume leads thus to the one-dimensional problem

$$\max_{t \in \mathbb{R}} \tilde{\nu}(t) = (r_1 - f_1(x) - td_1) \times (r_2 - f_2(x) - td_2), \tag{15}$$

where $d_{III} = (d_1, d_2)^T$, which has an analytic solution in case the weight vector α has no entries equal to zero.

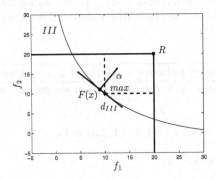

Fig. 3. Movement along linearized Pareto front in order to improve the hypervolume in Region *III*

Proposition 1. *Let $\alpha >_p 0$, then the global maximizer of Problem 15 is given by*

$$t^* = \frac{d_1 r_2 + d_2 r_1 - d_1 f_2(x) - d_2 f_1(x)}{2 d_1 d_2}. \tag{16}$$

Proof. If $\alpha_p >_p 0$, then it follows by (14) that $d_1, d_2 \neq 0$. The first derivative of $\tilde{\nu}$ is given by

$$\tilde{\nu}'(t) = 2t d_1 d_2 + d_2 f_1(x) - r_1 d_2 + d_1 f_2(x) - d_1 r_2 \tag{17}$$

Setting this to zero leads to

$$t^* = \frac{d_1 r_2 + d_2 r_1 - d_1 f_2(x) - d_2 f_1(x)}{2 d_1 d_2}. \tag{18}$$

Further, the second derivative at t^* is given by

$$\tilde{\nu}''(t^*) = 2 d_1 d_2 < 0. \tag{19}$$

The negativity holds since $\alpha >_p 0$ and by construction of d_{III}, and the claim follows. □

We stress that the above solution holds for the linearized problem which is of course a simplification of the problem at hand. We have observed that the step size t^* leads to satisfying results in particular if (i) the Pareto front is almost linear, and (ii) if the reference point R and the current objective vector $F(x)$ are not too far away from each other. For practical implementations, it is advisable to define a maximal step size t_{max} to bound the search. Also note that the step size t^* is defined for a search in *objective* space while the new iterate $x_{new} = x + t_x \nu$ is obtained via a line search in *parameter* space. For this, we follow the suggestion made in [11] to make the match $t_x = t^*$ that works particularly well for small values of t^*. Finally, we note that the above consideration is made for KKT points. However, these computations work also well if the candidate solution x is near to the Pareto set. In particular, d_{III} points along the Pareto front.

Algorithm 1 summarizes the above discussion and presents the HVDS as standalone algorithm.

Algorithm 1. HVDS as standalone algorithm for one element archives

---

Require: x_0: starting point, a, b: values for region assignment; R: reference point
  $i := 0$
  repeat
    compute the angle θ of $\nabla f_j(x_i)$, $j = 1, 2$ as in Eq. (7)
    if $\theta > a$ **then**   ▷ $F(x_i) \in I$
      Compute ν_I as in Eq. (9)
      Compute $t_I \in \mathbb{R}_+$
      $x_{i+1} = x_i + t_I \nu_I$
    else if $\theta \in (b, a)$ **then**   ▷ $F(x_i) \in II$
      $d_{II} = F(x_i) - R$
      $\nu_{II} = J(x_i)^+ d_{II}$
      Compute $t_{II} \in \mathbb{R}_+$
      $x_{i+1} = x_i + t_{II} \nu_{II}$
    else   ▷ $F(x_i) \in III$
      get the convex weight α according to Eq. (13)
      $d_{III} = (-\alpha[2], \alpha[1])^T$
      $\nu_{III} = J(x_i)^+ d_{III}$
      Compute t_{III} as in Eq. (16)
      $x_{i+1} = x_i + t_{III} * \nu_{III}$
    end if
    $i := i + 1$
  until $t_{III} = 0$ or a maximum number of iterations is reached

---

3.2   General Archives

We now consider the general case where the archive contains l elements, i.e., $A = \{x_1, \ldots, x_l\}$. As we will see (and which fits our intuition) the 'optimal' search direction for a given point $x \in A$ depends in some cases on the location of the other elements of A. However, we can reduce all cases to the one element case with appropriate adjustments to the reference point.

In the following we consider the local search in all three distance regions.

Local Search in Region I. If a point $x \in A$ that is chosen for local search is far away from the Pareto front, a movement into its direction is desired regardless of the location of the other elements of A. Hence, we propose to proceed as for the one element case.

Local Search in Region II. Since we consider two-objective problems, the images in objective space of A can be sorted by one of the objective values which we assume in the following. Let $x_i \in A$ be given that is assigned for local search. Figure 4 that shows such a scenario suggests that the hypervolume contribution of x_{new} that is obtained via a modification of x_i is restricted to the region between $F(x_{i-1})$ and $F(x_{i-1})$. Hence, for $i \in \{2, \ldots, l-1\}$ we propose to choose the new reference point

$$R_{F(x_i)} = \begin{pmatrix} f_1(x_{i+1}) \\ f_1(x_{i-1}) \end{pmatrix} \tag{20}$$

and to proceed analog to the one element case using the direction

$$d_{II,x_i} = F(x_i) - R_{F(x_i)}. \tag{21}$$

For the extreme points (i.e., $i \in \{1, l\}$) we proceed again with $R_{F(x_i)} = F(x_i) - R$.

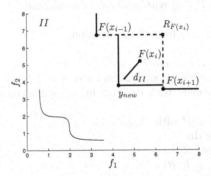

Fig. 4. Local search in Region II for multiple archive entries

Local Search in Region III. Analog to the above discussion we can proceed with points in the third distance region. To be more precise, we propose to use the reference point $R_{F(x_i)}$ for intermediate points (i.e., $i \in \{2, \ldots, l-1\}$) and the original point R for the extreme archive entries.

4   Integrating HVDS into SMS-EMOA

Here we make a first attempt to integrate the local search mechanism HVDS into an EMOA in order to obtain a fast and reliable algorithm to obtain hypervolume approximations of a given MOP. We have chosen to take the state-of-the-art algorithm SMS-EMOA ([2]), however, we stress that HVDS can in principle be hybridized with any other hypervolume based EMOA.

Each iteration step in SMS-EMOA is divided into three sections: First, one new element is generated through an evolutionary process that is inserted into the current population. In the second section, the population is partitioned into h separate groups $(S_1, \ldots, S_h)$ with respect to the degree of nondominance. Finally, the algorithm computes the contributions of the points according to hypervolume and the element with the least hypervolume contribution is discarded from the archive.

We propose to integrate the new local search mechanism as follows: After the update of the archive in iteration step i, m_i elements of the population P_i are chosen for local improvement via HVDS, X_{LS} will represents the set of the elements taken. Since it is assumed that HVDS actually improves the hypervolume value of a given element, no consideration of the hypervolume contributions is nesessary (which is a time-consuming task), but the new iterates replace the initial points. Algorithm 2 shows the pseudo-code of the new hybrid SMS-EMOA-HVDS.

Algorithm 2. SMS-EMOA-HVDS

Initialize a population $P \subset Q$ with μ elements at random
repeat
    generate offspring $x \in Q$ from P by variation
    $P := P \cup \{x\}$
    build ranking $S_1, \ldots, S_h$ from P
    compute the hypervolume contribution for each $x \in S_h$
    denote by x^* the element with the least hypervolume contribution
    $P := P \setminus \{x^*\}$
    choose the set $X_{LS} \subset P$ with $|X_{LS}| = m$
    for all $i = 1, \ldots, m$ **do**
        $x_{i,0} = i$th element of X_{LS}
        $\tilde{x}_i = $HVDS$(x_{i,0}, a, b, R)$
        $P := P \cup \{\tilde{x}_i\} \setminus \{x_{i,0}\}$
    end for
until stopping criterion fulfilled
return P

---

5   Numerical Results

5.1   HVDS as Standalone Algorithm

First we test the ability of the HVDS as standalone algorithm. For this, we will use the following two uni-modal problems:

$$F_1 : \mathbb{R}^{10} \to \mathbb{R}^2 \tag{22}$$

$$f_1(x) = ||x - a_1||_2^2, \quad f_2(x) = ||x - a_2||_2^2,$$

where $a_1 = (1, \ldots, 1)^T$, $a_2 = (-1, \ldots, -1)^T$, and

$$F_2 : \mathbb{R}^2 \to \mathbb{R}^2 \tag{23}$$

$$f_1(x) = \frac{1}{2}(\sqrt{1 + (x_1 + x_2)^2} + \sqrt{1 + (x_1 - x_2)^2} + x_1 - x_2) + \lambda \cdot e^{-(x_1 - x_2)^2}$$

$$f_2(x) = \frac{1}{2}(\sqrt{1 + (x_1 + x_2)^2} + \sqrt{1 + (x_1 - x_2)^2} - x_1 + x_2) + \lambda \cdot e^{-(x_1 - x_2)^2},$$

where $\lambda = 0.85$. MOP (22) ([16], denoted by 'Convex') has a convex Pareto front, and the front of MOP (23) ([18], 'Dent') is convex-concave (see Figure 5).

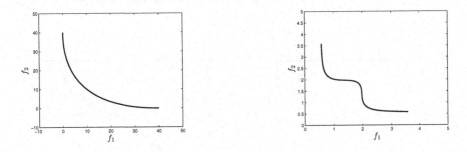

Fig. 5. Pareto fronts of MOP (22) (left) and MOP (23) (right)

First we test the HVDS for one element archives. For the sake of a comparison, we define a simple hill climber as follows: For a given point x, a further candidate solution y is taken from a neighborhood of x. As next iterate, the solution with the highest hypervolume value is taken and the search is continued in the same manner. We have chosen this strategy since it relates to a stochastic local search procedure within hypervolume-based MOEAs. Figures 6 and 7 show exemplary runs for both methods on each problem. Figure 8 shows the hypervolume against the number of function evaluations for both problems and methods. Here we count five function evaluations for the cost of one gradient evaluation which would be the case when using automatic differentiation [6]. In both cases, HVDS is able to get higher hypervolume values in the early stage of the algorithm. For Dent, the algorithm is even able to terminate after 130 function evaluations at the optimal hypervolume value.

Next, we make a first attempt to investigate the ability of the HVDS within set based search. For this, we have made the following adaption of the standalone HVDS as presented in Algorithm 1: Instead of one starting point x_0 we choose an initial population $X = \{x_0^{(1)}, \ldots, x_0^{(5)}\}$ consisting of five elements. The iteration step is then performed individually for all elements (i.e., $x_{i+1}^{(j)} = x_i^{(j)} + t\nu$ as described in Algorithm 1) using the choice of the reference point as proposed in

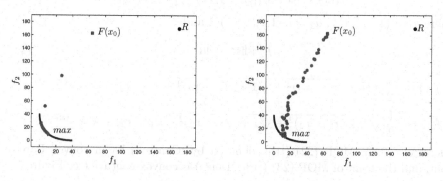

Fig. 6. Result of the HVDS and the hypervolume hill climber on Convex

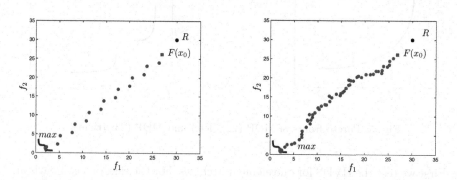

Fig. 7. Result of the HVDS and the hypervolume hill climber on Dent

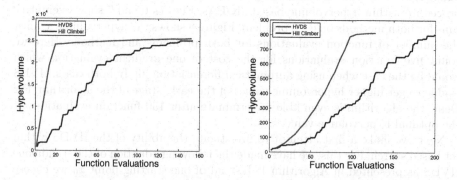

Fig. 8. Comparison the hypervolume hill climber on Convex (left) and Dent (right). The results are averaged over 20 test runs.

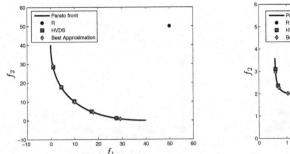
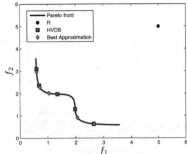

Fig. 9. Numerical results of the 5 element HVDS on Convex (left) and Dent (right)

Table 1. Comparison of the 5 element HVDS and the SMS-EMOA with $\mu = 5$. The results are averaged over 20 test runs.

	HVDS		SMS-EMOA		
	# Iterations	Hypervolume	# Iterations	Hypervolume	Best Value
Convex	1400	2100.1424	1400	1992.9788	2107.6523
Dent	885	16.6941	900	16.5721	16.8225

Table 2. HV results of SMS-EMOA with and without HVDS as local searcher after 2500 iterations of the algorithm (using the same number of function evaluations). The values are obtained from 20 test runs.

	SMS-EMOA			SMS-EMOA-HVDS		
	Average	Deviation	Median	Average	Deviation	Median
Convex	2003.867	68.956	2021.200	2161.668	18.039	2164.803
Dent	17.234	0.031	17.241	17.245	0.023	17.248
ZDT1	105.015	0.948	105.002	108.965	1.654	109.512
ZDT2	97.592	2.965	96.176	107.463	3.563	109.207
ZDT3	113.771	1.857	114.330	116.097	1.948	117.576
ZDT4	76.536	13.485	82.107	71.552	15.770	71.352

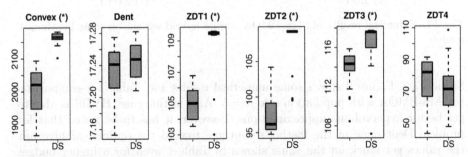

Fig. 10. Boxplots of the HV at the final iteration of the SMS-EMOA and its hybrid variant (DS) on the considered test problems. Statistically significant differences due to the Wilcoxon-Rank-Sum Test with $\alpha = 0.05$ are marked with (*).

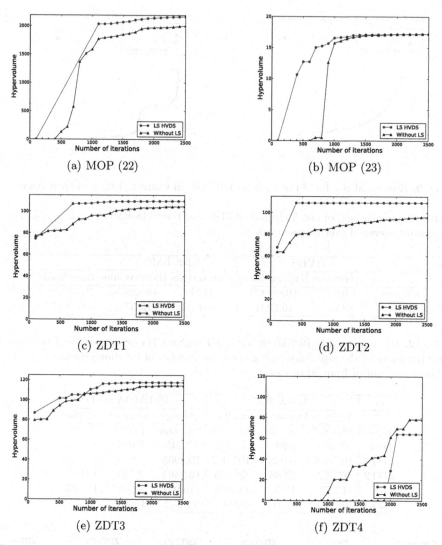

Fig. 11. Numerical results of SMS-EMOA and its hybrid variant on some benchmark models

Section 3.2. Figure 9 shows some numerical results and Table 1 a comparison to SMS-EMOA with population size $\mu = 5$. Also in this case, HVDS is able to get better hypervolume approximations. However, it has to be noted that for problem Dent none of the methods converge toward the optimal archive but the values get stuck on the value shown in Table 1 even for a higher budget of function evaluations. This might be due to the fact that only *one* point is iterated at each step. A possible remedy would be to modify *all* points in each iteration, however, to the sacrifice of a much higher computational burden.

5.2 HVDS within SMS-EMOA

Finally, we investigate the potential of HVDS as local searcher within SMS-EMOA. For the computations we have realized SMS-EMOA-HVDS as follows: To pull the current archive to the Pareto set, we have chosen to run two HVDS runs in the beginning of the search. Due to the relatively high cost of this search, we have omitted further HVDS calls in subsequent iterations (i.e., we have taken $m_1 = 2$ together with a budget of 50 iterations and $m_i = 0$ for $i > 1$).

Table 2 and Figure 11 show some numerical results on the above MOPs as well as on ZDT1-4 from [19]. Boxplots of the respective HV values after the final iteration are given in Figure 10. In 4 out of 6 cases the new hybrid is superior to its base EMOA while the differences in location of the HV values are not statistically significant for Dent and ZDT4. The latter is certainly due to the choice of the local search since the two runs got stuck in local minima, and hence, the effort was lost. Further variants of local search, e.g., the application of more but shorter HVDS runs, have to be tested which we leave for future research.

6 Conclusions and Future Work

In this paper, we have presented a new local search procedure for hypervolume approximations of a given multi-objective optimization problem. The new method, HVDS, is based on the Directed Search Method which is able to steer the search into any direction given in objective space and has been adapted to the given context. We have presented the HVDS both as standalone algorithm and as local search engine within the state-of-the-art hypervolume based algorithm SMS-EMOA. The benefit of the novel method has been shown on several numerical experiments.

For future work, there are many aspects that have to be considered. For instance, the current study was restricted to unconstrained bi-objective problems which has to be generalized for sake of a broader applicability. Further, by the same reason, it would be desirable to use the gradient free version of the Directed Search Method in the hybrid which needs a careful consideration of the neighborhood structure of the base EMOA [9]. Finally, it might be interesting to adapt the method to other indicators, e.g., to obtain Hausdorff approximations of the set of interest [5,13,17].

Acknowledgement. The first author acknowledges support from CONACyT through a scholarship to pursue undergraduate studies at the Computer Science Department of CINVESTAV-IPN. The second author acknowledges support from CONACyT project no. 128554. The third and fourth author acknowledge support from DFG, project no. TR 891/5-1.

References

1. Beume, N.: S-metric calculation by considering dominated hypervolume as Klee's measure problem. Evolutionary Computation 17(4), 477–492 (2009)
2. Beume, N., Naujoks, B., Emmerich, M.: SMS-EMOA: Multiobjective selection based on dominated hypervolume. European Journal of Operational Research (2006)
3. Brown, M., Smith, R.E.: Directed multi-objective optimisation. International Journal of Computers, Systems and Signals 6(1), 3–17 (2005)
4. Deb, K.: Multi-Objective Optimization using Evolutionary Algorithms. John Wiley & Sons, Chichester (2001) ISBN 0-471-87339-X
5. Emmerich, M.T.M., Deutz, A.H., Kruisselbrink, J.W.: On quality indicators for black-box level set approximation. In: Tantar, E., Tantar, A.-A., Bouvry, P., Del Moral, P., Legrand, P., Coello Coello, C.A., Schütze, O. (eds.) EVOLVE- A Bridge between Probability, Set Oriented Numerics and Evolutionary Computation. SCI, vol. 447, pp. 153–184. Springer, Heidelberg (2013)
6. Griewank, A.: Evaluating Derivatives: Principles and Techniques of Algorithmic Differentiation. Frontiers in Appl. Math., vol. 19. SIAM, Philadelphia (2000)
7. Hillermeier, C.: Nonlinear Multiobjective Optimization: A Generalized Homotopy Approach. Springer (2001) ISBN-13:9783764364984
8. Jaszkiewicz, A., Ishibuchi, H., Zhang, Q.: Multiobjective memetic algorithms. In: Neri, F. (ed.) Handbook of Memetic Algorithms. SCI, vol. 379, pp. 201–217. Springer, Heidelberg (2011)
9. Lara, A., Alvarado, S., Salomon, S., Avigad, G., Coello Coello, C.A., Schütze, O.: The gradient free directed search method as local search within multi-objective evolutionary algorithms. In: Schütze, O., Coello Coello, C.A., Tantar, A.-A., Tantar, E., Bouvry, P., Del Moral, P., Legrand, P. (eds.) EVOLVE - A Bridge Between Probability, Set Oriented Numerics, and Evolutionary Computation II. AISC, vol. 175, pp. 153–168. Springer, Heidelberg (2012)
10. Lara, A., Coello Coello, C.A., Schütze, O.: A painless gradient-assisted multi-objective memetic mechanism for solving continuous bi-objective optimization problems. In: 2010 IEEE Congress on Evolutionary Computation (CEC), pp. 1–8. IEEE, IEEE Press (2010)
11. Mejia, E., Schütze, O.: A predictor corrector method for the computation of boundary points of a multi-objective optimization problem. In: International Conference on Electrical Engineering, Computing Science and Automati Control (CCE 2010), pp. 1–6 (2007)
12. Schäffler, S., Schultz, R., Weinzierl, K.: A stochastic method for the solution of unconstrained vector optimization problems. Journal of Optimization Theory and Applications 114(1), 209–222 (2002)
13. Schütze, O., Esquivel, X., Lara, A., Coello Coello, C.A.: Using the averaged Hausdorff distance as a performance measure in evolutionary multi-objective optimization. IEEE Transactions on Evolutionary Computation 16(4), 504–522 (2012)
14. Schütze, O., Lara, A., Coello Coello, C.A.: The directed search method for unconstrained multi-objective optimization problems. Technical Report TR-OS-2010-01 (2010)
15. Schütze, O., Lara, A., Coello Coello, C.A.: The directed search method for unconstrained multi-objective optimization problems. In: Proceedings of the EVOLVE – A Bridge Between Probability, Set Oriented Numerics, and Evolutionary Computation (2011)

16. Schütze, O., Lara, A., Coello Coello, C.A.: On the influence of the number of objectives on the hardness of a multiobjective optimization problem. IEEE Transactions on Evolutionary Computation 15(4), 444–455 (2011)
17. Trautmann, H., Rudolph, G., Dominguez-Medina, C., Schütze, O.: Finding evenly spaced Pareto fronts for three-objective optimization problems. In: Schütze, O., Coello Coello, C.A., Tantar, A.-A., Tantar, E., Bouvry, P., Del Moral, P., Legrand, P. (eds.) EVOLVE - A Bridge Between Probability, Set Oriented Numerics, and Evolutionary Computation II. AISC, vol. 175, pp. 89–105. Springer, Heidelberg (2012)
18. Witting, K.: Numerical Algorithms for the Treatment of Parametric Optimization Problems and Applications. PhD thesis, University of Paderborn (2012)
19. Zitzler, E., Deb, K., Thiele, L.: Comparison of Multiobjective Evolutionary Algorithms: Empirical Results. Evol. Comput. 8(2), 173–195 (2000)
20. Zitzler, E., Thiele, L.: Multiobjective Optimization Using Evolutionary Algorithms - A Comparative Case Study. In: Eiben, A.E., Bäck, T., Schoenauer, M., Schwefel, H.-P. (eds.) PPSN 1998. LNCS, vol. 1498, pp. 292–301. Springer, Heidelberg (1998)
21. Zitzler, E., Thiele, L.: Multiobjective evolutionary algorithms: a comparative case study and the strength Pareto approach. IEEE Transactions on Evolutionary Computation 3(4), 257–271 (1999)

A Hybrid Algorithm for the Simple Cell Mapping Method in Multi-objective Optimization

Yousef Naranjani[1], Carlos Hernández[3], Fu-Rui Xiong[2],
Oliver Schütze[3], and Jian-Qiao Sun[1]

[1] School of Engineering
University of California at Merced
Merced, CA 95344, USA
{ynaranjani,jqsun}@ucmerced.edu
[2] Department of Mechanics
Tianjin University
Tianjin, 300072, China
xfr90311@gmail.com
[3] CINVESTAV-IPN, Depto de Computacion
Mexico City, 07360 Mexico
chernandez@computacion.cs.cinvestav.mx,
schuetze@cs.cinvestav.mx

Abstract. This paper presents a hybrid gradient free-gradient (GFG) algorithm for the simple cell mapping (SCM) method for multi-objective optimization problems (MOPs). The SCM method is briefly reviewed in the context of the multi-objective optimization problems (MOPs). We present a mixed application of gradient free directed search and gradient search algorithms for the SCM method and discuss its potentials for higher dimensional MOPs. We present several numerical exmaples to demonstrate the effectiveness of the proposed hybrid algorithm. The examples include two simple geometric MOPs, an example with five design parameters, and a proportional-integral-derivative (PID) control design for a second order linear system.

Keywords: Simple cell mapping, Multi-objective optimization, Feedback control.

1 Introduction

The solution of MOPs does not consist of a single point in the design space, but rather forms a set, called *Pareto set*. The corresponding objective function values are called *Pareto front*. Many numerical methods for MOPs have been studied. Lately, we have found that the simple cell mapping (SCM) method can obtain the global optimal solution in a quite effective manner for low and moderate dimensional problems. This paper presents a hybrid algorithm of the SCM method for MOPs to enhance its efficiency. The algorithm has a potential to reduce the computing time of the SCM method for MOPs.

M. Emmerich et al. (eds.), *EVOLVE - A Bridge between Probability, Set Oriented Numerics, and Evolutionary Computation IV*, Advances in Intelligent Systems and Computing 227,
DOI: 10.1007/978-3-319-01128-8_14, © Springer International Publishing Switzerland 2013

A hybrid gradient free directed search algorithm has been studied in conjunction with evolutionary algorithms [1]. Evolutionary algorithms are most widely used for MOPs [2]. The underlying idea in evolutionary computation is to steer (or evolve) an entire set of solutions (population) toward the set of interest during the search. Evolutionary algorithms specialized for multi-objective problems have been shown their strength in many applications. Due to the global and stochastic nature of evolutionary algorithms, the Pareto set can be approximated quite well in most cases, although there is always an uncertainty left regarding whether the global Pareto set has indeed been found. There also exist attempts to compute nearly optimal solutions with evolutionary algorithms [3]. However, this approach suffers the drawback that only a subset of the nearly optimal solutions can be stored. In particular, a significant fraction of the set of interest may be ignored by the archiving strategy in certain cases [3].

Another approach to approximate the Pareto set is to use the set oriented methods with subdivision techniques [4–6]. The advantage of the set oriented methods is that they generate an approximation of the global Pareto set in one single run of the algorithm. Further, they are applicable to a wide range of optimization problems and are characterized by a great robustness. Hence, these methods are interesting alternatives to 'classical' mathematical programming techniques in particular for the thorough investigation of low or moderate dimensional MOPs. The cell mapping method in this study is the predecessor of the set oriented methods, and was proposed by Hsu [7] for global analysis of nonlinear dynamical systems. In the cell mapping method for MOPs, the dynamical systems are derived from multi-objective optimization search algorithms. The cell mapping method can obtain a finite size approximation $\mathcal{A}$ of the Pareto set $\mathcal{P}$ such that the distance between $\mathcal{P}$ and $\mathcal{A}$ is less or equal to a given threshold value in the Hausdorff sense. That is, the approximation quality of $\mathcal{A}$ can in principle be determined. A first implementation of the multi-objective cell mapping technique is used that already yields promising results.

Two cell mapping methods have been extensively studied, namely, the simple cell mapping (SCM) and the generalized cell mapping (GCM) to study the global dynamics of nonlinear systems [7,8]. The cell mapping methods have been applied to optimal control problems of deterministic and stochastic dynamic systems [9–11]. Other interesting applications of the cell mapping methods include optimal space craft momentum unloading [12], single and multiple manipulators of robots [13], optimum trajectory planning in robotic systems by [14], and tracking control of the read-write head of computer hard disks [15]. Sun and his group studied the fixed final state optimal control problems with the simple cell mapping method [16,17], and applied the cell mapping methods to the optimal control of deterministic systems described by Bellman's principle of optimality [18]. MOPs represent new application domains of the cell mapping methods.

In this paper, we present a hybrid gradient free-gradient (GFG) algorithm of the SCM method for MOPs in order to improve the efficiency of the method. We present several examples to demonstrate the algorithm including the time domain design of PID controls for linear systems.

2   Multi-Objective Optimization

A multi-objective optimization problem (MOP) can be expressed as follows:

$$\min_{\mathbf{k} \in Q}\{\mathbf{F}(\mathbf{k})\}, \tag{1}$$

where $\mathbf{F}$ is the map that consists of the objective functions $f_i : Q \to \mathbf{R}^1$ under consideration.

$$\mathbf{F} : Q \to \mathbf{R}^k, \;\; \mathbf{F}(\mathbf{k}) = [f_1(\mathbf{k}), \ldots, f_k(\mathbf{k})]. \tag{2}$$

$\mathbf{k} \in Q$ is a q-dimensional vector of design parameters. The domain $Q \subset \mathbf{R}^q$ can in general be expressed by inequality and equality constraints:

$$Q = \{\mathbf{k} \in \mathbf{R}^q \mid g_i(\mathbf{k}) \le 0, \; i = 1, \ldots, l, \tag{3}$$
$$\text{and } h_j(\mathbf{k}) = 0, \; j = 1, \ldots, m\}.$$

Next, we define optimal solutions of a given MOP by using the concept of *dominance* [19].

Definition 1

(a) Let $\mathbf{V}, \mathbf{W} \in \mathbf{R}^k$. The vector $\mathbf{V}$ is said to be *less than* $\mathbf{W}$ (in short: $\mathbf{V} <_p \mathbf{W}$), if $V_i < W_i$ for all $i \in \{1, \ldots, k\}$. The relation $\le_p$ is defined analogously.
(b) A vector $\mathbf{v} \in Q$ is called *dominated* by a vector $\mathbf{w} \in Q$ ($\mathbf{w} \prec \mathbf{v}$) with respect to the MOP (1) if $\mathbf{F}(\mathbf{w}) \le_p \mathbf{F}(\mathbf{v})$ and $\mathbf{F}(\mathbf{w}) \ne \mathbf{F}(\mathbf{v})$, else $\mathbf{v}$ is called non-dominated by $\mathbf{w}$.

If a vector $\mathbf{w}$ dominates a vector $\mathbf{v}$, then $\mathbf{w}$ can be considered to be a 'better' solution of the MOP. The definition of optimality or the 'best' solution of the MOP is now straightforward.

Definition 2

(a) A point $\mathbf{w} \in Q$ is called *Pareto optimal* or a *Pareto point* of the MOP (1) if there is no $\mathbf{v} \in Q$ which dominates $\mathbf{w}$.
(b) The set of all Pareto optimal solutions is called the *Pareto set* denoted as

$$\mathcal{P} := \{\mathbf{w} \in Q : \mathbf{w} \text{ is a Pareto point of the MOP (1)}\}. \tag{4}$$

(c) The image $\mathbf{F}(\mathcal{P})$ of $\mathcal{P}$ is called the *Pareto front*.

Pareto front typically forms $(k-1)$-dimensional manifolds under certain mild assumptions on the MOP [20]. Recent studies with the SCM method seem to suggest that the Pareto front may have fine structures for MOPs of complex dynamical systems.

3 Simple Cell Mapping Method

The cell mapping methods describe system dynamics with cell-to-cell mappings by discretizing both the phase space and time. The point-to-point mapping obtained from the gradient search algorithm can be written as

$$\mathbf{x}(i) = \mathbf{G}(\mathbf{x}(i-1)), \tag{5}$$

where $\mathbf{x}(i) \in \mathbf{R}^n$ is the state vector at the ith mapping step. In the SCM, the dynamics of an entire cell denoted as Z is represented by the dynamics of its center. The center of Z is mapped according to the point-to-point mapping. The cell that contains the image point is called the image cell of Z. The cell-to-cell mapping is denoted by C,

$$Z(i) = C(Z(i-1)). \tag{6}$$

For the gradient free search, the image cell of Z is selected by comparing the objective function values of all its neighboring cells. If there is only one dominant cell in the neighborhood, it becomes the image of Z. If there are more than one dominant cells, we select the one that has the highest objective function value decrease per unit distance. Such a choice mimics the steepest gradient decent algorithm.

There are, however, other gradient free approaches that could be realized efficiently in the context of SCM. If the cells are small enough, one could, for instance, use the center points of the neighboring cells to obtain a finite difference approximation of the gradient at a given cell. This would open the door for the usage of in principle all gradient based iterations, but without explicitly computing the gradient. If the function values for the center points of neighboring cells in all q directions are already known, the approximation of the gradient comes for free in terms of the additional function calls in computation.

We should note that the exact image of the center of Z is approximated by the center of its image cell. This approximation can cause significant errors in the long term solution of dynamical systems [9,10,14]. The cell mapping with a finite number of cells in the computational domain will eventually lead to closed groups of cells of the period same as the number of cells in the group. The periodic cells represent invariant sets, which can be periodic motion and stable attractors of the system. The rest of the cells form the domains of attraction of the invariant sets. For more discussions on the cell mapping methods, their properties and computational algorithms, the reader is referred to the book by Hsu [7].

In the rest of the paper, whenever a gradient decent algorithm is used, we adopt the method due to Fliege and Svaiter to generate the cell-to-cell mapping [21]. For a comprehensive survey of various search algorithms, see the book [22].

4 Directed Search Algorithm

In this work, we shall also consider the directed search (DS) algorithm [23,24] that has the benefit of needing less information to perform the local search.

The DS algorithm allows to steer the search into any direction $\mathbf{d} \in \mathbf{R}^k$ given in the objective space. To apply the gradient free version of this method within SCM we can proceed as follows: Choose $r \geq q$ neighboring cells where the images of the center points are known. Define a unit vector as

$$\nu_i = \frac{\mathbf{x}_i - \mathbf{x}_0}{\|\mathbf{x}_i - \mathbf{x}_0\|_2}, \tag{7}$$

where $\mathbf{x}_0$ is the center of the cell under consideration, and $\mathbf{x}_i$ is the center of the ith cell in the neighborhood of $\mathbf{x}_0$. Define the matrix $\mathcal{F} = \{m_{i,j}\} \in \mathbf{R}^{k \times r}$ as

$$m_{i,j} = \frac{f_i(\mathbf{x}_j) - f_i(\mathbf{x}_0)}{\|\mathbf{x}_j - \mathbf{x}_0\|_2}, \tag{8}$$

which is an approximation of the directional derivative of f_i in direction ν_j, and compute

$$\lambda = \mathcal{F}^+ \mathbf{d}, \tag{9}$$

where $\mathcal{F}^+$ denotes the pseudo inverse of $\mathcal{F}$ [25]. Then a line search (in phase space) in the direction

$$\nu = \sum_{i=1}^{r} \lambda_i \nu_i \tag{10}$$

leads to a movement along $\mathbf{d}$-direction in the objective space. Note that only the function values for $r \geq k$ test points have to be known to get such a search vector while a gradient approximation via finite differences requires q or even $2q$ test points. We refer to [24] for details about the gradient free DS.

It remains to discuss the choice of $\mathbf{d}$. In the context of multi-objective optimization two movements are of particular interest: Toward and along the Pareto set. In order to steer toward the Pareto set, we follow the suggestion made in [26], namely that a search orthogonal to the convex hull of individual minima (CHIM) is most promising. An approximation of the CHIM can be obtained by taking all the k minimal values found during the search of the SCM so far. If a current iterate is already very close to the Pareto front (which can be checked numerically by the 2-condition of $\mathcal{F}$), no further improvements can be obtained but a movement along the Pareto set may be beneficial. This is due to the fact that the SCM may lose parts of the Pareto set in case the cell structure is not fine enough. To move along the set of interest, one can choose a direction $\mathbf{d}$ that points along the CHIM.

5   Numerical Examples

The numrical results for the following examples are obtained using laptop personal computers.

Example 1

Consider two objective functions defined as

$$\mathbf{F}(\mathbf{x}) = [(x_1 - 1)^2 + (x_2 - 1)^2, (x_1 + 1)^2 + (x_2 + 1)^2], \qquad (11)$$

and

$$Q = \{\mathbf{x} \in [-2, 2] \times [-2, 2] \subset \mathbf{R}^2\}. \qquad (12)$$

We consider two implementations of the SCM method for this problem with the numbers of initial divisions along x_1 and x_2 are $\mathbf{N} = [30, 30]$. The sub-division is 3×3. We intend to compare the hybrid approach with the gradient search algorithm with the SCM method.

With the hybrid approach, we first use the gradient free algorithm to generate an approximate solution of the Pareto set that contains the true Pareto set. We then apply the gradient search algorithm to construct the cell mapping on the approximate Pareto set obtained in the first run. Finally, we apply the sub-division to the set of periodic cells obtained in the second run.

As a baseline for comparison, we also apply the gradient search algorithm with the SCM method to compute the Pareto set and Pareto front on the entire region Q.

The Pareto set and Pareto front in two implementations are shown in Figures 1 and 2. The top left of Figure 1 shows the set of 81 periodic cells obtained by the gradient free algorithm. This set certainly contains the true Pareto set. The bottom left of Figure 1 shows the refined solution with the gradient search algorithm for the SCM method. The Pareto set consisting of 86 periodic cells obtained with the hybrid algorithm is identical to the one obtained with the gradient search algorithm applied to the entire region Q.

The computing time with the gradient search algorithm for the SCM method over the entire domain Q is 10.19 seconds and with the hybrid algorithm 7.78 seconds, 24% saving. When we choose $\mathbf{N} = [60, 60]$, the computing time with the gradient search algorithm is 51.3 seconds, while with the hybrid algorithm 34 seconds, 34% saving. Since the Pareto set usually occupies a small percentage of volume in Q, it is anticipated that the higher dimensional MOPs are, the more computational savings can be achieved.

Example 2

We conduct the same comparison studies for the MOP defined by the following objective functions

$$\mathbf{F}(\mathbf{x}) = \left\{ 1 - \exp\left[-(x_1 - 1/\sqrt{2})^2 - (x_2 - 1/\sqrt{2})^2\right], \qquad (13) \right.$$
$$\left. 1 - \exp\left[(x_1 + 1/\sqrt{2})^2 + (x_2 + 1/\sqrt{2})^2\right] \right\},$$

and

$$Q = \{\mathbf{x} \in [-4, 4] \times [-4, 4] \subset \mathbf{R}^2\}. \qquad (14)$$

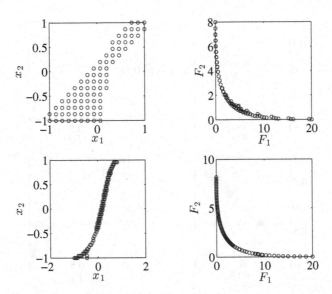

Fig. 1. The Pareto set and Pareto front of Example 1 obtained with the SCM method using the hybrid algorithm for the construction of the cell mapping. Top row: The first run on 30 × 30 grid. Bottom row: 3 × 3 sub-division only applied to the Pareto set obtained in the first run.

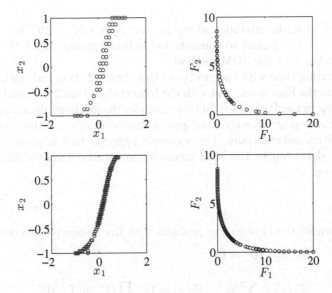

Fig. 2. The Pareto set and Pareto front of Example 1 obtained with the SCM method using the gradient based search algorithm for the construction of the cell mapping on the entire region Q. Top row: The first run on 30 × 30 grid. Bottom row: 3 × 3 sub-division only applied to the Pareto set obtained in the first run.

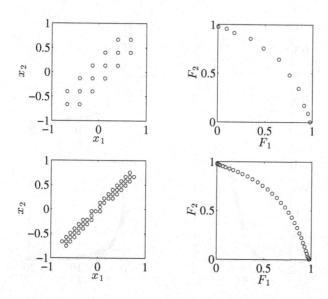

Fig. 3. The Pareto set and Pareto front of Example 2 obtained with the SCM method using the hybrid algorithm for the construction of the cell mapping. Top row: The first run on 30×30 grid. Bottom row: 3×3 sub-division only applied to the Pareto set obtained in the first run.

The numbers of initial divisions along x_1 and x_2 are $\mathbf{N} = [30, 30]$. The subdivision is 3×3. We intend to compare the hybrid approach with the gradient search algorithm with the SCM method.

The computing time with the gradient algorithm is 244 seconds yielding 3052 refined cells in the Pareto set, and with the hybrid algorithm 2.2 seconds yielding 45 refined cells in the Pareto set. In this example, the gradient algorithm applied to the entire region Q picks up many pseudo-periodic cells and thus slows down the computation substantially. This example suggests that a proper choice of the gradient algorithm for the SCM method is not a trivial matter and deserves much attention.

Example 3

Next, we consider the bi-objective problem with five design parameters [27]:

$$F_1, F_2 : \mathbf{R}^5 \to \mathbf{R}$$

$$F_1(x) = \sum_{j=1}^{5} x_j, \quad F_2(x) = 1 - \prod_{j=1}^{5}(1 - w_j(x_j)), \tag{15}$$

where

$$w_j(z) = \begin{cases} 0.01 \cdot \exp\left[-\left(\frac{z}{20}\right)^{2.5}\right] & \text{for } j = 1, 2 \\ 0.01 \cdot \exp\left[-\frac{z}{15}\right] & \text{for } j = 3, \ldots, 5 \end{cases}$$

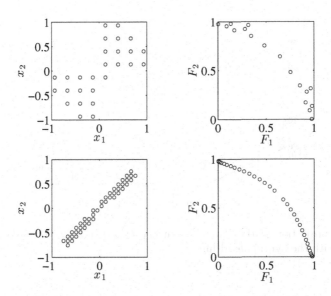

Fig. 4. The Pareto set and Pareto front of Example 2 obtained with the SCM method using the gradient based search algorithm for the construction of the cell mapping in the entire region Q. Top row: The first run on 30×30 grid. Bottom row: 3×3 sub-division only applied to the Pareto set obtained in the first run.

Objective F_1 is related to the cost of a given product and F_2 to its failure rate. For the domain $Q = [0, 40]^5$ the Pareto set consists of four connected components. The domain is divided into 10 segments, leading to 10^5 cells. The hybrid algorithm for the SCM method is implemented.

The first run of the gradient free algorithm finds 517 cells in periodic groups. The second run with the DS algorithm is carried out over the periodic cells only. Note that because the dimension of the parameter space is 5, a sub-division by 3 along each direction at the same time would lead to $125631 = 517 \times 3^5$ cells. This would require more memory and slow down the program. A strategy to save memory and to gain speed is to sub-divide the design parameter one at a time. We divide one design parameter at a time into 5 sub-divisions, and apply the gradient free descent direction method 5 times to each of the cells found. The total CPU time of the entire computing is 372.01 seconds, while the first run takes 351.12 seconds, and the refinement takes 89.74 seconds.

Example 4

Finally, we consider a second order oscillator subject to a proportional-integral-derivative (PID) control.

$$\ddot{x} + 2\zeta\omega_n\dot{x} + \omega_n^2 x = \omega_n^2 u(t), \tag{16}$$

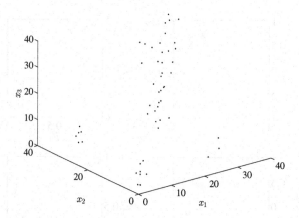

Fig. 5. The projection of the Pareto set in (x_1, x_2, x_3)-subspace on a coarse grid of Example 3 with the hybrid algorithm

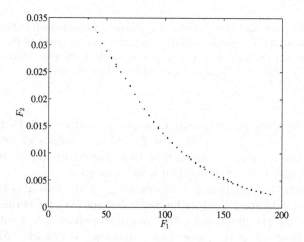

Fig. 6. The Pareto front on a coarse grid of Example 3 with the hybrid algorithm

where $\omega_n = 5$, $\zeta = 0.01$,

$$u(t) = k_p \left[r(t) - x(t) \right] + k_i \int_0^t \left[r(\hat{t}) - x(\hat{t}) \right] d\hat{t} - k_d \dot{x}(t), \qquad (17)$$

$r(t)$ is a step input, k_p, k_i and k_d are the PID control gains. We consider the MOP with the control gains $\mathbf{k} = [k_p, k_i, k_d]^T$ as design parameters. The design space for the parameters is chosen as follows,

$$Q = \{ \mathbf{k} \in [10, 50] \times [1, 30] \times [1, 2] \subset \mathbf{R}^3 \}. \qquad (18)$$

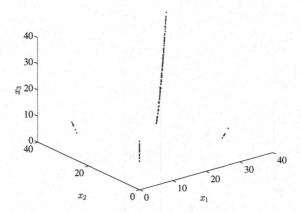

Fig. 7. The projection of the Pareto set in (x_1, x_2, x_3)-subspace on the refined grid of Example 3 with the hybrid algorithm

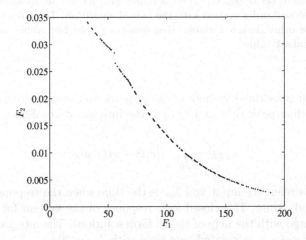

Fig. 8. The Pareto front on the refined grid of Example 3 with the hybrid algorithm

Initially, we select the number of divisions in the three control gain intervals as $\mathbf{N} = [10, 10, 10]$.

Peak time and overshoot are common in time domain control design objectives [28–30]. We consider the multi-objective optimization problem to design the control gain $\mathbf{k}$,

$$\min_{\mathbf{k} \in Q}\{t_p, M_p, e_{IAE}\}, \tag{19}$$

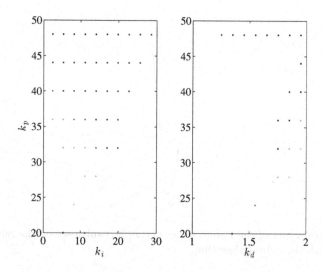

Fig. 9. The Pareto set of $[k_p, k_i, k_d]$ on a coarse grid for the multi-objective optimal control of the second order system with the hybrid algorithm. The color code indicates the level of the other design variable. Red denotes the highest value, and dark blue denotes the smallest value.

where M_p stands for the overshoot of the response to a step reference input, t_p is the corresponding peak time and e_{IAE} is the integrated absolute tracking error

$$e_{IAE} = \int_0^{T_{ss}} \left| r(\hat{t}) - x(\hat{t}) \right| d\hat{t}. \tag{20}$$

where $r(t)$ is a reference input and T_{ss} is the time when the response is close to be in the steady state. The closed-loop response of the system for each design trial is computed with the help of closed form solutions. The integrated absolute tracking error e_{IAE} is calculated over time with $T_{ss} = 20s$.

We first discuss the hybrid algorithm. In the first run of the SCM program, we use the gradient free algorithm. We then apply the gradient search algorithm to the periodic cells with $3 \times 3 \times 3$ subdivisions. The Pareto set and the corresponding Pareto front are shown in Figures 9 and 10. 142 cells are in the Pareto set. After the sub-division, 714 cells are found in the Pareto set. The total CPU time is 90 seconds.

Next, we consider the gradient search algorithm applied to the entire region Q. The first run of the algorithm identifies 325 periodic cells in the Pareto set. After the sub-division, 578 cells are identified in the Pareto set. To save space, we only show the Pareto front on the refined grid in Figure 13. The Pareto front exhibits the same fine structure as in Figure 12. The total CPU time for this example with the gradient search algorithm applied to the entire region Q is 244 seconds. The computational saving of the hybrid algorithm is remarkable.

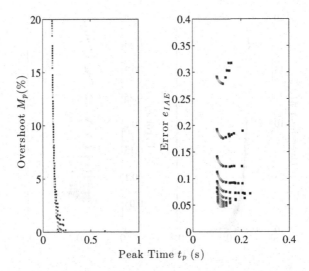

Fig. 10. The Pareto front of $[t_p, M_p, e_{IAE}]$ on the coarse grid for the multi-objective optimal control of the second order system with the hybrid algorithm corresponding to the Pareto set in Figure 9. The color code indicates the level of the other objective function. Red denotes the highest value, and dark blue denotes the smallest value.

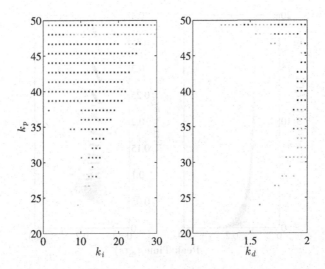

Fig. 11. The Pareto set of $[k_p, k_i, k_d]$ on the refined grid for the multi-objective optimal control of the second order system with the hybrid algorithm. The color code indicates the level of the other design variable. Red denotes the highest value, and dark blue denotes the smallest value.

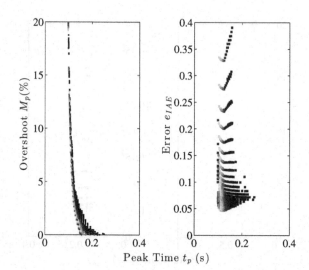

Fig. 12. The Pareto front of $[t_p, M_p, e_{IAE}]$ on the refined grid for the multi-objective optimal control of the second order system with the hybrid algorithm corresponding to the Pareto set in Figure 9. The color code indicates the level of the other objective function. Red denotes the highest value, and dark blue denotes the smallest value.

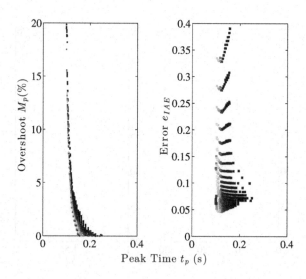

Fig. 13. The Pareto front of $[t_p, M_p, e_{IAE}]$ on the refined grid for the multi-objective optimal control of the second order system with the gradient search algorithm applied to the entire region Q. The color code indicates the level of the other objective function. Red denotes the highest value, and dark blue denotes the smallest value.

6   Concluding Remarks

We have presented a hybrid gradient free-gradient (GFG) algorithm for the simple cell mapping (SCM) method applied to MOPs. We have studied several examples of the multi-objective optimization problems including an example with five design parameters and the time domain design of PID controls for linear systems. In the first run of the hybrid algorithm, the DS search for the SCM method delivers a set of cells that cover the Pareto set and Pareto front. In the second run, a gradient search is applied to the covering set of cells and delivers much fine resolution of the global optimal solutions of Pareto set and Pareto front. It has been shown through all the numerical examples that the hybrid algorithm for the SCM method delivers substantial computational savings while obtaining comparably accurate solutions for the Pareto set and Pareto front. It appears that the hybrid algorithm may be a key element when we extend this global optimal solution method to higher dimensional MOPs.

Acknowledgement. The work is supported by the UC MEXUS-CONACyT through the project "Cell-to-cell Mapping for Global Multi-objective Optimization", by the Natural Science Foundation of China through the grant 11172197 and by the Natural Science Foundation of Tianjin through a key-project grant. The second author also acknowledges support from CONACyT through a scholarship to pursue graduate studies at the Computer Science Department of CINVESTAV-IPN.

References

[1] Lara, A., Alvarado, S., Salomon, S., Avigad, G., Coello Coello, C.A., Schütze, O.: The gradient free directed search method as local search within multi-objective evolutionary algorithms. In: Schütze, O., Coello Coello, C.A., Tantar, A.-A., Tantar, E., Bouvry, P., Del Moral, P., Legrand, P. (eds.) EVOLVE - A Bridge Between Probability, Set Oriented Numerics, and Evolutionary Computation II. AISC, vol. 175, pp. 153–168. Springer, Heidelberg (2012)

[2] Coello Coello, C.A., Lamont, G., Van Veldhuizen, D.A.: Evolutionary Algorithms for Solving Multi-Objective Problems. Springer, New York (2007)

[3] Schütze, O., Vasile, M., Coello Coello, C.A.: Computing the set of epsilon-efficient solutions in multiobjective space mission design. Journal of Aerospace Computing, Information, and Communication 8(3), 53–70 (2011)

[4] Dellnitz, M., Schütze, O., Hestermeyer, T.: Covering Pareto sets by multilevel subdivision techniques. Journal of Optimization Theory and Applications 124, 113–155 (2005)

[5] Jahn, J.: Multiobjective search algorithm with subdivision technique. Computational Optimization and Applications 35(2), 161–175 (2006)

[6] Schütze, O., Vasile, M., Junge, O., Dellnitz, M., Izzo, D.: Designing optimal low thrust gravity assist trajectories using space pruning and a multi-objective approach. Engineering Optimization 41(2), 155–181 (2009)

[7] Hsu, C.S.: Cell-to-Cell Mapping, A Method of Global Analysis for Nonlinear Systems. Springer, New York (1987)

[8] Hsu, C.S.: A theory of cell-to-cell mapping dynamical systems. Journal of Applied Mechanics 47, 931–939 (1980)

[9] Hsu, C.S.: A discrete method of optimal control based upon the cell state space concept. Journal of Optimization Theory and Applications 46(4), 547–569 (1985)

[10] Bursal, F.H., Hsu, C.S.: Application of a cell-mapping method to optimal control problems. International Journal of Control 49(5), 1505–1522 (1989)

[11] Crespo, L.G., Sun, J.Q.: Stochastic optimal control of nonlinear dynamic systems via bellman's principle and cell mapping. Automatica 39(12), 2109–2114 (2003)

[12] Flashner, H., Burns, T.F.: Spacecraft momentum unloading: the cell mapping approach. Journal of Guidance, Control and Dynamics 13, 89–98 (1990)

[13] Zhu, W.H., Leu, M.C.: Planning optimal robot trajectories by cell mapping. In: Proceedings of Conference on Robotics and Automation, pp. 1730–1735 (1990)

[14] Wang, F.Y., Lever, P.J.A.: A cell mapping method for general optimum trajectory planning of multiple robotic arms. Robotics and Autonomous Systems 12, 15–27 (1994)

[15] Yen, J.Y.: Computer disk file track accessing controller design based upon cell to cell mapping. In: Proceedings of the American Control Conference (1992)

[16] Crespo, L.G., Sun, J.Q.: Solution of fixed final state optimal control problems via simple cell mapping. Nonlinear Dynamics 23, 391–403 (2000)

[17] Crespo, L.G., Sun, J.Q.: Optimal control of target tracking via simple cell mapping. Journal of Guidance and Control 24, 1029–1031 (2000)

[18] Crespo, L.G., Sun, J.Q.: Fixed final time optimal control via simple cell mapping. Nonlinear Dynamics 31, 119–131 (2003)

[19] Pareto, V.: Manual of Political Economy. The MacMillan Press, London (1971); original edition in French (1927)

[20] Hillermeier, C.: Nonlinear Multiobjective Optimization - A Generalized Homotopy Approach. Birkhäuser, Berlin (2001)

[21] Fliege, J., Svaiter, B.F.: Steepest descent methods for multicriteria optimization. Mathematical Methods of Operations Research 51(3), 479–494 (2000)

[22] Liu, G.P., Yang, J.B., Whidborne, J.F.: Multiobjective Optimisation and Control. Research Studies Press, Baldock (2002)

[23] Schütze, O., Lara, A., Coello Coello, C.A.: The directed search method for unconstrained multi-objective optimization problems. In: Proceedings of the EVOLVE – A Bridge Between Probability, Set Oriented Numerics, and Evolutionary Computation (2011)

[24] Lara, A., Alvarado, S., Salomon, S., Avigad, G., Coello Coello, C.A., Schütze, O.: The gradient free directed search method as local search within multi-objective evolutionary algorithms. In: Schütze, O., Coello Coello, C.A., Tantar, A.-A., Tantar, E., Bouvry, P., Del Moral, P., Legrand, P. (eds.) EVOLVE - A Bridge Between Probability, Set Oriented Numerics, and Evolutionary Computation II. AISC, vol. 175, pp. 153–168. Springer, Heidelberg (2012)

[25] Nocedal, J., Wright, S.: Numerical Optimization. Springer Series in Operations Research and Financial Engineering. Springer (2006)

[26] Das, I., Dennis, J.: Normal-boundary intersection: A new method for generating the Pareto surface in nonlinear multicriteria optimization problems. SIAM Journal of Optimization 8, 631–657 (1998)

[27] Schäffler, S., Schultz, R., Weinzierl, K.: A stochastic method for the solution of unconstrained vector opimization problems. Journal of Optimization, Theory and Application 114(1), 209–222 (2002)

[28] Liu, G.P., Daley, S.: Optimal-tuning PID controller design in the frequency domain with application to a rotary hydraulic system. Control Engineering Practice 7(7), 821–830 (1999)

[29] Liu, G.P., Daley, S.: Optimal-tuning nonlinear PID control of hydraulic systems. Control Engineering Practice 8(9), 1045–1053 (2000)

[30] Panda, S.: Multi-objective PID controller tuning for a FACTS-based damping stabilizer using non-dominated sorting genetic algorithm-II. International Journal of Electrical Power & Energy Systems 33(7), 1296–1308 (2011)

Planning and Allocation Tasks in a Multicomputer System as a Multi-objective Problem

Apolinar Velarde[1], Eunice Ponce de León[2], Elva Diaz[2], and Alejandro Padilla[2]

[1] Technological Institute of Ciudad Valles, S.L.P., México
avelardem@gmail.com
[2] Autonomous University of Aguascalientes, Ags. México
{eponce,elvitad,apadilla}@correo.uaa.mx

Abstract. In this article we address the task planning and assignment problem in a multicomputer system using architectural 2D mesh. The problem of planning and allocation of tasks to a group of computers consists of several sub-problems that can be made to correspond to functions to optimize. The proposed solution to this problem is; first: establish the identification of distinct parts that are involved, such as; maximizing processor usage, minimize task wait time in the queue and avoid indefinite task delay (starvation). Second: a planning algorithm and an allocation algorithm are implemented through the search engine within the queue, the first algorithm makes a previous planning to the allocation to identify the task lists that fit in the mesh, and the second is a sole variant distribution algorithm to identify the best allocations in the processor mesh through a dynamic quadratic allocation. Finally, our final results are presented; they allow us to see that a previous allocation in the queue and a search engine allocation of the tasks best positions in the available (free) sub meshes, are determining factors for bettering the longevity of the processors and optimize answer time in a multicomputer system.

Keywords: Estimation Distribution Algorithm, Univariate Marginal Distribution Algorithm, Multicomputer system, Multi-objective optimization, NP-Hard problem, 2D Mesh Architecture.

1 Introduction

Parallel computation has had a tremendous impact on a variety of areas ranging from computation simulations for scientific applications, and engineering to commercial applications in data mining and transaction processing. The benefits of parallelism, together with yield requirements and application speed present convincing arguments in favor of informatics parallelism [1].

In Parallel Systems, the processors are connected through network communications, the most widely used architectures are Multicomputer systems connected through 2D and 3D meshes. Multicomputer architectures have proven to be a viable alternative to parallel computing. Different from traditional supercomputers, Multicomputer architectures present a simple architecture, regular

M. Emmerich et al. (eds.), *EVOLVE - A Bridge between Probability, Set Oriented Numerics, and Evolutionary Computation IV*, Advances in Intelligent Systems and Computing 227, DOI: 10.1007/978-3-319-01128-8_15, © Springer International Publishing Switzerland 2013

and scalable, and have been used in parallel commercial and experimental computers such as the IBM Blue-Gene/L and the Intel Paragon [2].

Different from sequential system programming, parallel systems should divide problems (tasks) into sub-problems (sub-tasks) in order to permit the distribution of the computing load between processors [1]. It is not easy to break down all of the big problems into sub-problems, due to the data dependence between the sub-problems. Due to this dependence the processors should communicate between themselves. It is important to point out that the consumed communication time between processors is very high compared to processing time. Due to this, the communication squeme should be carefully planned in order to obtain an efficient parallel algorithm [3, 4].

Once this division has been done, the tasks along with the sub-tasks are put into the queue where the planning algorithm determines the execution order among these functions. Then a task allocation algorithm becomes responsible for placing said tasks into the processor mesh, where they are held until they are executed [5–8].

Allocating tasks in the sub-mesh of available processors implies a joining of objectives such as [9]:

1. Minimize the number of allocations to the processor mesh that carries out the task allocation algorithm.
2. Minimize the task wait time in the queue
3. Maximize usage of processor meshes, or in other words, diminish the percentage of processors that remain available after the allocation algorithm places one or more tasks in the processor mesh (external fragmentation) [10].
4. Minimize task starvation. Avoid a task allocation discrimination that would require a large amount of processors (large tasks), that in which is provoked by tasks that require small amounts of processors (small tasks), that are being continually allocated.
5. Maximize adjacency between processors (joint assignment of available processors that are closest in proximity), in order to minimize the communication route distance and avoid interference; the purpose of this is to obtain a good parallel algorithm, that will diminish communication time and maximize processing time [3, 4, 11].

Considering the aforementioned objectives, the task allocation within the processor mesh is outlined as a multi-objective optimization problem NP-hard. A Meta-heuristic could see itself as a general work setting, referring to algorithms that could be applied to diverse optimization problems (combinatorial) with few significant changes, if a previous specific heuristic method already exists for the problem. In fact, Meta-heuristics are widely known as one of the best approximations in attacking combinatorial optimization problems [14, 15].

In this investigation we present an Estimation Distribution Algorithm (EDA) in especially an UMDA (Univariate Marginal Distribution Algorithm), in order to model the dynamic quadratic task allocation to the 2D processor. We do this through a joint probability distribution calculation of selected tasks in the queue, which are susceptible of being assigned to the processor mesh.

This paper is organized in the following manner: The second section explains the planning process and task assignment in a 2D mesh multicomputer system. The third section outlines past works, in specific non-adjacent allocation strategies that over the course of many years carry out task allocations in a Multicomputer System. The fourth section outlines a joining of utilized and referenced definitions used throughout the entirety of this document. The fifth section, presents a description of the algorithm that is used in this investigation. The sixth section, houses the description of the developed method for task allocation in a processor mesh using an UMDA algorithm. And finally, summing up the seventh section, we discuss the results obtained experimenting with the developed method.

2   Planning Allocation Tasks in a 2D Mesh

Task planning and allocation in a processor mesh is constituted by a list of tasks that remain in the queue. Tasks that are waiting to be assigned to a joining of processors to then start their execution. A planning algorithm of tasks is responsible for the specific order, in which each task or tasks will enter into the processor mesh. The allocation algorithm informs the available sub-meshes that exist within the processor mesh in time t [11, 13].

The process of task planning before its execution implies the determination in time t the sub-list of tasks that will enter into the processor mesh.

Once a task or tasks have been selected by a planning algorithm (whose code is shown in section 6), they are assigned within the mesh to start their execution without interruption until they have finished. They leave the processor mesh and upon continuing, the queue contracts and the subsequent task remains at the head of the queue line awaiting its entrance into the processor mesh. For example, a six task queue, where each task consists of a joining of subtasks that are between two and five tasks, and an eight by eight processor mesh from processor $< 1, 1 >$ to processor $< 8, 8 >$.

With time t, the planning process determines which and how many tasks will enter the mesh depending on the number of available processors. In this case we suppose that task 1, at the head of the queue, has been selected to be allocated in the mesh. After it enters, the queue contracts, the second task passes to the head of the queue and all of the remaining tasks in the mesh wait for the following planning, and allocation to be executed. The sixth task will go onto the fifth position of the queue, and a space will become available for a new task to enter that will compete for entrance into the mesh for the following planning.

In this way task 1 was uniquely contemplated for its execution. Other tasks can be planned to enter the mesh depending on the number of unoccupied processors that exist, and the type of planning that the algorithms will use to place the tasks.

In order to carry out the task allocation with in the processor mesh two allocation strategies have been developed [12, 13]. Their classification obeys the type of recognition that is observed in the processor mesh: Adjacent Processor Assignment Strategies and Non-adjacent Processor Assignment Strategies.

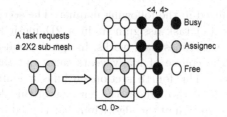

Fig. 1. Adjacent allocation strategy when a task requests an available sub-mesh of four processors

Adjacent Processor Assignment Strategies, appear when a partial system recognition is had and it is possible to be assigned to the work execution when adjacent processor sub-meshes are what the task requests. Figure 1 shows an adjacent allocation strategy when a task requests an available sub-mesh of four processors, within a four by four mesh. The identification of processors comes from processor $< 1, 1 >$ to processor $< 4, 4 >$. Given that there is adjacency in the available processor sub-mesh $< 1, 1 >$ to the processor $< 2, 2 >$; the algorithm can carry out the task allocation without any problems. Non-adjacent processor allocation strategies come about to eliminate the presented deficiencies that are involved with adjacent allocations and they imply that a total recognition algorithm within the mesh is hard. In this way if a mesh with a size of n x n can not be allocated into an available m x m sub-mesh it will occupy different available sub-meshes when said sub-meshes are not found to be adjacent. Figure 2 shows a non-adjacent processor allocation within the mesh. In this example the tasks requests a 2×2 sub-mesh, given that the processors $< 1, 3 >< 2, 3 >< 3, 2 >$ y $< 4, 2 >$ are available in the mesh and the allocation strategy permits the non-adjacent of the processors so that a task can be executed, the allocation then will be possible. To improve the performance of a paralleled computing system, the objective is to look for a strategy that permits the processors to be adjacent, and in case available sub-meshes can not be found that satisfy the previous ones, one should make sure that processors are as close as possible to avoid communication overloading in the mesh. Also one must look for available sub-meshes with few un-added processors in the mesh, that can find

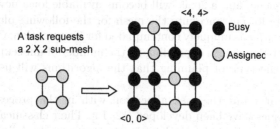

Fig. 2. Non-adjacent allocation in a 4×4 processor mesh

tasks with few processor requirements in the queue during all the planning that happen, this could provoke that tasks that require high quantities of processors could become discriminated, it even could mean that available sub-meshes with a large number of processors could become existent, due to this, the strategy should avoid that large tasks fall into starvation within the system.

To exemplify the previous paragraph consider a queue of six tasks that is waiting to enter, and the un-added available sub-meshes within the mesh, we suppose that the first five tasks T_0 to T_4 solicit a reduced number of processors: 5, 4, 5, 6 and 3 respectively, and the task T_{14} that solicits twenty nine processors; once the planning has been carried out the first five tasks could be allocation objects within the mesh, but given the number of resources of task T_{14} that solicits twenty nine processors, it should wait if the conditions of the planning algorithm do not establish a strategy that is able to attend tasks with large resource solicitations and avoid task starvation. Being as the task is not assigned, it should wait for the following planning operation in order to compete for its allocation, with the condition that small tasks that would directly compete for the same assignment do not exist because if this happens, the task would end up waiting indefinitely.

The processor allocation policy FIFO (first input, first output) is an open policy of starvation. It permits that all tasks be assigned according to their arrival time in the queue but these politics lack planning. One big problem that is found is that available processors can exist within the mesh but they can not be allocated due to the required number of processors does not satisfy the number of available processors within the mesh. For example, suppose that task T_{14} remains at the head of the queue and solicits 30 processors, the five tasks that have arrived cannot be allocated due to the fact that the order does not permit the operation, and the number of available processors that solicits that the task remain at the head of the queue cannot be completed. In this way twenty nine processors would become useless in the mesh, and twenty nine processors would be required for the five tasks that wait to enter at the same time, and would not be attended until the fourteen tasks have been allocated into the mesh.

The previous paragraph gives us an established series of objectives, that interfere with designing a task planning strategy within the mesh: on one hand, the system could have available resources that it would not be able to use, and on the other hand it would have tasks that run the risk of waiting indefinitely in the queue, if not adequately planned.

In opposition of politics concerning available allocation starvation, the algorithms that plan the tasks permit that they be allocated in a faster way, due to that in each allocation more than one task can be placed at one time, a non-existent task number relation with allocation numbers within the mesh, for example, using the a task queue and considering that task planning has existed, the allocation method in non-adjacent, then in a single assignment may directly enter the first 5 tasks, and task 14 may go to the head of the queue, freeing up 5 spaces for future tasks entering the mesh. So in just one allocation they could enter the mesh, a choosing process based on different criteria is necessary: task location in relation to the number of processors that are required (adjacency) and allocate the highest number of tasks in the available nodes.

One method of task planning and allocation for a parallel system, that models the aforementioned could diminish the number of allocations that occur in the processor mesh, diminish task wait time in the queue, and maximize the use of the processors within the mesh. In this way, minimize and maximize the objectives in the task allocation of a 2D mesh, its seen as a multi-objective problem.

One way of handling a multi-objective problem is through meta-heuristic algorithms that establish strategies to cover the solution space of the problem, transforming themselves from the recurrent form to the final form. In our proposal, once visualizing a joining of available meshes, and a joining of possible tasks to be allocated a meta-heuristic search engine is ran to specify recurrently the best solution, in order to arrive at an optimization in the allocation, or in other words, find the cheapest allocation having to do with communication, adjacency and number of processors that are going to be allocated. The necessary allocations that should take place are considered to be dynamic and quadratic in the mesh, due to the possible combinations that are based on the number of available processors and the number of tasks that are waiting to enter.

Now to counter arrest the risks run by using meta-heuristics as a search engine in a multi-objective problem, we have implemented a partial systematic search engine to avoid repetitions maintaining a high grade of randomness. We implement this search engine through an exhaustive process, using a halting method without necessity of covering all solution space to avoid falling into optimal local solutions. In the following sections the algorithm application form is defined.

3   Previous Works

In this section the investigation methods used in non-adjacent allocation strategy areas that have been found in modern literature are described, and are related by their allocation type along with the method that is proposed in this article.

The non-adjacent allocation strategies that have been developed are: Random [16], Paging [16], Multiple Buddy Strategy MBS [16], Adaptive Scan and Multiple Buddy ANCA [17], Adaptive Scan and Multiple Buddy AS & MB [18], and recent paging variants [19]. In Random [16], the internal fragmentation and external fragmentation are eliminated but high communication interference between these works does exist. In the paging method a certain level of adjacency exists between processors assigned to a parallel task, which increases if large numbers of pages are used. However, fragmentation can exist internally within the processor when large numbers of pages are assigned to tasks that do not require them. MBS [16], betters the performance compared to the previous strategies but problems are presented upon assigning an adjacent sub-meshes to available processors, because this can increase the communication overload. ANCA subdivides the request into equal parts in the i_{th} recurrence and requires that the partitionment, and the allocation occur in the same recurrence that in

which provokes that a previous recurrency is not to be assigned, large sub-meshes to one part of the request; this could also increase the communication overload. The performance of the AS & MB [18], in respond and service times are identical to those of the MBS [16], however AS & MB have a high allocation overloading for large meshes. In paging variants the allocation unit is just one processor that requires more time to make one decision in larger meshes, whereas in the MBS [16] and ANCA [17], the allocation unit increases for what is to be a longer time to carry out an allocation in larger systems [20].

4   Basic Concepts

The proposal is a multicomputer system connected to a 2D mesh, with a task queue that awaits its admission into the mesh and where the assignments are established by a dynamic quadratic allocation. The following definitions formally describe a system of this type.

Definition 1. *An n-dimensional mesh has $k_0 \times k_1 \times ... \times k_{n-2} \times k_{n-1}$ nodes, where k_i is the number of nodes along the i-th dimension and $k_i \geq 2$. Each node is identified by n coordinates, $\rho_0(a), \rho_1(a), ... , \rho_{n-2}(a), \rho_{n-1}(a)$, where $0 \leq \rho_i(a) < k_i$ for $0 \leq i < n$. Two nodes a and b are neighbours if and only if $\rho_i(a) = \rho_i(b)$ for all dimensions except for one dimension j, where $\rho_j(b) = \rho_j(a) \pm 1$. Each node in a mesh refers to a processor and two neighbours are connected by a direct communication link.*

Definition 2. *A 2D mesh, which is referenced as $M(W, L)$ consists of $W \times L$ processors, where W is the mesh width and L is mesh height. Every processor is represented by a pair of coordinates (x, y), where $0 \leq x < W$ and $0 \leq y < L$. A processor is connected by a bidirectional communication link to each one of its neighbours. For each 2D mesh $a = P_{ij}$.*

Definition 3. *In a 2D mesh, $M(W, L)$, a sub-mesh $S(w, l)$ is a mesh of 2 dimensions that belongs to $M(W, L)$ with a width w and a height l, where $0 < w \leq W$ and $0 < l \leq L$. $S(w, l)$ they are represented by the coordinates (x, y, x', y'), where (x, y) is the bottom left corner of the mesh and (x', y') is the upper right corner. The bottom left corner node is called the sub-mesh base node and the upper right corner is called the final node. In this case $w = x' - x + 1$ and $l = y' - y + 1$. The size $S(w, l)$ is $w \times l$ processors.*

Definition 4. *In a 2D mesh $M(W, L)$, an available sub-mesh $S(w, l)$ is a sub-mesh that satisfies the conditions: $w \geq \alpha$ and $w \geq \beta$ assuming that the allocation of $S(\alpha, \beta)$ required, where the allocation is referred to by selecting linked processors for an arriving task.*

Definition 5. *Let ϑ be a joining of system tasks, as here in $\vartheta = J_1, , J_n$ where n is the number of tasks in time t and ϑ_k is the joining of sub-tasks of task K where: $\vartheta_k = j_{k1}, j_{k2}, , j_{kf(k)}$ and f(k) is the number of sub-tasks of task K. For* *each sub-task $i \in s$ a k task is had as seen here in $S_i \in J_k$ and a processor $m_i \in P$ that in which must execute consuming time $t \in \mathbb{N}$ uninterrupted.*

Definition 6. *Given the size of matrix $n x n$: fluid mesh F whose (i,j)ith elements represent the fluidness between tasks i and j and an arrangement of distances D whose (i,j)ith elements represent the distance between sites i and j. An allocation is represented by vector p, which is a permutation of numbers $1, 2n$. $p(j)$ is where the task j is assigned. Like this, the quadratic assignment task can be written as the following:*

$$min_{p \in} \sum_{i=1}^{n} \sum_{i=1}^{n} f_{ij} dp(i)p(j)$$

Definition 7. *An optimizing problem is that in which whose solution applies finding a joining of alternate solution candidates that best satisfies some objectives. Formally, the problem is made up of solution space S and of the objective function f. Resolving the optimization problem (S, f) consists of determining an optimal solution, or in other words, a feasible solution $x^* \in S$ as in $f(x^*) \leq f(x)$, for either $x \in S$. The alternative solutions can be expressed by the value allocation to a finite joining of variables $X = X_i : i = 1, 2, , n$. For U_i the dominance is indicated or universe (joining of possible values) of each on of these n variables, the problem consists in sectioning the x_i value assigned to each variable X_i of the dominance U_i, submissive to certain restrictions, it optimizes an objective function f. The universe of solutions is identified with the joining of $U = x = (x_i : i = 1, 2, , n) : x_i \in Ui$. The problem restrictions reduce the solutions universe to a sub-joining of solutions $S \subseteq U$, denominating feasible space.*

Definition 8. *Minimize task wait times in the queue is given by:*

$$t_{sj} = \sum_{i=1}^{s_j} t_{ij}$$

Whose objective function is established as:

$$minTQ = \left(\sum_{i=1}^{s_j} t_{i1} + \sum_{i=1}^{s_j} t_{i2} + ... + \sum_{i=1}^{s_j} t_{ij} \right)$$

Subject to:

$$t_i \in \mathbb{R} \text{ and } 1 \leqslant j < |J|$$

5 Multi-objective Univariate Marginal Distribution Algorithm

Multi-objective EDA are evolutionary algorithms that use a collection of solutions as candidates to carry out, search engine trajectories avoiding local minimizations [14]. These algorithms use the distribution estimation and simulation joined probability as an evolution mechanism, instead of directly manipulating the individuals that represent problem solutions. An EDA algorithm begins by generating randomly an individual population that represents problem solutions. Three types of operations are recurrently used with the population. The first type of operation consists in the generation of a sub-joining of the best individuals of the population. Then secondly, a learning process is used as a distribution model of probability form, those selected individuals. In third place, new individuals are generated simulating the obtained distribution model. The algorithm stops itself when it reaches a certain number of generations, or when the output of the population stops significantly improving.

In order to estimate each distribution generation of the probability, in conjunction with the selected individuals we use an UMDA. So that the probability distribution coordinated is factored, as the product of independent univariate distributions.

$$p_l(x) = p(x|D_{l-1}^{Se}) = \prod_{i=1}^{n} p_l(x_i)$$

Each univariate probability distribution is estimated from marginal frequencies:

$$p_l(x_i) = \frac{\sum_{j=1}^{N} \delta_j(X_i = x_i | D_l - 1_e^S)}{N}$$

where:

$$\delta_j\left(X_i = x_i | D_{l-1}^{Se}\right) = \begin{cases} 1 \; if \; in \; the \; j-th \; event \; of \; D_{l-1}^{Se}, \; X_i = x_i \\ 0 \qquad\qquad\quad in \; another \; case. \end{cases}$$

The pseudocode for MO UMDA can be found in Table 1.

Table 1. Pseudocode for the UMDA

UMDA
$D_0 \leftarrow$ Generate M individuals (initial population) randomly
Repeat for $l = 1, 2, ...$ until the stopping criterion check
$\quad D_{l-1}^{Se} \leftarrow$ Select $N \leq M$ individuals of D_{l-1} according to a selection method
$\quad p_l(x) = p(x|D_{l-1}^{Se}) = \prod_{i=1}^{n} p_l(x_i) =$
$\quad p_l(x_i) = \frac{\sum_{j=1}^{N} \delta_j(X_i = x_i | D_{l-1}^{Se})}{N} \leftarrow$ Estimate the joined probability distribution
$\quad D_l \leftarrow$ Sampling (M individuals, the new population) from $p_l(x)$)

---

6   Method Description

The planting of the proposed method, part of the assumed that has a previous understanding of the communication grade measurement between the main task and the sub-tasks that it consists of all the tasks that are found in the queue, like the relation between the same sub-tasks. To exemplify the relationship between

Fig. 3. Message paths between tasks for a task with three sub-tasks

the main task and the sub-tasks, consider having a task T_1 with three sub-tasks T_{11}, T_{12} and T_{13}, the interaction that could arise between them is exemplified in Figure 3 through lines that show message transfer, in this way T_1 can send and receive messages about its sub-tasks and in its time the sub-tasks can do the same with the main task amongst themselves. Also a symmetrical distance matrix is had between processors that specify the movements that a message should make in order to be carried out between one processor and another. Given the aforementioned the process, of assigning the processors is based on the carrying out of a processor allocation calculation based on the availableness of the same ones, within the mesh and the tasks that remain in the queue. This quadratic assignment is allowed to be determined, through calculations and the apparition frequencies that represent the most feasible solution. The pseudo-code for the task search engine (planning algorithm) in the queue is shown in Table 2. After that the planning algorithm builds the list of selected tasks, the allocation algorithm receives the list to assign tasks to the mesh. The allocation algorithm is shown in Table 3 and explained in each of the following sections by an example. In the following example the method function is shown. With time as t a 4×4 processor mesh is had where 1 represents the occupied processors that were assigned to a task with a time $t-1$, a zero represents the available processors that have not been assigned to a task or sub-task. Table 4 represents the queue that contains four tasks pending execution, these tasks await to be executed in the mesh and one cost communication matrix is represented in Table 5, T_1 and T_2 with sub-tasks and within the same sub-tasks.

Table 2. Pseudo-code for the task search engine (planning algorithm) within the queue

Allocation algorithm informs free Processors in the mesh
Repeat $((FreeProcessorsinthemesh > 0)$ or
    (Total tasks in the mesh = Tasks Verified))
    Select randomly one task from the queue.
    $If(thenumberofprocessorsrequestingtask <= Freeprocessorsinthemesh)$
        If (the first selected task) Create the list of selected tasks
        Else Add task to the list of selected tasks
        Decrement free processors in submesh
    Else Select other task
    Add 1 to TasksVerified
End repeat

Table 3. Pseudocode for the allocation algorithm

Receive the list of selected tasks
Repeat until a minimum in the total cost of solutions
    Allocation generation, Population evaluation, Estimating the probabilistic model
    Saving the best individual
End repeat
Sort the solutions and Inform the best solution and assign the tasks in the mesh
Verify and Inform free processors in the mesh to the planning algorithm

Table 4. Task queue shown in time t

T_1	T_{11}	T_{12}	T_{13}	T_{14}
T_2	T_{21}	T_{22}	T_{23}	T_{24}
T_3	T_{31}	T_{32}	T_{33}	T_{34}
T_4	T_{41}	T_{42}	T_{43}	T_{44}

Table 5. Communication costs matrix for tasks T_1, T_2

	T_1	T_{11}	T_{12}	T_{13}	T_2	T_{21}	T_{22}
T_1	0	3	0	3	0	0	0
T_{11}	2	0	1	4	0	0	0
T_{12}	0	1	0	2	0	0	0
T_{13}	3	5	3	0	0	0	0
T_2	0	0	0	0	1	3	0
T_{21}	0	0	0	0	1	3	0
T_{22}	0	0	0	0	2	0	4

6.1   Generation of Allocations

In order to generate the first allocation, the state matrix is taken from the mesh by making a random task selection from the queue, and whose number of processors that request are available in the free sub-meshes. To carry out the random selection, a random number is generated as the task queue base number, if the selected task fits into the sub-mesh or free sub-meshes, then it is considered in calculating its allocation and cost, in case of the opposite another task would be chosen.

Each individual of the population is represents by a task allocation and its sub-tasks in the free spaces that are shown agglutinated in Table 6. This representation dominates the task allocation matrix or state matrix in time t, in this way we would generate a random initial population of size 2, constituted by the tasks T_1 and T_2. In order to obtain the first value of the objective function, the

Table 6. Task allocation matrix according to the state matrix of the mesh shown in time t, which represents a first solution to the dynamic quadratic allocation problem

1	1	1	1
1	1	1	1
T_{11}	T_{12}	T_{21}	1
T_1	T_{13}	T_2	T_{22}

Table 7. Transference cost calculation of messages for task T_1

$T_1 \to T_{11}$	$T_{11} \to T_1$	$(3+2)*1$	5
$T_1 \to T_{12}$	$T_{12} \to T_1$	$(0+0)*2$	0
$T_1 \to T_{13}$	$T_{13} \to T_1$	$(3+3)*1$	6
$T_{11} \to T_{12}$	$T_{12} \to T_{11}$	$(1+1)*1$	2
$T_{11} \to T_{13}$	$T_{13} \to T_{11}$	$(4+5)*2$	18
$T_{12} \to T_{13}$	$T_{13} \to T_{12}$	$(2+3)*1$	5
		TOTAL	35

allocation cost calculation is made for each task based on the communication costs, between tasks and the distances between processors, considering the message path of a processor from one to another and vice versa. Upon considering the message path between processors one should also calculate the transference cost of the same, form the origin to the destination and vice versa, or in other words, in the case of exemplification the transfer rate of task T_1 to the sub-task T_{11} is different form the sub-task T_{11} to task T_1, even though it may be that both weights are the same, the values of the distances between the processors remain the same. Like this the values upon calculating are given in the shown operations in Table 7 for task T_1. The totals of the respective individuals are added together to obtain the total cost of the solution.

The representations of the previous calculations are represented by: C_{ij} is the communication cost of i and j. Where: i is the task or sub-task of the task. j is the task or sub-task of the task. C_{ij} is the distance between processors where the task is assigned with sub-tasks.

6.2 Population Evaluation

Once the allocation costs have been calculated, the obtained totals are added together to make an evaluation of the individual through the obtained values in the objective function. The population evaluation pseudocode is shown in Table 8.

Table 8. Pseudocode for population evaluation

Evaluate population
Repeat from Tasks = 1 to M Tasks that keep in the mesh
    Make the sum of the amounts obtained in each calculation
        of the cost of transfer of messages for tasks
End repeat
Report the total value of the objective function

6.3 Estimating the Probabilistic Model

In this part of the document we are going to use the simplest probabilistic model, that in which all of the variables that describe the product are independent, for this, we calculate the task apparition frequencies in every empty cell of the mesh in time t of one part of the population that are the best individuals, through a selection done by truncation and the percent of the truncation. The pseudocode for estimate the probabilistic model is shown in Table 9

Table 9. Pseudocode for estimating the probability model

Probabilistic Model Estimate
Repeat for Assigned Tasks = 1, 2, ..., M to the end of the tables that
    contain the tasks that can be assigned to the free sub-meshes
    Verify that the next position is assigned each task, counting by 1
    each assignment and store it in the Matrix of Probabilistic Model Estimation
End repeat
Report the values obtained at frequencies allocation

6.4 Saving the Best Individual

This process is carried out by taking every generated population, the best individual at the moment in which the current population is ordered. The ordering process from least to greatest is generated in order to obtain allocation minimization.

7 Obtained Results

In this section we describe the obtained results gathered by performed experiments using proposed methods. The experiments are based on a method comparison with two task allocation methods: the lineal allocation and Hilberts curves method. The lineal allocation method is the most used method in processor allocation, its implementation simplicity allows that the tasks become quickly placed within the available sub-meshes, but it presents difficulties when it processors become freed. The nature of this method allows allocations to be done from the bottom left to top right positions.

The recognition that the algorithm makes is base on a linear mesh route. The main problem with this method is that it can not assign available sub-meshes in a different order than the established linear method, that in which produces high task segmentation and it increases message transfer in the mesh processors.

Hilberts curve method establishes a filling of spaces in the form of curves that visit each point in the squared meshes of $2 \times 2, 4 \times 4, 8 \times 8, 16 \times 16$ dimensions, or in whichever order of potential 2. This method was written by David Hilbert in 1892. The method applications have existed in image processing especially in the comprehension and dilatation of images.

7.1   Utilized Experimental Parameters

Before specifying each of the carried out experiments, in this section we define each one by the involved experimental parameters, as possible values that can be taken. In this section the specific term references, for example; free mesh, are used without explanation. Their significance is assumed to be known by the reader having understood the previous sections of this document.

1. Mesh sizes. Definition 2 given that the reference dimension as M (W, L) consisting of $W \times L$ processors, where W is the width of the mesh and L is the height of the mesh.
2. Number of tasks at the start of the test. It is considered as an initial set of tasks to perform the first direct allocation of processors in the mesh, once the tasks begin to remove from the mesh, the allocation method which uses meta-heuristic is enabled.
3. Number of sub-task by task. A variable parameter defined by the user which indicates to the system the maximum number of sub-tasks of a task. In terms of a parallelizable computable problem, subtasks represent the number of sub-problems in which the problem can be divided.
4. Average waiting time of tasks. The calculation of the sum of the waiting times of tasks between the total numbers of tasks.
5. Total number of tasks. The total number of tasks that the system responded.
6. Verification time task completion. It is a period of time set by the system for verification of completion times of the tasks, when the system verifies the completion of a task can occur:
   (a) The task finishes its time.
   (b) The task is not finished at the time and only decreases in relation to the time spent in the system.
7. Execution time. It is the total time the system takes to execute n tasks.
8. Number of assignments made. The number of assignments that the algorithm takes to process all the tasks.
9. Average of sub-tasks processed. Obtained by summing all sub-tasks of each task between the total number of tasks.

7.2   First Solution Proposal Considering Cost Allocation of Joined Tasks in a Free Sub-mesh

The first experiment done with the proposed method establishes the use of allocated costs in joined tasks with their respective sub-tasks to one or many available sub-meshes. The procedure is explained in the following points:

1. The allocation processor algorithm informs the task planner of the number and position of free n processors in the mesh.
2. The task planner activates the method used to generate the quadratic allocation with n as possible combinations.
3. For each generated combination a calculation of communication costs is made for every proposed solution.
4. The bubble sort algorithm is activated to order proposed solution, from least to greatest.
5. The best solution (lowest cost) is considered to be assigned in the processor mesh through a task allocation algorithm.

7.3   Number of Allocations vs. Number of Tasks

The allocation of one or many tasks to joint processors is done once the allocation algorithm has informed the planning algorithm of the number of free processors with-in the mesh; with the quantity of available processors the planning algorithm can determine which task or tasks are to be assigned.

This experiment compares two methods that do not use a previous planning upon entering the tasks into the mesh, against the method that is done by planning. The objective is to minimize the time that the system consumes in assigning tasks for processing through the minimization of the number of carried out allocations.

The numbers of allocations are obtained by adding together the carried out allocations during the processing of n tasks of each experiment. The obtained results are shown in Figure 4, using 360, 819, 988, 1338, 2657, 5120, 5444, 7230 and 10115 tasks in the system. In these results we have found that the proposed method once carrying out the planning diminishes the number of allocations to the mesh by 50% in relation to the other two methods, those in which, upon not using planning, the number of allocations corresponds with the number of tasks that are processed.

The previous paragraph shows that diminishing the number of allocations to the mesh significantly betters multicomputer system processing time.

7.4   Average Task Wait Time (Expressed in Seconds)

The average task wait time, is the time that a task spends waiting in the queue to be assigned to linked available processors within the mesh. One of the objectives in parallel computing is to minimize the time that a task remains in the queue, if it has to do with a task that requires a large amount of processors.

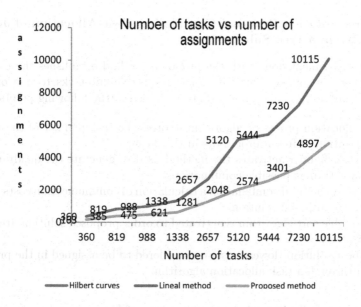

Fig. 4. Number of tasks vs. the number of allocations in the mesh that the three methods use during their execution

The objective of this experiment is to determine the average wait time of tasks within the queue when a load of 360, 819, 988, 1338, 2657, 5120, 5444, 7230 and 10115 tasks exists within the system. The Figure 5 shows that the proposed method produces a shorter wait time, even with heavy loads in the system (more than 10,000 tasks), in relation to the linear method and Hilberts curve method. The average wait time is expressed in seconds. The previous paragraph shows that anticipates task planning to the allocation minimizes task wait time.

7.5 External Fragmentation. Processor Occupation Percentage for Each Completed Allocation

External fragmentation occurs when there are available processors within the mesh and can not be assigned to a task that requires them, due to a utilized allocation strategy. This experiment calculates the external fragmentation that each method produces for each allocation within the task mesh. In Figure 6 we observe that the proposed method gives us better results than the other two methods, in this experiment minimizing the percentage of available in the mesh after each allocation. The allocations that the proposed method does are based on the calculations of the best produced allocations, by a Meta-heuristic algorithm, calculated on a communications cost base, and the best tasks location (calculating the apparition frequencies of the tasks within the cells) in the mesh, it is worth mentioning that the processor total is not used in each allocation, but the available processor percentage during the experiment remains in an average of .08% that in which is less when compared to the other two methods.

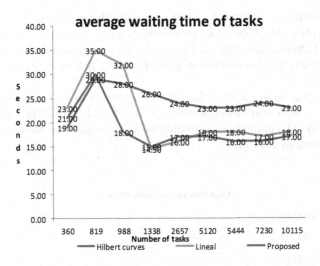

Fig. 5. Average time expressed in seconds that a tasks awaits its entrance into the processor mesh

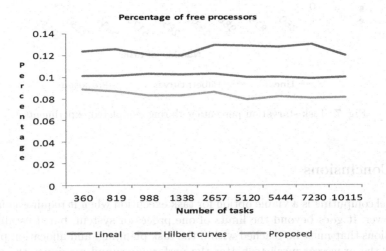

Fig. 6. Internal fragmentation. Percentage of available processors in each experiment done.

7.6 Task Starvation Percentage

In the proposed method, in order to detect task starvation, first we must identify the tasks with the largest amount of processor requests. This is monitored during the allocation phases. If more than 5 allocations have gone by and the largest tasks have not been assigned, then the acceptance process is detained in the queue until the starvation task or tasks have been assigned the number

of processor that they require for their execution. In our tests, starvation exists
when the number of system tasks grows significantly (more than 2500) as shown
in Figure 7. Different from the methods that we have used to make the comparison, a null starvation percentage in present due to not carrying out a previous
queue planning.

It is important to know that the proposed system has an advantage, in that
it presents a smaller starvation percentage, due to this, if a large task competes
with small tasks the allocation cost could be less, because the processor that will
be assigned remain adjacent.

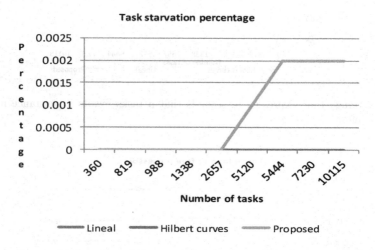

Fig. 7. Task starvation percentage during completed experiments

8   Conclusions

Parallel computing is a viable option for task execution when it requires computing power. It goes beyond the limits of one processor system, but it establishes
conditions that must be reached within the task planning, and allocation phases
within the processor mesh to better the performance and maximize the use of
resources. In this research we have presented objective detection that opposes
task planning and task allocation within a processor mesh and a method that
early task planning uses for the tasks that remain in the queue, that in which
minimizes the number of allocations to the mesh, diminishes the task wait time
and maximizes the number of available processors within the mesh, it diminishes
starvation of large tasks and maximizes adjacency between allocated processors
assigned to a task, helping it by use of a mechanism to avoid the starvation of
large tasks in the queue.

The obtained results from the experiments using the proposed method based
on planning, show that by diminishing the number of allocations within the mesh,

we are allowed to minimize task processing times diminish the time that a task waits in the queue and diminish internal fragmentation. The aforementioned also allows us to reduce the risks that saturate the communication network, helping with the allocation algorithm that makes recognition of the mesh, and maintains the available sub-mesh information existent through a dynamic data structure. Task starvation still is present in our system but there are fewer cases, and furthermore, the algorithm maintains a strict control over the tasks that would begin to fall into starvation.

References

1. Grama, A., Gupta, A., Karypis, G., Kumar, V.: Introduction to Parallel Computing, 2nd edn. Addison Wesley (January 16, 2003) ISBN: 0-201-64865-2
2. Bani-Ahmad, S.: Submesh Allocation in 2D Mesh Multicomputers: Partitioning at the Longest Dimension of Requests. In: ADVCOMP 2010: The Fourth International Conference on Advanced Engineering Computing and Applications in Sciences, pp. 99–104 (2010) ISBN: 978-1-61208-101-4
3. Torres, J., Rodriguez, E.: Conceptos de Computo Paralelo. Trillas, Mayo (2000) ISBN: 968-24-6223-3
4. Xavier, C., Iyengar, S.: Introduction to Parallel Algorithms. Wiley Inter-science, New York (1998) ISBN 0-471-25182-8
5. Sharma, D.D., Pradhan, K.: Job Scheduling in Mesh Multicomputers. IEEE Transactions on Parallel and Distributed Systems 9(1) (January 1998)
6. Chen, J., Taylor, V.E.: Mesh Partitioning for Efficient Use of Distributed Systems. IEEE Transactions on Parallel and Distributed Systems 13(1) (January 2002)
7. Dutot, P., Takpe, T.N., Suter, F.: Scheduling Parallel Task Graphs on (Almost) Homogeneous Multicluster Platforms. IEEE Transactions on Parallel and Distributed Systems 20(7) (July 2009)
8. Amoroso, A., Marzullo, K.: Multiple Job Scheduling in a Connection Limited Data Parallel System. IEEE Transactions on Parallel and Distributed Systems 17(2) (February 2006)
9. Velarde, A., Ponce de Leon, E., Diaz, E., Padilla, A.: Planning and Allocation of processors in 2D meshes. In: Doctoral Consortium. Mexican Internacional Conference on Artificial Intelligence, MICAI 2010, Pachuca Hidalgo, México (2010)
10. Heiss, H.U.: Dynamic Partitioning of Large Multicomputer Systems Department of Informatics. In: Proc. Int. Conf. on Massively Parallel Computing Systems (IEEE MPCS 1994), Ischia, May 2-6. University of Karlsruhe, Germany (1994)
11. Bani-Ahmad, S.: Processor Allocation with Reduced Internal and External Fragmentation in 2D Mesh-based Multicomputer. Journal on Applied Sciences 11(6), 943–952 (2011) ISSN 1812-5654, doi:10.3923 / jas. 2011.943.952, 2011 Asian Network for Scientific Information
12. Bani-Mohammad, S., Ould-Khaoua, M., Ababneh, I., Machenzie, L.: Noncontiguous Processor Allocation Strategy for 2D Mesh Connected Multicomputers Based on Sub-meshes Available for Allocation. In: Proc. of the 12th Int. Conference on Parallel and Distributed Systems (ICPADS 2006), Minneapolis, Minnesota, USA, vol. 2, pp. 41–48. IEEE Computer Society Press (2006)
13. Das Sharma, D., Pradhan, D.K.: Job Scheduling in Mesh Multicomputers. IEEE Transactions on Parallel and Distributed Systems 9(1) (January 1998)

A. Velarde et al.

14. Larrañaga, P., Lozano, J.A., Mühlenbein, H.: Estimation of Distribution Algorithms Applied To Combinatorial Optimization Problems. Inteligencia Artificial. Revista Iberoamericana de Inteligencia Artificial (2003)
15. Lozano, J.A., Larrañaga, P.: Estimation of Distribution Algorithms. A New Tool for Evolutionary Computation. Kluwer Academic
16. Lo, V., Windisch, K., Liu, W., Nitzberg, B.: Non-contiguous processor allocation algorithms for mesh-connected multicomputers. IEEE Transactions on Parallel and Distributed Systems 8(7), 712–726 (1997)
17. Chang, C.Y., Mohapatra, P.: Performance improvement of allocation schemes for mesh-connected computers. Journal of Parallel and Distributed Computing 52(1), 40–68 (1998)
18. Suzaki, K., Tanuma, H., Hirano, S., Ichisugi, Y., Connelly, C., Tsukamoto, M.: Multi-tasking Method on Parallel Computers which Combines a Contiguous and Non-contiguous Processor Partitioning Algorithm. In: Madsen, K., Olesen, D., Waśniewski, J., Dongarra, J. (eds.) PARA 1996. LNCS, vol. 1184, pp. 641–650. Springer, Heidelberg (1996)
19. Bunde, D.P., Leung, V.J., Mache, J.: Communication Patterns and Allocation Strategies. Sandia Technical Report SAND2003-4522 (January 2004)
20. Liu, P., Hsu, C., Wu, J.J.: I/O Processor Allocation for Mesh Cluster Computers. In: IEEE Proceedings of the 2005 11th International Conference on Parallel and Distributed Systems, ICPADS 2005 (2005)

Real World System Architecture Design Using Multi-criteria Optimization: A Case Study

Rajesh Kudikala[1], Andrew R. Mills[1], Peter J. Fleming[1],
Graham F. Tanner[2], and Jonathan E. Holt[2]

[1] Department of Automatic Control and Systems Engineering,
The University of Sheffield,
Mappin Street, S1 3JD, Sheffield, UK
`{r.kudikala,a.r.mills,p.fleming}@sheffield.ac.uk`
[2] Rolls-Royce plc, PO Box 31, Derby, UK
`{graham.tanner,jonathan.holt}@rolls-royce.com`

Abstract. System architecture design using multi-criteria optimization is demonstrated using a case study of an aero engine health management (EHM) system. A design process for optimal deployment of EHM system functional operations over physical architecture component locations, e.g., on-engine, on-aircraft and on-ground, is described. The EHM system architecture design needs to be optimized with respect to many qualitative criteria in terms of operational attributes within the constraints of resource limitations. In this paper the system architecture design problem is formulated as a multi-criteria optimization problem. Considering the large discrete search space of decision variables and many-objective functions and constraints, an evolutionary multi-objective genetic algorithm along with a progressive preference articulation technique, is used for solving the optimization problem. The optimization algorithm found a family of Pareto solutions which provided valuable insight into design trade-offs. Using the progressive preference articulation technique, the optimization search can be focused for the industrial decision maker on to a region of interest in the objective space. Performance of the proposed method is evaluated using various test metrics. Using this approach it was possible to identify the most significant design constraints ("hot spots") and the opportunities afforded by either the relaxation or the tightening of these constraints, along with their performance implications.

Keywords: System architecture design, multi-criteria optimization, many-objective optimization, preference articulation, genetic algorithms.

1 Introduction

The architecture of a complex system can be described in terms of functional requirements, physical elements, and element interrelationships. It is often seen as a generic framework or blueprint of the overall system. Designing a complex system architecture can be a difficult task involving multi-faceted trade-off decisions. The design process often needs to consider experience, models and

M. Emmerich et al. (eds.), *EVOLVE - A Bridge between Probability, Set Oriented Numerics,* 245
and Evolutionary Computation IV, Advances in Intelligent Systems and Computing 227,
DOI: 10.1007/978-3-319-01128-8_16, © Springer International Publishing Switzerland 2013

data from many design disciplines. A typical architecture design process starts by identifying the main functional requirements and follows a process of decomposition [4]. The top level system functional requirements (use cases) are divided into several sub-functions. The physical form of the system is also divided into sub-systems and components. Designers try to map the functionalities onto the physical hardware components. Designers then iterate between the upper and lower levels of the system decomposition to optimize the deployment of functional operations onto physical systems. However, due to the large and discontinuous design search space, many qualitative and quantitative criteria, it is almost impossible for designers to find optimal architecture designs. Designers have considered various methods and tools for exploring these trade-offs in the design space [13, 18].

Since they are efficient, global, parallel search methods, evolutionary algorithms (EAs) are very popular for solving such complex design problems [6, 12]. These algorithms apply generic operators of variation and thus are applicable to search non-standard combinatorial search spaces taking into account domain specific characteristics of the problem. Thompson et al. [20] explored the potential of the increasing use of embedded intelligence through deployment of smart sensors and actuators in future distributed control system (DCS) architectures for aero-engines. They have used a multi-objective evolutionary algorithm (MOEA) to generate and assess competing architectures. A model-driven architecture design approach, SESAME, developed in [16], which explores the design space of embedded system architectures using an MOEA. Similarly, the Palladio component modelling (PCM) framework was developed to automatically evaluate software architectures with an MOEA [15]. Armstrong et al. [2] developed a tool set for the function based architecture design exploration, which included function decomposition, adaptive function mapping and complex inter relations between architecture elements. This information aids in architectural definition and trading of architecture alternatives. Other frameworks [18] adopted multiple stages and commercial life-cycle viewpoints in terms of qualitative and quantitative analysis for evaluating the architecture trade-off decisions.

Many of the architecture design frameworks concentrated on only multi-objective problems having 2 to 3 objective functions. However, in general real-world design problems will have many-objective functions to be optimized [8]. Fonseca and Fleming [9] introduced the first Pareto-optimal MOEA; it was named MOGA - multi-objective genetic algorithm. There are many different MOEAs available now for solving multi-objective optimization problems such as, NSGAII [7], SPEA2 [22] etc., However, their search ability can significantly deteriorate when solving many-objective optimization problems [14, 17]. The optimal trade-off surface of a multi-objective optimization problem can contain a potentially infinite number of Pareto-optimal solutions. The task of an MOEA is to provide an accurate and useful representation of the trade-off surface to the decision-maker (DM). With the increased number of objectives, the number of non-dominant solutions increases exponentially, decreasing the selection pressure and convergence towards true Pareto optimal surface. Visualization of the

Pareto surface also becomes difficult for many-objective functions. The preference based MOEAs are one of the proven techniques in such scenarios e.g. [3,5]. Fonseca and Fleming's Progressive Preference Articulation (PPA) [10] was first preference articulation technique for an MOEA, and it remains an important approach in tackling many-objective optimization problems [1]. The core of this PPA technique is based on a preferability operator, that incorporates goal and priority information about the objectives and which consequently modifies the dominance definition.

In this paper, we present a design process for an aero engine EHM system architecture for an aero engine. It uses a multi-criteria optimization approach, for the deployment of EHM functional operations over physical architecture component locations. In this study, four physical component architecture locations have been considered. The EHM functional operations have to be deployed in order to satisfy many qualitative criteria such as, criticality, output data security and flexibility etc., within the constraints of resource limitations. System architecture models in Systems Modeling Language (SysML)/Unified Modeling Language (UML) are successfully integrated into optimization platform. An MOEA is used for solving the optimization problem. Using the PPA technique, the optimization search can be focused on to a region of interest (ROI) for the DM who is able to experiment with changing goals for the objectives, in order to arrive at a satisfactory compromised solutions that takes account of domain knowledge.

The remainder of the paper is organized as follows. The next section describes the EHM system for an aero engine. In section 3 the multi-criteria optimization problem for the EHM system architecture design is formulated. Section 4 presents the MOEA with progressive preference articulation for solving the many-objective EHM system architecture design optimization problem and test metrics for performance evaluation. In section 5 various future design scenarios are explored. Finally, section 6 provides concluding remarks.

2   Engine Health Management System

Engine health management (EHM) system has become an essential part of aero engines. The main aim of an EHM system is to perform real-time parameter analysis and anomaly detection of the aero engine. Output from on-board analysis can be passed to an on-ground computer resource for further analysis to predict, classify and locate developing engine faults and anomalies [19]. The optimum combination of the on-engine and on-ground computational resources for the EHM system will deliver the benefits of reducing the engine through life operating and maintenance costs.

The EHM system in aero engines forms part of the engine electronic control (EEC) system, with an engine monitoring unit (EMU) at its centre. The EMU along with the EEC are mounted on the fan case of the aero engine. Figure 1 shows an EHM system for an aero engine. The EMU collects data from a number of sensors mounted on the various engine components and sub-systems of the

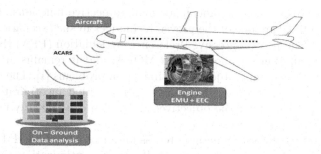

Fig. 1. Engine health management system

engine. The EEC along with the EMU uses data from the sensors on engines to monitor numerous critical engine characteristics such as temperatures, pressures, speeds and vibration levels to ensure they are within known tolerances and to highlight when they are not. The EMU transmits applicable EHM data to an on-aircraft system for reporting and storage on maintenance server. From the aircraft system potentially useful EHM data can be compressed and transmitted real-time to an on-ground station using an aircraft communications addressing and reporting system (ACARS) over satellite or radio. Further detailed and complex data analysis necessary for monitoring the engine health status can be performed at the on-ground station using a substantial computational resource and a knowledge base accumulated from all the other engines in the fleet. The EHM data analysis will highlight any changes in the engine component characteristics. Expert knowledge is used for diagnosis and prognosis of the developing engine faults and to generate necessary maintenance reports.

For designing the EHM functional architecture for the aero engines, all primary EHM functional requirements and stake holder requirements are captured using requirements analysis and flowdown techniques, and represented as EHM system use cases. The EHM system has significant data limits and cost limits for development. A SysML model for a baseline EHM system has been developed by system engineers. The EHM system primary use cases are decomposed into several EHM functions and these functions are further decomposed into 74 EHM functional operations (OPs). For each functional operation (OP) several operational attributes are defined, which indicate the specific requirements of that operation in terms of its input data flowrate, processing power, immediacy and security etc., In the SysML model for all the 74 EHM functional operations, these attribute requirements are expressed by system designers mainly in terms of different qualitative levels: 'high', 'medium' and 'low'. The operational attributes are described below:

Input Data Flow Rate: An estimate of the input information / data flow rate (in Kbits/sec) required for the operation.

Processing Power: A measure of the processing power (in MFLOPS: mega floating-point operations per second) necessary to perform the operation.

Criticality: A judgement of the level of safety criticality (Level A to E) of the function output to the engine safety.

Immediacy: A measure of how quickly the operation output needs to be acted upon by downstream OPs to fulfil the functional requirements.

Coupling: A design judgement concerning whether there is a high sequence dependency on a preceding functional operation or external system events.

Security: A judgement on data security level required for the output of a functional operation.

IP Sensitivity: A judgement on the intellectual property sensitivity of a functional operation.

Flexibility: A judgement on level of modifiability of a functional operation in the future upgrades.

The current EHM system architecture design process seeks to find an optimal deployment of the 74 EHM functional operations over the physical architecture component locations. In the present study, four physical architecture component locations for the EHM system have been considered. They are:

   (i) engine monitoring unit (EMU) (on-engine),
   (ii) engine electronic controller (EEC) (on-engine),
   (iii) on-aircraft, and
   (iv) on-ground.

These physical architecture component locations have certain resource limitation in terms of the operational attributes. The 74 EHM functional operations are to be deployed in order to satisfy the operational (OP) attribute requirements within the constraints of the resource limitations.

3   Multi-criteria Optimization Problem

In this section, the EHM system architecture design problem has been formulated as a multi-criteria optimization problem. In general, to formulate an optimization problem it is necessary to identify the following key components: a model of the system to be optimized, its decision variables, the objective functions to be optimized and system constraints. Figure 2 shows the EHM system multi-criteria optimization problem components. In the current study, the deployment of the functional operations in various physical architecture components has been considered as the model of the optimization problem. The deployment locations of EHM functional operations are considered as decision variables. As there are 74 EHM functional operations, 74 decision variables are binary coded

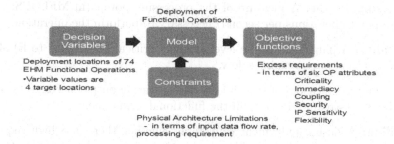

Fig. 2. EHM system multi-criteria optimization problem components

to represent the deployed location number of each functional operation. Each of these decision variables can have values of 1, 2, 3 or 4 representing a physical architecture location: 'EMU', 'EEC', 'on-aircraft', and 'on-ground', respectively.

The EHM system has certain resource limitations for the physical architecture components in terms of their capability to handle the functional operations attribute requirements. The EHM functional deployment solutions should both satisfy all the limitations of these physical resources and enable full functionality of the EHM system. Hence, the hardware limitations of 'data flow rate' and 'processing capacity' on each of the four physical architecture component locations are imposed as eight constraints in the optimization process. In the optimization process the functional operations deployed on each location are separated and grouped. The total data flow rate and processing power required for all the operations deployed in each location are computed and compared with the corresponding resource limitations/ constrains. If these requirements are satisfied, then the solution is considered as a feasible solution, otherwise it is treated as an infeasible solution.

In the current design process, in order to facilitate the search for finding the best deployment solutions that satisfy EHM functional OP attribute requirements, violations are permitted in terms of the remaining six OP attributes on physical architecture components. These attribute requirement violations are treated as "excess requirements" in that OP attribute. The total excess requirements in terms of the six operational attributes: **'Criticality'**, **'Immediacy'**, **'Coupling'**, **'Security'**, **'IP sensitivity'** and **'Flexibility'**, are considered as six individual objective functions to be minimized in the optimization process. Since the design problem has more than 3 objective functions, it is classified as a many-objective optimization problem.

The multi-criteria optimization problem for the EHM system architecture design can be formulated as below:

$$\min_{\mathbf{x}} \ \mathbf{F}(\mathbf{x}) = (f_1(\mathbf{x}), f_2(\mathbf{x}), ..., f_6(\mathbf{x})), \quad \text{where} \quad f_k(\mathbf{x}) = \sum_{i=1}^{74} E^2{}_{ik}(\mathbf{x}) \qquad (1)$$

$$\text{subject to} \quad \sum_{i=1}^{74} d_{ir} \le D_r, \qquad r = \{1, 2, 3, 4\} \tag{2}$$

$$\sum_{i=1}^{74} p_{ir} \le P_r, \qquad r = \{1, 2, 3, 4\} \tag{3}$$

$$\mathbf{x} = [x_1, ..., x_i, ..., x_{74}], \qquad x_i \in \{1, 2, 3, 4\} \tag{4}$$

where, $\mathbf{F}(\mathbf{x})$ are six objective functions representing the sum of squares of total excess requirements $E_{ik} = (OP\ requirement\ -\ resource\ limit)$ of 74 EHM operations. D_r and P_r are the constraint limitations of data flowrate and processing capacity on the four physical architecture locations, and, d_{ir} and p_{ir} are individual attribute requirement measures of each operation deployed at the corresponding location. x_i are the 74 decision variables, which can have values $\{1, 2, 3, 4\}$ to represent the deployment locations of the corresponding operation.

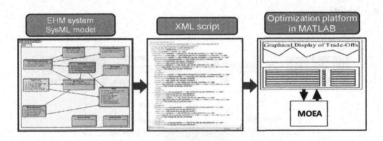

Fig. 3. Integrating the system architecture SysML model to optimization platform

The baseline EHM system for the aero engine is developed as a SysML model, which has information concerning all the EHM functions and their operational attribute requirements. The information available in the SysML model has to be transferred to the platform used by the optimization algorithm. In order to integrate the SysML model into the optimization process, the SysML model is exported as an XML script file. From this XML script file all the attributes information of the functional operations is decoded and converted into a format suitable for use within optimization platform. The process of integrating the SysML model and importing data to the optimization platform in MATLAB is depicted in Figure 3. Data for all the functional operations and their attribute levels are exported from the SysML model to the optimization platform in MATLAB. Then, for ease of manipulation in the optimization, the qualitative attribute levels for most of the operations, i.e. 'High', 'Medium', and 'Low', are transformed to numerical values 9, 4, and 1 ($3^2, 2^2 and\ 1^2$), respectively. This data is further used for the evaluation of objective functions values for the candidate architecture alternatives and for finding the optimal architecture solutions for the EHM system.

4   Implementing the Optimization Approach

The proposed EHM system many-objective optimization problem is solved using an MOEA, MOGA [9], in the MATLAB environment. The algorithm uses a non-dominated classification of the population of solutions and rank-based fitness sharing techniques in evolving multi-objective optimization solutions. MOGA is further enhanced to incorporate a unique progressive preference articulation technique [11]. This denotes the process of introducing, incorporating and modifying designer preferences in an interactive and progressive way at any time during the optimization search process; this is a key feature for multi-criteria decision making (MCDM). This enables the DM to set goal values (design requirements or aspirations) for the objectives being optimized and to introduce and change priorities for objectives in a progressive fashion at any time during the optimization process. This enhanced version of MOGA also allows the designer to specify constraints together with a set of priority levels for design objectives in a transparent and consistent manner. A "parallel coordinates" graph in the MOGA software suite facilitates visualization of the interplay between the different objectives.

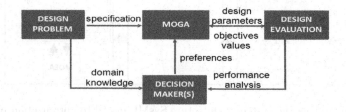

Fig. 4. Interactive multi-criteria optimization design framework

The MOGA design framework with progressive preference articulation technique is shown in Figure 4.

- Initially, the design problem is formulated as a multi-criteria optimization problem. The specification is passed to MOGA.
- MOGA generates initial candidate solutions and passes them to the model for evaluation of the design objective functions.
- MOGA iterates the optimization process and creates new solutions using the genetic operators and displays the non-dominated trade-off solutions.
- While MOGA evolves new non-dominated solutions, the decision maker can observe progress of the optimization process and analyse the performance improvements.
- The decision maker can interact with MOGA using the progressive preference articulation approach. Using domain knowledge the decision maker can express preferences and steer the optimization into the design region of interest (ROI).

For solving the current EHM system architecture optimization problem, the following parameters are selected in MOGA:

- population size = 200,
- number of generations = 500,
- probability of cross-over = 0.7, and
- probability of mutation = 0.01.

Stochastic universal sampling selection, single point binary crossover and mutation genetic operators are used in the MOGA algorithm. In this study elitism is incorporated in MOGA by maintaining an archive of non-dominated solutions of fixed size, which will keep the best non-dominated solutions found so far in all the generations. The archive size is set to 200 non-dominated solutions. For maintaining diversity and fixed size for the archive solutions, a density estimation operator and archive truncation technique similar to SPEA2 [22] are employed. With a view to increasing the confidence in the Pareto solutions, MOGA was run 50 times using a different seed for the random number generator in each run and various performance metrics were also evaluated.

In the optimization process, a candidate architecture solution has 74 decision variables each representing the deployment resource location number for its corresponding operation. Then the operations are separated and grouped according to their deployment location. The solution should satisfy all the constraints. The total data flow rate and processing resource requirements on each location, i.e., values of eight constraints, are estimated and compared with the corresponding constraint limitations. For the six qualitative criteria: criticality, immediacy, coupling, security, IP sensitivity and flexibility, if the requirements of operations are not satisfied, then the excess requirements are estimated for all OPs, using numerical transformation described in last section. The total excess requirements in each criteria are considered as six individual objective functions to be minimized in the optimization process.

Due to the many objective functions and constraints in the optimization problem, non-dominated solutions predominate in each generation. This decreases the selection pressure towards the true Pareto optimal surface, thus reducing the performance of the optimization algorithm. Using the PPA technique, the decision maker's preferences are expressed progressively to reduce the objective search space, and steer the optimization into region of interest. The additional preference operator used by PPA helps restore selection pressure. Several Pareto optimal solutions are obtained for the EHM system architecture design using MOGA.

The Pareto optimal solutions for the EHM system architecture design are shown in Figure 5. In MOGA the trade-off values of all objective functions are shown in a "parallel coordinates" graph. In the case of many-objective optimization, the parallel coordinates graph aids the visualization of all constraints and objective functions values on a single plot. This approach places all the axes parallel to each other thus allowing any number of axes to be shown in a 2-D representation. In Figure 5, the top window shows a "trade-off graph" for the 14 criteria of the current EHM system architecture design. Criteria 1 to 8 are

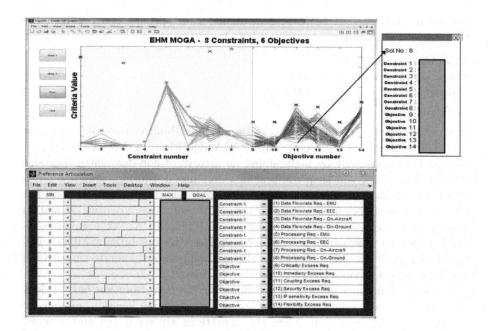

Fig. 5. MOGA parallel coordinates trade-off graph with preference articulation

constraints on each physical architecture location in terms of data flow rate and processing power. Criteria 9 to 14 are the six objective functions: criticality, immediacy, coupling, security, IP sensitivity and flexibility excess requirements. In order to clearly distinguish, the constrains are shown in shaded region and objectives are shown in unshaded region. On the 'x-axis' constraints and objective numbers are displayed and on the 'y-axis' corresponding criteria values are displayed. Each connected line in the trade-off graph represents a Pareto optimal solution for the EHM system architecture design. With a mouse click on a solution, a pop-up notification window will display the corresponding solution number and all the constraints and objective functions values for that Pareto solution.

The bottom window shows the "preference articulation" facility. In this window, all the constraints and objective functions of the EHM system architecture design are listed. The DM can set goal values for each objective by moving the sliders between maximum and minimum bounds at any time during the optimization process. (Note: Due to data confidentiality requirements of Rolls-Royce plc the goal values are masked in the plot.) Goal points for each of the objectives are marked with an "x" in the trade-off graph. As the decision maker exercises progressive articulation of preferences, the optimization process will steer the search in that preferred region of interest in the feasible objective space and try to minimize the objective functions values within the specified goals to find Pareto optimal solutions.

In the trade-off plot, it can be observed that, crossing lines between criteria 9 and 10, demonstrate that the objectives 'criticality' and 'immediacy' are in conflict with each other, while concurrent lines between criteria 12 and 13 demonstrate that the objectives 'security' and 'IP sensitivity' are in relative harmony with each other. A limitation of the parallel coordinates representation is that only adjacent objectives can be easily compared. However, in MOGA, there is a facility to interactively switch the order of representation of the objectives.

It can be seen from the trade-off graph, that the data flow requirements of deployed OPs on EMU (1) and on-aircraft (3), the processing resource requirements on-aircraft (7) and off-board (8) are far below the goals/limitations. These constraints are thus satisfied very easily. Whereas, the data flow requirements on EEC (2) and on-ground (4), the processing resource requirements on EMU (5) and EEC (6) are close to the goals/limitations. These constraints are tightly satisfied. These constraints can be identified as the most significant design constraints ("hot spots"). There are several Pareto solutions with zero values for the objective functions: criticality (9), immediacy (10) and IP sensitivity (13). Hence the requirements of OPs in criticality, immediacy and IP sensitivity attributes can be completely satisfied by these deployment solutions. It is shown that other OP attributes, coupling (11) and flexibility (14) requirements, cannot be satisfied completely. By adjusting the sliders, the DM can isolate a single Pareto optimal solution from the set of trade-off solutions.

The test metrics we employed to measure the performance of the proposed method are described below:

- **Inverted Generational Distance (IGD)**, introduced in [21],

$$D(A, \mathcal{P}^\star) = \frac{\sum_{s \in \mathcal{P}^\star} \min\{\|A_1 - s\|_2, \ldots, \|A_N - s\|_2\}}{|\mathcal{P}^\star|}, \tag{5}$$

where $|\mathcal{P}^\star|$ is the cardinality of the set $\mathcal{P}^\star$, which is true Pareto front (PF) and A is an approximated non-dominated set of the PF obtained from optimization algorithm. The IGD metric measures the distance of the elements in the set A from the nearest point of the true Pareto front $\mathcal{P}^\star$. In this paper, since the true Pareto front for the current problem is unknown, we have considered the best non-dominated solutions found so far out of all combined solutions from 50 runs to be the Pareto front $\mathcal{P}^\star$ and A as non-dominated set obtained in each run.

- **Mean Nearest Neighbour Distance**,

$$S(A) = \frac{\sum_{i=1}^{|A|} d_i}{|A|}, \tag{6}$$

where d_i is,

$$d_i = \min_j\{\|F(x_i) - F(x_j)\|_2\}.$$

This metric can serve as a measure of the density of solutions. The ratio of this metric is used as,

$$S_R(A, \mathcal{P}^\star) = \frac{S(A)}{S(\mathcal{P}^\star)}. \tag{7}$$

and the coverage metric as defined in [23] is described below,

- **Coverage Metric (C-Metric)**

$$C(A, B) = \frac{|\{u \in B | \exists v \in A : v \preceq u\}|}{|B|}, \tag{8}$$

$C(A, B) = 0$ is interpreted as: there is no solution in A that completely dominates any solution in B. And $C(A, B) = 1$ is interpreted as all the solutions in B are dominated by at least one solution in A.

Table 1. $D(A, \mathcal{P}^\star)$, $S_R(A, \mathcal{P}^\star)$ and C-Metric values for the solutions obtained by the optimizer, where A is the obtained set in each run, $\mathcal{P}^\star$ is the considered PF

	$D(A, \mathcal{P}^\star)$			$S_R(A, \mathcal{P}^\star)$			$C(\mathcal{P}^\star, A)$			$C(A, \mathcal{P}^\star)$		
Problem	min	mean	std	min	mean	std	min	mean	std	min	mean	std
EHM	0.013	0.031	0.003	0.863	0.933	0.038	0.004	0.090	0.054	0.00	0.00	0.00

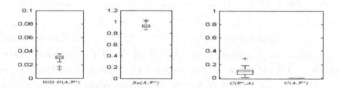

Fig. 6. Box plots of the metrics computed for the solutions in 50 runs

Table 1, shows the IGD index $D(A, \mathcal{P}^\star)$, mean nearest neighbour distance ratio $S_R(A, \mathcal{P}^\star)$ and C-metrics $C(\mathcal{P}^\star, A)$ and $C(A, \mathcal{P}^\star)$ values for the EHM system architecture design problem. The low scores of IGD index $D(A, \mathcal{P}^\star)$ for the solutions indicate that the obtained non-dominated sets in each run are very close the true Pareto front considered. Therefore, for this EHM system architecture design problem, the proposed methodology is consistent in producing solutions that are close to the PF in every run. The $S_R(A, \mathcal{P}^\star)$ ratio values indicate that the obtained solutions in each run have a good distribution and diversity comparable to the PF. Low values of the C-metrics imply that, there are very few solutions in A that are being dominated by solutions in the PF. Figure 6, shows the box plots for the above metrics computed for the solutions obtained by the optimizer in 50 runs. These plots indicate that the PPA method is able to find well distributed non-dominated solutions with good repeatability in the DM's preferred region close to the true Pareto front. The PPA method can be combined with other MOEA to improve the convergence of the algorithm in many-objective optimization.

5   Exploring What-If Design Scenarios

Progressive preference articulation in MOGA enables the decision maker to explore different architecture design scenarios, such as improved processor technology on the EMU and improved wireless transmission rate between on-board and on-ground systems. This is possible by interacting with constraints and specifying different goal/limitation settings. It can be observed from Figure 5, that the constraints on - data flow rate to on-ground (4), processing capacity on EMU (5) and processing capacity on EEC (6) - are narrowly satisfied. By increasing the goal values for these constraints, the decision maker can explore future ("what-if") architecture design scenarios and analyse prospective performance improvements. Here, in the first attempt, the constraint goal value for the ACARS data flow rate to the on-ground system is increased (50%). The trade-off solutions obtained after running the MOGA for a further 50 generations are shown in Figure 7 (a). It is observed that solutions are obtained with zero values (completely satisfied) for coupling objective (11) and with a slight improvement in flexibility objective (14). In the second attempt, the constraint goal value for the processing capacity on the EMU is increased (20%), and the data flowrate for the on-ground system decreased to the actual constraint value. Figure 7 (b) shows the trade-off solutions obtained after running the MOGA for a further 50 generations. It can be seen that solutions are obtained with a zero value for the security objective (12) and no significant improvement in the coupling objective (14). Out of these solutions one interesting solution is obtained with zero values

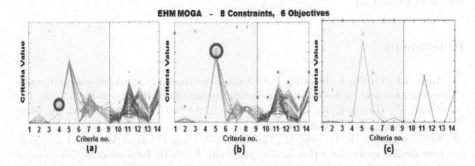

Fig. 7. Optimal non-dominated solutions with (a) increased dataflow rate to the on-ground, (b) increased processor capacity on the EMU and (c) four objectives completely satisfied (zero values)

for four objectives: criticality (9), immediacy (10), security (12) and IP sensitivity (14), as shown in Figure 7 (c). This means that in this deployment solution all OP requirements in criticality, immediacy, security and IP sensitivity attributes are completely satisfied, when the EMU processing capacity is increased.

6   Conclusions

In this paper a system architecture design process using multi-criteria optimization technique for a real world application, aero-engine EHM system, has been presented. The EHM system architecture design problem has been formulated as a many-objective optimization problem. System architecture models in SysML are successfully integrated into optimization platform. A strategy for optimal deployment of the functional operations on physical architecture locations has been successfully developed using an MOEA. The optimizer supports the decision maker by providing a facility for progressive preference articulation, empowering closely coupled user and optimization process interaction.

The MOEA optimizer is able to find several Pareto optimal trade-off solutions for the given model specifications. Candidate solutions are found through expression of explicit sets of preferences, as demonstrated in the work. Through this design process, it is revealed that it is not possible to fully satisfy all attributes for the EHM system, while observing the given constraints, thereby highlighting the value of a multi-criteria approach. Performance of the proposed method is assessed using test metrics. Various architecture design scenarios, such as hardware upgrades to data input rates and processor capacities, are explored by changing the goal values of constraints. It is shown that improved system performance achieved for the new specification. This strategy is deemed to be easily applied to other systems architecture design studies.

Acknowledgements.   The first author wishes to acknowledge the financial support of Rolls-Royce plc and EPSRC through a Dorothy Hodgkin Postgraduate Award (DHPA).

References

1. Adra, S.F., Griffin, I., Fleming, P.J.: A comparative study of progressive preference articulation techniques for multiobjective optimisation. In: Obayashi, S., Deb, K., Poloni, C., Hiroyasu, T., Murata, T. (eds.) EMO 2007. LNCS, vol. 4403, pp. 908–921. Springer, Heidelberg (2007)
2. Armstrong, M., de Tenorio, C., Garcia, E., Mavris, D.: Function based architecture design space definition and exploration. In: 26th International Congress of Aeronautical Sciences
3. Branke, J., Deb, K.: Integrating user preferences into evolutionary multi-objective optimization. In: Jin, Y. (ed.) Knowledge Incorporation in Evolutionary Computation. STUDFUZZ, vol. 167, pp. 461–478. Springer, Heidelberg (2004)

4. Crawley, E., de Weck, O., Eppinger, S., Magee, C., Moses, J., Seering, W., Schindall, J., Wallace, D., Whitney, D.: The influence of architecture in engineering systems. Engineering Systems Monograph (2004)
5. Cvetkovic, D., Parmee, I.: Preferences and their application in evolutionary multiobjective optimization. IEEE Transactions on Evolutionary Computation 6(1), 42–57 (2002)
6. Deb, K.: Multi-objective optimization using evolutionary algorithms, vol. 16. John Wiley & Sons, Hoboken (2001)
7. Deb, K., Pratap, A., Agarwal, S., Meyarivan, T.: A Fast and Elitist Multiobjective Genetic Algorithm: NSGA-II. IEEE Transactions on Evolutionary Computation 6(2), 182–197 (2002)
8. Fleming, P.J., Purshouse, R.C., Lygoe, R.J.: Many-objective optimization: An engineering design perspective. In: Coello Coello, C.A., Hernández Aguirre, A., Zitzler, E. (eds.) EMO 2005. LNCS, vol. 3410, pp. 14–32. Springer, Heidelberg (2005)
9. Fonseca, C., Fleming, P.: Genetic Algorithms for Multiobjective Optimization: Formulation, Discussion and Generalization. In: Proceedings of the Fifth International Conference on Genetic Algorithms, San Mateo, California, vol. 1, pp. 416–423 (1993)
10. Fonseca, C., Fleming, P.: Multiobjective Genetic Algorithms Made Easy: Selection Sharing and Mating Restriction. In: First International Conference on Genetic Algorithms in Engineering Systems: Innovations and Applications, pp. 45–52. IET (1995)
11. Fonseca, C., Fleming, P.: Multiobjective Optimization and Multiple Constraint Handling with Evolutionary Algorithms. I. A Unified Formulation. IEEE Transactions on Systems, Man and Cybernetics, Part A: Systems and Humans 28(1), 26–37 (1998)
12. Goldberg, D.: Genetic algorithms in search, optimization, and machine learning (1989)
13. Gries, M.: Methods for evaluating and covering the design space during early design development. Integration, the VLSI Journal 38(2), 131–183 (2004)
14. Ishibuchi, H., Tsukamoto, N., Nojima, Y.: Evolutionary Many-Objective Optimization: A Short Review. In: IEEE Congress on Evolutionary Computation, pp. 2419–2426 (June 2008)
15. Martens, A., Koziolek, H., Becker, S., Reussner, R.: Automatically improve software architecture models for performance, reliability, and cost using evolutionary algorithms. In: Proceedings of the First Joint WOSP/SIPEW International Conference on Performance Engineering, pp. 105–116. ACM (2010)
16. Pimentel, A., Erbas, C., Polstra, S.: A systematic approach to exploring embedded system architectures at multiple abstraction levels. IEEE Transactions on Computers 55(2), 99–112 (2006)
17. Purshouse, R., Fleming, P.: On the Evolutionary Optimization of Many Conflicting Objectives. IEEE Transactions on Evolutionary Computation 11(6), 770–784 (2007)
18. Selva, D., Crawley, E.: Integrated assessment of packaging architectures in earth observing programs. In: IEEE Aerospace Conference, pp. 1–17. IEEE (2010)
19. Tanner, G., Crawford, J.: An integrated engine health monitoring system for gas turbine aero-engines. IEE Seminar on Aircraft Airborne Condition Monitoring 2003(10203), 5 (2003)
20. Thompson, H., Chipperfield, A., Fleming, P., Legge, C.: Distributed aero-engine control systems architecture selection using multi-objective optimisation. Control Engineering Practice 7(5), 655–664 (1999)

21. Veldhuizen, D.A.V.: Multiobjective evolutionary algorithms: Classifications, analyses, and new innovations. Tech. rep., Air Force Institute of Technology (1999)
22. Zitzler, E., Laumanns, M., Thiele, L.: SPEA2: Improving the Strength Pareto Evolutionary Algorithm. Tech. Rep. 103, Computer Engineering and Networks Laboratory (TIK), ETH Zurich, Zurich, Switzerland (2001)
23. Zitzler, E., Thiele, L.: Multiobjective evolutionary algorithms: A comparative case study and the strength pareto approach. IEEE Transactions on Evolutionary Computation 3(4), 257–271 (1999)

A Self-adaptive Genetic Algorithm Applied to Multi-Objective Optimization of an Airfoil

John M. Oliver, Timoleon Kipouros, and A. Mark Savill

Cranfield University, School of Engineering,
College Rd, Cranfield, MK43 0AL, UK
j.m.oliver@cranfield.ac.uk

Abstract. Genetic algorithms (GAs) have been used to tackle non-linear multi-objective optimization (MOO) problems successfully, but their success is governed by key parameters which have been shown to be sensitive to the nature of the particular problem, incorporating concerns such as the numbers of objectives and variables, and the size and topology of the search space, making it hard to determine the best settings in advance. This work describes a real-encoded multi-objective optimizing GA (MOGA) that uses self-adaptive mutation and crossover, and which is applied to optimization of an airfoil, for minimization of drag and maximization of lift coefficients. The MOGA is integrated with a Free-Form Deformation tool to manage the section geometry, and XFoil which evaluates each airfoil in terms of its aerodynamic efficiency. The performance is compared with those of the heuristic MOO algorithms, the Multi-Objective Tabu Search (MOTS) and NSGA-II, showing that this GA achieves better convergence.

Keywords: Genetic, Algorithm, MOGA, GA, Multi-Objective, Optimization, MOO, Self-Adaptive, Parameters, MOOP, Airfoil.

1 Introduction

GAs, originally proposed by Holland [1] and expanded upon by Goldberg [2] and Schaffer [3], are heuristic, stochastic methods of searching very large non-linear problem spaces in order to attempt to obtain near optimal solutions [4] for problems upon which classical optimization methods do not perform well. GAs are characterized by populations of potential solutions that converge towards local or global optima through evolution by algorithmic selection as inspired by neo-Darwinian [5] evolutionary processes. An initial population of random solutions is created and through the evaluation of their fitnesses for selection for reproduction, and by the introduction of variation through mutation and recombination (crossover), the solutions are able to evolve towards the optima.

Research into GAs over more recent years by Fleming [6], Fonseca [7], Deb [8], Zitzler [9] and others, has extended their use to multi-objective optimization problems, in which two or more conflicting objectives, each with their own criteria, are optimized simultaneously, yielding a Pareto-optimal trade-off attainment surface from which a solution can be chosen by a higher-level decision maker.

M. Emmerich et al. (eds.), *EVOLVE - A Bridge between Probability, Set Oriented Numerics,* 261
and Evolutionary Computation IV, Advances in Intelligent Systems and Computing 227,
DOI: 10.1007/978-3-319-01128-8_17, © Springer International Publishing Switzerland 2013

GA performance on a given problem has been shown, since De Jong [10], to be extremely sensitive to the settings of its parameters, these being the probabilities of mutation and crossover occurring, the population size and the number of players in a tournament selection (when this selection method is used). Moreover, for certain real/continuous encoded GAs, it is necessary to consider operators' polynomial distribution indices [11].

Real-encoded GAs can be thought of as being similar to Evolutionary Strategies (ES) introduced by Rechenberg and Schwefel as described by Bäck [12], except that ESs also are able to self-adapt their control parameters (or strategy parameters as they call them). The GA described in this work adopts this extra capability. The term *self-adaptive* used here is meant in the sense of that coined by Eiben *et al.* [13], to indicate control parameters of the GA that are encoded in the chromosome along with the problem definition parameters applying to the objective functions (the *main* parameters), and that these control parameters are subject to change along with the *main* parameters due to mutation and crossover. This is different from a purely *adaptive* control parameter strategy as in that case the change is instigated algorithmically by some feedback at the higher level of the GA rather than the lower level of each chromosome/solution in the population. The *deterministic* approach is rule-based and is not considered adaptive.

Eiben *et al.* [14] showed how population size and tournament selection size can be made to be self-adaptive, although in the former case to the detriment of performance of the optimization. Nonetheless, the latter case was shown to improve performance, and the method by which a parameter whose context is the population can be set through the aggregation of its representation at the individuals within the population, can be extended to other parameters having the same high-level context. However, the above work only uses mutation to affect each self-adapting parameter gene, rather than including the parameter genes in the crossover of the chromosome as a whole, and the model used is a steady-state GA (SSGA) with relatively low replacement strategy rather than a generational one (GGA). This work uses a fixed tournament size in order to keep all the self-adaptation occurring at the level of the individual, rather than by aggregation, since this is the focus of the work.

Zhang and Sanderson [15], [16] describe differential evolution (DE) algorithms that use self-adaptation, including their multi-objective (MO) JADE2 and JADE algorithms, that generate new values for mutation factors and crossover probabilities based on probability distributions governed by self-adapting *means*. DEs [17] are similar to GAs but new solutions are produced by adding the weighted difference of two population vectors to a third, to create a new donor vector which is recombined (crossover) with a target (parent) vector to produce the trial (child) vector. Differences between GAs and DEs, both algorithmic and from a performance perspective, are discussed in [18]. In a DE scheme, the mutation factor is a weight rather than a probability as in a GA, and notably crossover acts on whole parameters (the genes in a GA) rather than parts of parameters (Holland's *schemas*).

Sareni *et al.* [19] describe self-adaptation in a multi-objective genetic algorithm (MOGA) in which there is a self-adaptive choice between three different crossover operators for crossover, and in which mutation is self-adapted by the standard deviation of the amount of perturbation applied to a gene. Both of these mechanisms are different to the ones employed by the MOGA in this work.

Tan *et al.* [20] expounded their binary MOGA in which the mutation rate is deterministically assigned as a function of time, and Tan *et al.* [21] discuss a deterministic binary MOGA in which rules assign values for mutation and crossover probabilities. Ho *et al.* [22] used a binary GA for single objective optimisation in which sub-population groups adapted their mutation or crossover rates based on feedback from average fitness increase, while Li *et al.* [23] investigated diversity-guided mutation and deterministically adaptive mutation and crossover rates in a binary single-objective GA. These works all found their implementations of the various adaptive methods provided advantages on mathematically based benchmark problems, but differ from this work which is concerned with self-adaptation in a real-encoded multi-objective GA that addresses a real-world optimisation problem having time-consuming function evaluations.

This work presents a real-encoded generational MOGA employing elitism in which each solution has an evolving self-adaptive mutation-rate and self-adaptive crossover-rate, together with their own perturbation factors, encoded in its chromosome and which are subject to both mutation and crossover themselves, along with the *main* problem parameters. The MOGA is used on a real engineering multi-objective optimization problem (MOOP), that of airfoil optimization, and its performance on the problem is compared with two other leading heuristic algorithms, Multi-Objective Tabu Search (MOTS) and NSGA-II. The MOGA uses a novel crossover mechanism in order to recombine both the mutation rate and crossover rate control parameters at the level of the chromosome, and unlike other GAs, controls the number of duplicate chromosomes in each generation. Whereas DE algorithms have used probability distributions governed by self-adapting *means*, this MOGA uses its own mutation and crossover operators to control its self-adapting control parameters in the same way that they change the *main* genes.

2   The Genetic Algorithm Used

The MOGA developed in this work adopted much of the NSGA-II algorithm [24] with some modifications to the non-domination sorting method and to the construction of the new generation, and is called Ganesh (as it is a GA using non-dominated sorting and elitism). Constraints are implemented similarly to NSGA-II, requiring the constraint definition to return an increasingly negative number indicating the increasing degree of violation, and where 0 indicates no violation. Internally the MOGA is constructed to minimize, requiring objective functions that maximize to return a negative number, by the principle of duality [8].

A tournament selection method of degree two is used, polynomial mutation [25] is used along with a simulated binary crossover (SBX) [11] for real parameters, and the crossover strategy used in the problem discussed here is uniform crossover. Self-adaptive crossover requires extra consideration and is defined further, below.

The non-domination sorting is amended from NSGA-II to ensure that each solution is compared with every other one once in a simple and efficient manner with the number of comparisons equivalent to that of the *continuously updated* method [8], and the method of updating dominated-by count and dominated-solutions lists are modified accordingly. The new generation is produced by pruning one solution at a time from the merged parent and child populations and re-calculating the distance/crowding metric each time, giving a more accurate estimate of best-to-remove based on crowding (and non-dominated ranking) [26].

This MOGA additionally provides the ability to choose the cardinality of duplicate solutions in each generation, meaning that 0, 1 or many duplicates may be kept, with the default being many. Zero duplicates means one solution having no duplicates, and so on, where a duplicate is defined as all corresponding genes in both chromosomes having the same values. The ability to control the existence of duplicates is achieved here through the use of a linked hash map data structure where the key is the chromosome and the value includes a count and list of chromosomes having the same genes.

Bäck et al. [27], uses a random mutation rate initially, each solution being initialized to a random number in the range (0.001, 0.25), however he suspected that this randomness slowed down convergence to some extent. Ganesh allows mutation and crossover rates to be specified for the initial population, or to be set to random values in a uniform distribution, or to default to certain values. The default mutation rate of each solution would be set to $1/n$ where n is the number of variables of the objective functions (OFs), and the default crossover rate would be 0.6, both as probability of occurrence. This MOGA (henceforth referred to as a GA for brevity) also allows alternative initializers to be written and specified per problem, allowing for different probability distributions, such as the uniform or Gaussian, however this work uses the uniform distribution.

Similarly to Bäck [12] and Smith & Fogarty [28] (a steady-state GA), mutation first occurs to the gene encoding the mutation rate and then the new mutation rate is applied to the *main* genome, but unlike the previous studies, this is based on a generational GA, that is one in which the entire population is in theory able to be replaced by fitter solutions, and for which the variables, and operator parameters, are encoded as real numbers in the genes.

2.1   Self-adaptation

The GA control parameters undergoing self-adaptation are the mutation probability pM (per gene) and the crossover probability pC (per chromosome), and also the associated polynomial distribution indices, [25] & [11], for each, ηM and ηC respectively, which are all real values. Each solution has a chromosome encoding its objective function parameters and its control parameters. Mutation occurs to all of the parameters including the control ones and their indices, but mutation occurs first to the control ones at the current rate of mutation, and then the *main* ones using the newly mutated values.

The uniform crossover specifies that each gene has a 50% chance of crossing over if the chromosome is to undergo crossover at all, and the probability of chromosome crossover occurring is given by pC. However since crossover occurs between chromosomes but each chromosome has its own pC, the pC to be used is chosen stochastically at 50% probability from either of the parent chromosomes selected for breeding, and the ηC is taken from the same chromosome. The ηC value is then used in the crossover of the respective controls from each chromosome (pC, ηC, pM and ηM) and the *main* chromosome genes, with the new control values being written to the recombined chromosomes, as shown in Fig. 1.

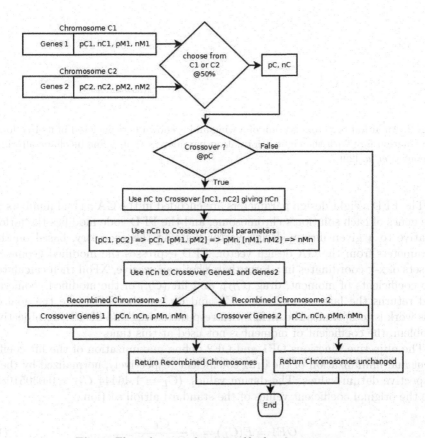

Fig. 1. Flow-chart explaining self-adaptive crossover

3   Airfoil Optimization

The real-world engineering problem to which this GA is applied is airfoil optimization, using the NACA 0012 airfoil section, as previously carried out by Kipouros et al. [29] using their Multi-Objective Tabu Search (MOTS) software [30] and NSGA-II. NACA 0012 [31] is a standard symmetric airfoil having a 12% thickness to chord length ratio, defined originally by the U.S. National Advisory Committee for Aeronautics, now part of the National Aeronautics and Space Administration (NASA). Airfoil shape modification is carried out by free-form deformation (FFD) [32] code and the shape is evaluated for aerodynamic efficiency by the XFoil tool [33] which calculates moment, drag and lift coefficients of flow, based on eight parameters as illustrated in Fig. 2.

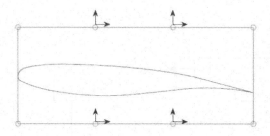

Fig. 2. An airfoil is a cross-section of a wing and is shown here enclosed in its free-form deformation hull with the eight deformation parameters that define its shape altering. Kipouros et al. [29].

The FFD's eight design parameters are encoded in the GA as real numbers in the genes of each solution's chromosome, and the FFD code modifies the airfoil relative to a given datum design vector defining the geometry, based on the parameters from the GA design vector. FFD expresses the modified geometry as sets of x-y coordinates in a form that XFoil can receive, XFoil then calculates the coefficients of moment, drag (C_D), and lift (C_L) of the modified geometry and returns the latter two results, C_D and C_L, to the GA. Since the goal of this work is to optimize the airfoil with respect to drag and lift as a bi-objective problem, the coefficient of moment is not used at this time.

The objective functions OF1 and OF2 define maximization of the lift coefficient and minimization of the drag co-efficient respectively, normalized by their respective datum values. The datum values, $(C_L = 1.46444, C_D = 0.0305108)$, are the original coefficient values of the standard airfoil section.

$$OF1 = F(C_L) = -\frac{C_L}{C_{L,\,\text{datum}}} \tag{1}$$

$$OF2 = F(C_D) = -\frac{C_D}{C_{D,\,\text{datum}}} \tag{2}$$

Ganesh is set to perform 6,000 function evaluations in each run with a population size of 120 and being allowed to run for 50 generations, as was performed in [29]. The probability of crossover pC is initially set to 0.9 and that of mutation pM to 0.5 for each member of the initial population, and their respective polynomial indices ηC to 10 and ηM to 20, as was the case for NSGA-II, but in the succeeding generations these values self-adapt. The probabilities may self-adapt in the interval [0, 1] while the polynomial indices may self-adapt in the interval [1, 100], noting that for the latter, larger values cause smaller perturbations in the original gene values, and vice-versa.

The problem definition used by the GA also defines the range by which the design vector is allowed to be modified, and thus how much the geometry of the airfoil may change, specified as follows: ±0.3, ±0.4, ±0.6 and ±1.0, with a larger range enabling larger variation in the free-form deformation. A given run of the GA uses one of those ranges, and the corresponding results are compared with those of MOTS and NSGA-II.

The airfoil is subject to two hard geometrical constraints, these being implemented inside XFoil, parametrically: the thickness of the airfoil section at (a) 25% and (b) 50% of the chord. Soft constraints are also specified against the objective function results to attempt to keep the solutions within a feasible region, since the current version of XFoil does not in all circumstances unambiguously return an indication as to whether it successfully converged. The constraints are as follows:

$$C_L \geq -3 \tag{3}$$

$$C_L \leq 0 \tag{4}$$

$$C_D \leq 2 \tag{5}$$

4   Results

Figures 3-7 show scatter plots of non-dominated solutions in the objective space obtained by the GA, in which OF1 and OF2 give the normalized values of C_L and C_D as previously described, plotted along the x and y axes respectively. Solutions which create unfeasible design vectors have been removed, and dominated solutions are not shown. At the wider ranges, Ganesh had not finished converging and so if left to run longer, more non-dominated solutions are available, giving a better spread of solutions. The values of C_L are shown as negative since it is being maximized and the GA is constructed internally to assume minimization. Figures contain results from Ganesh and the other optimizers MOTS and NSGA-II (where available). All results are for generation 50 (numbered as 0 to 49) unless stated otherwise in the figure caption, to provide a direct comparison with the MOTS and NSGA-II results obtained previously.

Having obtained a result for the ±0.4 range which generated unfeasible results at extreme minima, constraints were applied to the problem definition. This run produced more reasonable points but almost all design vectors had the same values, so more runs were executed but specifying no duplicates allowed in each generation, enabling a much better spread to be achieved, even if most points were dominated by a small number of extreme optima relative to the constraint boundaries.

Results obtained for the range ±1.0, albeit after many more generations, included solutions that had extreme values in both objectives and which represented design vectors that were unfeasible, therefore the constraints were redefined to the following, to exclude the obviously unfeasible regions:

$$C_L \geq -2 \tag{6}$$

$$C_L \leq 0 \tag{7}$$

$$C_D \leq 1.5 \tag{8}$$

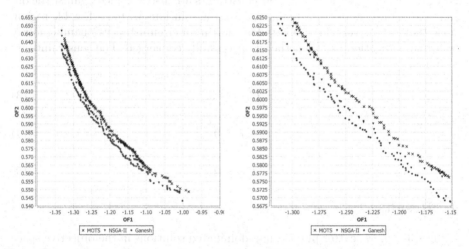

Fig. 3. Results for range ±0.3. MOTS, NSGA-II, & Ganesh, using constraints given by equations (3), (4) & (5), duplicate solutions allowed.

Fig. 4. Shows a zoom-in of Fig. 3, for clarity

4.1 Discussion

It was apparent that as the range increased, the GA was able to find attainment surfaces that were better approximations of the Pareto-optimal front, as does MOTS, as it is able to explore more of the search space; yet without constraints

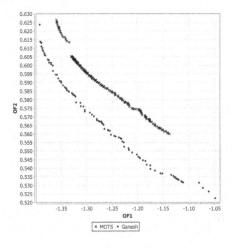

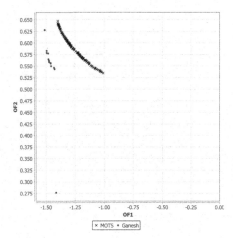

Fig. 5. Results for range ±0.4. MOTS & Ganesh, using constraints (3), (4) & (5), with no duplicates permitted.

Fig. 6. Results for range ±0.6. MOTS & Ganesh, using constraints (3), (4) & (5), with no duplicates permitted.

the GA would find extreme minima that are not feasible airfoil shapes due to a lack of feedback from XFoil under certain conditions about whether it was converging successfully. These problems were not seen with XFoil previously as such extreme points had not been found before. MOTS works better with XFoil because it explores more locally, and XFoil takes longer to run with larger variations in range and is less likely to converge successfully.

In Fig. 3 and Fig. 4 we see results for MOTS, NSGA-II and Ganesh for range ±0.3 showing that the Pareto attainment surface of Ganesh has converged more than both the others, and while not having quite as good a density of solutions as MOTS, is not significantly worse, and has a better spread than NSGA-II.

The original study had more NSGA-II results but for lower ranges (±0.1, ±0.2) which are not compared with here as their convergences are less than that of ±0.3, and a result for ±0.4 which had no better convergence than of ±0.3 and with only a few distinct points. NSGA-II results were not obtained for ranges ±0.6 and ±1.0 as no convergence was achieved, despite many runs, thus Ganesh out-performs in this aspect too.

Fig. 5 shows MOTS and Ganesh results for the range ±0.4, and again Ganesh converges more than MOTS and with a wider spread overall, but with a less dense front than MOTS. There is one extreme datum at approximately (-2.113, 0.330), which although dominating all other points, represents an unfeasible design, therefore it has been excluded by effectively zooming in to the region of interest. Were the point to be included, the resultant figure would need to be zoomed-out and would therefore contain much white space and make the region of interest more difficult to examine.

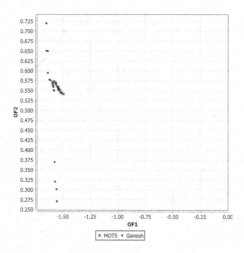

Fig. 7. Results for range ±1.0. MOTS & Ganesh, using constraints (6), (7) & (8), with no duplicates permitted.

Fig. 6 shows results for the range ±0.6, in which Ganesh has better convergence but with a less dense spread than MOTS. Some points ranked 1 and 2 have been removed as they were not feasible geometries, leaving some points ranked 3 which are nonetheless good solutions and showing better convergence than MOTS. It can be seen that one solution at around (-1.4, 0.275) has been found with high lift and very low drag which is nonetheless a feasible geometry. If Ganesh is left running for more generations, this solution generates more solutions around it, due to elitism and its high ranking.

Fig. 7 shows results for the range ±1.0, with Ganesh again having a better convergence than MOTS and having a wide spread of solutions although not as dense as MOTS. Some of the points found by Ganesh have both high lift (more negative) and low drag and would be of much interest to an airfoil designer. Allowing the GA to continue to generation 155 enabled it to find three regions clearly separated by apparent discontinuities.

Specifying that zero duplicates are permitted is beneficial as although many of the points may be dominated by extreme minima, if those minima are really unfeasible solutions, then the remaining dominated points may actually be useful real-world solutions. It also prevents the GA from prematurely converging to just a few solutions which are duplicated many times.

The constraints applied were not intended to be exact remedies, but they could perhaps be narrowed further to eliminate more unfeasible solutions and prevent premature convergence to inappropriate minima which distort the GA performance, since each minima produced will tend to breed further points around it, pulling the results away from more feasible regions. The result of range ±1.0 shows that the GA is producing results close to the constraints specified, however if XFoil was able to feedback the success or otherwise of its own convergence, the constraints could be eliminated, which would be beneficial as they naturally inhibit the exploration of the search space.

Fig. 8 is a parallel coordinates plot [34], [35] showing the eight parameters of the design vector of all solutions in the last generation of a run for range ±0.6, and their corresponding objective function values, C_L and C_D. Parallel lines between (p3 & p4), (p4 & p5), (p6 & p7), and (p7 & p8) indicate that those parameters, for certain values, are positively correlated. The lines crossing between p5 and p6 show that these are negatively correlated. This inter-dependency between parameters would be a good reason to use a multi-point crossover, rather than the uniform one currently used (or allow the GA to evolve which one of the two is used), since the uniform one is much more likely to disrupt the covariance ('epistasis'). That parameters p4, p7 and to a lesser extent p5, have most of their points at the extreme of their ranges may be because of the limits imposed on the values of the objective function values. Parameter p3 has most values passing through a relatively narrow band of values indicating that it is important and sensitive.

In Figs. 9 & 10 the means of the GA control parameters are plotted since each of the 120 solutions in each generation has its own value for each of these parameters. It can be seen that as the GA progresses through its generations, both pM and pC become smaller, hence the disturbance to good solutions is lessened, while their respective polynomial distribution indices become larger, which decreases the perturbation to the section geometry, thus at the start the GA is better at exploring the search space while towards the end it is better at converging to good solutions.

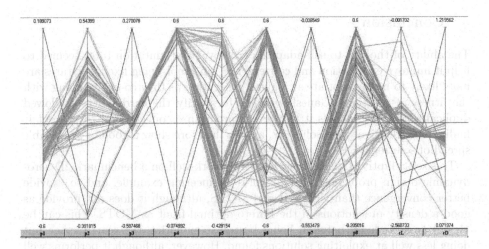

Fig. 8. Parallel coordinates plot for Ganesh final generation of range ±0.6, showing the 8 design parameters and the 2 objectives

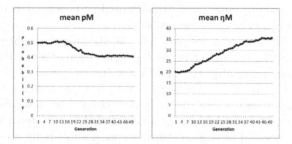

Fig. 9. Trends of the means of the pM & ηM GA control parameters against generation number for all ranges

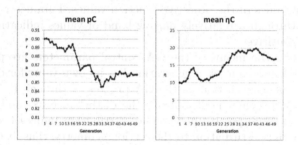

Fig. 10. Trends of the means of the pC & ηC GA control parameters against generation number for all ranges

5   Conclusion

The ability of the GA to self-adapt its crossover and mutation rates seemed to help it improve exploration and convergence, since not being fixed, the rates are more likely to be appropriate at a given generation as they co-evolve along with the fitness of solutions. Ganesh's ability to specify the cardinality of allowed duplicate solutions enables it to prevent premature convergence, to prevent it finding only relatively few solutions, and to therefore show a good or reasonable spread of solutions.

This self-adaptive GA has been shown to work well on a benchmark 2D aerodynamic design problem, as a real-world engineering example, and to provide better convergence than NSGA-II and MOTS, although it does not provide as good a density of solutions in the Pareto-optimal front as MOTS. This can be viewed as a trade-off between being better at exploring the search space but doing less well at exploiting solutions found. However, although it performs well at convergence, MOTS can be said to be better at arriving at feasible solutions through its local search, which works better with XFoil as it currently is.

6   Future Work

Although 80 runs (20 per range) of the GA have been performed, automatic removal of unfeasible designs is not available, therefore a lack of time and space has precluded the analysis of the results and the inclusion of the performance statistics here. The PISA package [36] will be used with the results obtained to acquire standard metrics such as hyper-volume indicator [37] and ϵ-indicator [9] to understand the performance better. At the same time an enhanced version of XFoil will be developed, in order to return all convergence failures and indications of the breaking of the hard constraints, which will enable unfeasible solutions to be automatically discarded and which should improve convergence. There are also the following improvement options to explore:

1. Enable the GA to switch between modes centered on either convergence or distribution, since these have been shown generally to be mutually exclusive. Like most GGAs, this GA is distribution-centric as driven by its ranking algorithm, but allowing it to adopt a convergence-centric mode when convergence halts, by only allowing new solution points that dominate at least one of the existing points to enter the next generation, should improve convergence performance. Any optimization algorithm that exhibits these characteristics is expected to be favorable for the exploration and exploitation of such complex engineering design problems. DE algorithms, in which a new solution vector only enters the population if it is better than the parent, can be only considered as convergent-centric.
2. Employ an additional multi-point crossover in which the number and location of crossover loci are chosen stochastically, within reasonable intervals, and allow each solution to self-adapt the crossover scheme it uses, choosing between uniform and the cardinality of multi-point crossover. The uniform crossover used in this version, although good for exploring the search space, can be disruptive to good solutions, but low cardinality multi-point crossover can be better at preserving good solutions and overcome linkage (epistasis) problems.
3. Explore the potential of using Epanechnikov kernels [38] to define an alternative density/crowding metric for the ranking part of the GA algorithm.
4. Enable duplicate solutions to be controlled across generations for the entire history of the GA.

References

1. Holland, J.H.: Adaptation in Natural and Artificial Systems: An Introductory Analysis with Applications to Biology, Control and Artificial Intelligence, 2nd edn. MIT Press, Massachusetts (1992)
2. Goldberg, D.E.: Genetic Algorithms in Search, Optimization & Machine Learning. Addison-Wesley, Reading (1989)
3. Schaffer, J.D.: Some experiments in machine learning using vector evaluated genetic algorithms (1984)

4. Jones, D.F., Mirrazavi, S.K., Tamiz, M.: Multi-objective meta-heuristics An overview of the current state-of-the-art. European Journal of Operational Research 137(1), 1–9 (2002)
5. Coello Coello, C.A.: Evolutionary multi-objective optimization: a historical view of the field. IEEE Computational Intelligence Magazine 1(1), 28–36 (2006)
6. Fleming, P.J., Purshouse, R.C., Lygoe, R.J.: Many-objective optimization: An engineering design perspective. In: Coello Coello, C.A., Hernández Aguirre, A., Zitzler, E. (eds.) EMO 2005. LNCS, vol. 3410, pp. 14–32. Springer, Heidelberg (2005)
7. Fonseca, C.M., Fleming, P.J.: Multiobjective optimization and multiple constraint handling with evolutionary algorithms - part i: A unified formulation. IEEE Transactions on Systems, Man and Cybernetics, Part A: Systems and Humans 28(1), 26 (1998)
8. Deb, K.: Multi-Objective Optimization using Evolutionary Algorithms. John Wiley, Chichester (2001) ID: 2
9. Zitzler, E., Thiele, L., Laumanns, M., Fonseca, C.M., da Fonseca, V.G.: Performance assessment of multiobjective optimizers: An analysis and review. IEEE Transactions on Evolutionary Computation 7(2), 117–132 (2003)
10. Jong, K.A.D.: An analysis of the behavior of a class of genetic adaptive systems (1975) AAI7609381
11. Deb, K., Agrawal, R.B.: Simulated binary crossover for continuous search space. Complex Systems, 115–148 (1995)
12. Bäck, T.: Self-adaptation in genetic algorithms. In: Proceedings of the First European Conference on Artificial Life, pp. 263–271. MIT Press (1992)
13. Eiben, A.E., Schut, M.C., Wilde, A.R.D.: Boosting genetic algorithms with self-adaptive selection. In: Proceedings of the IEEE Congress on Evolutionary Computation, pp. 1584–1589 (2006)
14. Eiben, A.E., Schut, M.C., de Wilde, A.R.: Is self-adaptation of selection pressure and population size possible? a case study. In: Runarsson, T.P., Beyer, H.-G., Burke, E.K., Merelo-Guervós, J.J., Darrell Whitley, L., Yao, X. (eds.) PPSN 2006. LNCS, vol. 4193, pp. 900–909. Springer, Heidelberg (2006)
15. Zhang, J., Sanderson, A.C.: Self-adaptive multi-objective differential evolution with direction information provided by archived inferior solutions. In: IEEE Congress on Evolutionary Computation, CEC 2008 (IEEE World Congress on Computational Intelligence), pp. 2801–2810 (2008) ID: 1
16. Zhang, J., Sanderson, A.C.: Jade: Self-adaptive differential evolution with fast and reliable convergence performance. In: IEEE Congress on Evolutionary Computation, CEC 2007, pp. 2251–2258 (2007) ID: 1
17. Storn, R., Price, K.: Differential evolution - a simple and efficient heuristic for global optimization over continuous spaces. Journal of Global Optimization 11(4), 341–359 (1997)
18. Tušar, T., Filipič, B.: Differential evolution versus genetic algorithms in multiobjective optimization. In: Obayashi, S., Deb, K., Poloni, C., Hiroyasu, T., Murata, T. (eds.) EMO 2007. LNCS, vol. 4403, pp. 257–271. Springer, Heidelberg (2007)
19. Sareni, B., Regnier, J., Roboam, X.: Recombination and self-adaptation in multiobjective genetic algorithms. In: Liardet, P., Collet, P., Fonlupt, C., Lutton, E., Schoenauer, M. (eds.) EA 2003. LNCS, vol. 2936, pp. 115–126. Springer, Heidelberg (2004)
20. Tan, K.C., Goh, C.K., Yang, Y.J., Lee, T.H.: Evolving better population distribution and exploration in evolutionary multi-objective optimization. European Journal of Operational Research 171(2), 463–495 (2006)

21. Tan, K.C., Chiam, S.C., Mamun, A.A., Goh, C.K.: Balancing exploration and exploitation with adaptive variation for evolutionary multi-objective optimization. European Journal of Operational Research 197(2), 701–713 (2009)
22. Ho, C.W., Lee, K.H., Leung, K.S.: A genetic algorithm based on mutation and crossover with adaptive probabilities. In: Proceedings of the 1999 Congress on Evolutionary Computation, CEC 1999, vol. 1, p. 775 (1999)
23. Li, M., Cai, Z., Sun, G.: An adaptive genetic algorithm with diversity-guided mutation and its global convergence property. Journal of Central South University of Technology 11(3), 323–327 (2004)
24. Deb, K., Agrawal, S., Pratap, A., Meyarivan, T.: A fast elitist non-dominated sorting genetic algorithm for multi-objective optimization: Nsga-ii. In: Schoenauer, M., Deb, K., Rudolph, G., Yao, X., Lutton, E., Merelo, J., Schwefel, H.P. (eds.) PPSN 2000. LNCS, vol. 1917, pp. 849–858. Springer, Heidelberg (2000)
25. Deb, K., Goyal, M.: A combined genetic adaptive search (geneas) for engineering design. Computer Science and Informatics 26, 30–45 (1996)
26. Kukkonen, S., Deb, K.: Improved pruning of non-dominated solutions based on crowding distance for bi-objective optimization problems. In: IEEE Congress on Evolutionary Computation, CEC 2006, pp. 1179–1186 (2006) ID: 1
27. Bäck, T., Eiben, A.E., van der Vaart, N.A.L.: An emperical study on GAs "without parameters". In: Schoenauer, M., Deb, K., Rudolph, G., Yao, X., Lutton, E., Merelo, J., Schwefel, H.P. (eds.) PPSN 2000. LNCS, vol. 1917, pp. 315–324. Springer, Heidelberg (2000)
28. Smith, J.E., Fogarty, T.C.: Self adaptation of mutation rates in a steady state genetic algorithm. In: Proceedings of the 1996 IEEE Conference on Evolutionary Computation, pp. 318–323. IEEE (1996)
29. Kipouros, T., Peachey, T., Abramson, D., Savill, M.: Enhancing and developing the practical optimisation capabilities and intelligence of automatic design software AIAA 2012-1677. In: 8th AIAA Multidisciplinary Design Optimization Specialist Conference (MDO). American Institute of Aeronautics and Astronautics (April 2012)
30. Jaeggi, D.M., Parks, G.T., Kipouros, T., Clarkson, P.J.: The development of a multi-objective tabu search algorithm for continuous optimisation problems. European Journal of Operational Research 185(3), 1192–1212 (2008) ID: 3
31. Abott, I., von Doenhoff, A.: Theory of wing sections: including a summary of airfoil data. Dover, New York (1959)
32. Sederberg, T.W., Parry, S.R.: Free-form deformation of solid geometric models. SIGGRAPH Comput. Graph. 20(4), 151–160 (1986)
33. Drela, M.: Xfoil - an analysis and design system for low reynolds number airfoils. In: Low Reynolds Number Aerodynamics Conference, Germany, Notre Dame, pp. 1–12 (1989)
34. Inselberg, A.: Parallel Coordinates: Visual Multidimensional Geometry and Its Applications. Springer (2009)
35. Kipouros, T., Mleczko, M., Savill, M.: Use of parallel coordinates for post-analyses of multi-objective aerodynamic design optimisation in turbomachinery. AIAA-2008-2138. In: 4th AIAA Multi-Disciplinary Design Optimization Specialist Conference. Structures, Structural Dynamics, and Materials and Co-located Conferences, Schaumburg, Illinois. American Institute of Aeronautics and Astronautics (April 2008)

36. Bleuler, S., Laumanns, M., Thiele, L., Zitzler, E.: Pisa - a platform and programming language independent interface for search algorithms. In: Fonseca, C.M., Fleming, P.J., Zitzler, E., Deb, K., Thiele, L. (eds.) EMO 2003. LNCS, vol. 2632, pp. 494–508. Springer, Heidelberg (2003)
37. Zitzler, E., Thiele, L.: Multiobjective optimization using evolutionary algorithms - a comparative case study. In: Eiben, A.E., Bäck, T., Schoenauer, M., Schwefel, H.-P. (eds.) PPSN 1998. LNCS, vol. 1498, pp. 292–301. Springer, Heidelberg (1998)
38. Epanechnikov, V.: Non-parametric estimation of a multivariate probability density. Theory of Probability & Its Applications 14(1), 153–158 (1969)

Genetic Programming with Scale-Free Dynamics

Hitoshi Araseki

Graduate School of Social and Cultural Studies, Nihon University
4-25, Nakatomi-Minami, Tokorozawa, Saitama, Japan
araseki.hitoshi@nihon-u.ac.jp

Abstract. This paper describe a new selection method, named SFSwT (Scale-Free Selection method with Tournament mechanism) which is based on a scale-free network study. A scale-free selection model was chosen in order to generate a scale-free structure. The proposed model reduces computational complexity and improves computational performance compared with a previous version of the model. Experimental results with various benchmark problems show that performance of the SFSwT is higher than with other selection methods. In various fields, scale-free structures are closely related to evolutionary computation. Further, it was found through the experiments that the distribution of node connectivity could be used as an index of search efficiency.

1 Introduction

In complex network study, the scale-free network is a universal phenomenon in various networks such as the World Wide Web, social networks, metabolic networks, and so on [1, 2]. Scale-free networks exhibit a power-law distribution of node connectivity such that the probability that a given node has k connections is governed by the relationship:

$$P(k) \propto k^{-\gamma} \tag{1}$$

where γ is referred to as the scaling exponent.

Although a large body of research has proposed various theoretical models to generate a scale-free structure, little is known about their applicability to real problems.

In evolutionary computation, various selection methods have been applied to many problems. The selection method controls the performance of evolution computation. Recently, the selection method in genetic algorithms (GA) has been considered relevant to the study of a scale-free network [3–5]. The network-related selection methods have been called the mating network, the network of reproduction that occurs in a GA population where the individuals in a population are the network nodes. Two nodes are connected by an edge if these nodes represent individuals that have mated during the course of the evolutionary process and have jointly contributed genetic material towards at least one offspring.

M. Emmerich et al. (eds.), *EVOLVE - A Bridge between Probability, Set Oriented Numerics, and Evolutionary Computation IV*, Advances in Intelligent Systems and Computing 227, DOI: 10.1007/978-3-319-01128-8_18, © Springer International Publishing Switzerland 2013

A previous selection method named the scale-free selection method (SFS) with scale-free properties was developed in order to create a scale-free network [6]. Although this selection mechanism is very different, the performance of the SFS is similar to other selection methods. However, the performance of the SFS model was not better than other selection models in a specific problem when many individuals have similar evaluation values. It is likely that selection pressure was missing in the previous selection method. As it is also necessary to evaluate the fitness of all individuals in the population, it is hard to apply SFS to real problems. Why then has the real world generated the scale-free structure in various fields?

The purpose of this study is to investigate whether or not a scale-free structure in a mating network is effective for evolutionary computation. A new selection method was developed in order to address this problem to examine the effectiveness of the propose selection model in a population containing many individuals with similar evaluation values. The new selection method has a controllable selection pressure mechanism; moreover, computational complexity is reduced compared with SFS. Experimental results with some benchmark problems show that performance of the new method is higher than other selection methods.

The paper is organized as follows. Section 2 describes the details of the new selection method. Section 3 gives experimental results. Section 4 discusses the search space and computation performance. Finally, concluding remarks are given in Section 5.

2   Scale-Free Selection with Tournament Mechanism

We had previously proposed an SFS model [6] with a scale-free properties by using a non-growing network [7–10]. In the SFS model, a mating pair is preferentially chosen by Eq. (2), where k_i is the number of edges connected to individual i, and η_i is a fitness value of individual i.

$$\Pi_i = \frac{\eta_i(k_i + 1)}{\sum_j \eta_j(k_j + 1)} \tag{2}$$

However, the performance of the SFS was not better than other selection models in a specific problem where many individuals have similar fitnesses. In other words, in a statistical method such as roulette selection or SFS, it is difficult to choose a higher fitness individual from a population with similar fitness values because these selections do not have a mechanism to control selection pressure [6].

It is well known that the tournament selection controls selection pressure in evolutionary computation. Tournament selection has the following features compared with other selection methods [11, 12]:

1. Its selection pressure can be adjusted.
2. It is simple code, efficient for both non-parallel and parallel architectures.
3. It does not require sorting the whole population first. It has the time complexity $O(N)$ where N is the population size.

The new selection method, named SFSwT (Scale-Free Selection method with Tournament mechanism), incorporates the tournament mechanism in SFS in order to improve the performance.

In an evolving population, the mating network is defined as follows [3–6]. The individuals in a population are the network nodes; two nodes are connected by an edge if these nodes represent individuals that have mated during the course of the evolutionary process and have jointly contributed fragments towards at least one offspring.

The SFSwT and SFS model are rewiring models that generate a scale-free network from a random network. In an initial network such as SFSwT and SFS, these models start with a random network [15] with N nodes and N×M edges. Accordingly, a random network is generated by doing random selection at the M times step. Therefore, the mating network is a random network where each individual has about M edges. In complex network study, M is about 5 [8, 9]. Similarly, M = 5 in these experiments.

The next steps are as follows:

1. Similar to tournament selection, a number of individuals (tournament size = T) are randomly chosen from the population. The evaluation values calculated by Eq. (3) are compared, and the best individual i is chosen to mate.

$$E_i = \eta_i k_i \tag{3}$$

where k_i is the number of edges connected to individual i, and η_i is the fitness value of individual i. It can be conjectured that an individual with a high E_i has generated an offspring with high fitness in the past. The selected individual i has edge l_{ij} connected to it. This evaluation value in Eq. (3) is a simplified evaluation value according to Bianconi and Barabási [13] compared to Eq. (4).

$$\Pi_i = \frac{\eta_i k_i}{\sum_j \eta_j k_j} \tag{4}$$

The calculation is simplified because there is no normalization coefficient in Eq. (4). Therefore, the computation complexity of SFSwT is $O(T)$.

2. The edge l_{ij} is removed in the least fit individual and replaced with a new $l_{ij'}$ that connects a parent i with parent j' where the j node with the lowest evaluation value is connected to the i node. The parent j' is chosen by tournament selection according to Eq. (5). Note that the evaluation value in Eq. (5) is chosen to be proportional to $(k_j + 1)$ so that when there is a non-zero value, isolated individuals $(k_j = 0)$ acquire a new edge [8, 9].

$$E'_{j'} = \eta_{j'}(k_{j'} + 1) \tag{5}$$

In operation 2 when there is already an edge $l_{ij'}$, the pair are reevaluated by the next step. Figure 1 illustrates an edge replacement mechanism using SFSwT in the mating network.

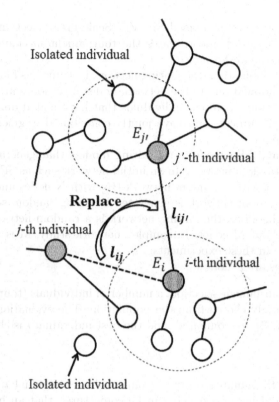

Fig. 1. Edge replacement mechanism using the SFSwT method with a tournament size of 5

3. Select an individual pair (i and j' nodes) and create two offspring from these parents by a crossover operation. Hence, two individuals are selected from four individuals to propagate to the next generation.
4. Select one elite individual and randomly select another individual from the rest of the individuals.
5. Replace the individual pair with one elite individual and one randomly selected individual. As a result of the replacement, the fitness of an individual i and/or individual j' is replaced by one or two offspring's fitness, but the fitness of individual i and/or individual j' does not change the number of edges k_i and/or individual $k_{j'}$. Furthermore, when two offspring are not selected for replacement, do not replace the individual pair. In this case, a new edge is not formed between i node and j' node. Figure 2 illustrates an individual inheritance mechanism in the mating network. Of course, the individual is constructed by the instruction set of the tree structure.
6. Repeat steps 1-5 until the terminal conditions are reached.

In particular, it is expected that steps 3-5 which are similar to MGG (Minimal Generation Gap) [14] and have been designed to maintain the diversity of individuals should avoid early convergence and suppression of evolutionary

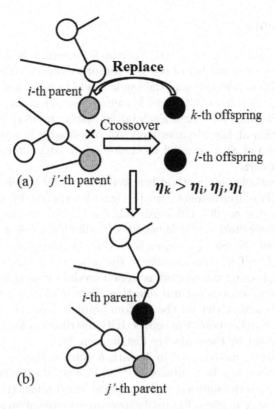

Fig. 2. Individual inheritance mechanism in the mating network. (a) Two parents produce two offspring by crossover. (b) A parent is replaced by one elite offspring which has higher fitness value than other individuals.

stagnation. In these experiments, steps 3-5 were applied after choosing parents by all selection methods. Furthermore, one generation at the N (= population) times step is defined by repeating steps 3-5.

Raw fitness (ε_j) is calculated as the sum of the absolute values of the difference in Eq. (7). However, this paper uses the fitness (η_j) in Eq. (6) which includes the raw fitness (ε_j).

$$\eta_i = \frac{1}{1 + \varepsilon_i} \tag{6}$$

$$\varepsilon_i = \sum_{j=1}^{n} |f_i(\boldsymbol{x_j}) - y_j| \tag{7}$$

where $f_i(\boldsymbol{x_j})$ is each i-th individual's fitness value on the j-th given target data $\{(\boldsymbol{x_1}, y_1), (\boldsymbol{x_2}, y_2), \cdots, (\boldsymbol{x_n}, y_n)\}$. Therefore, the goal is to achieve success with $\eta_i = 1$.

3 Experiments

Experiments used four different selection methods including the SFSwT method. In SFSwT and SFS, the number of edges is conserved during runs. On the other hand, the tournament selection and the random selection methods have another way to maintain the number of edges because these selections do not have a conservation mechanism for edges. Each initial mating network is generated by doing each selection at the M times step. An oldest edge is removed when an individual pair (i and j) is selected to create a new edge after generating the initial mating network.

The tournament selection method is a well known and widely used selection method in GP. This tournament size (5) was determined by experiments in order to optimize the results. Obviously when a large tournament size is used in evolutionary computation, the diversity of individuals is lost. However, the tournament size of the SFSwT depends on the specific problem because the performance of SFSwT is very sensitive to the problem to which it is applied.

A random graph using the random selection method is used to set a baseline for the stochastic process of natural selection. From Erdös and Rényi, the distribution of node connectivity for the random graph is represented by a Poisson distribution [15]. Furthermore, the connectivity distribution for the SFS can be expected to represent by the scale-free distribution [6].

In all experiments, many individuals with a subtree show a success fitness value, i.e., the subtree is a best subtree ($\eta_i^{sub} = 1$). For all experiments, the best subtree was chosen as the solution of the problem; other nodes (not belonging to the best subtree) were deleted. Figure 3 shows an example of an individual with the best subtree (gray nodes) which is recognized as a new individual [16, 17]. However, subtree evaluation does not require any new calculations because when an individual fitness value was obtained, all the subtree evaluation values for the individual were known as the fitness value of individual is the summation of subtree evaluations.

The GP parameters used for these experiments are shown in Table 1. The average of 100 different runs is calculated for each selection method on each problem in Figure 4 and Figure 5.

Table 1. General parameters for GP computation

Population Size	500
Max Generation	100
Max Depth	15
Mutation Ratio	0.1

3.1 Royal Tree Problem

The Royal Tree problem, proposed by Punch et al. [18], is a natural extension of the Royal Road problem. The function set consists of symbols "A", "B", "C",

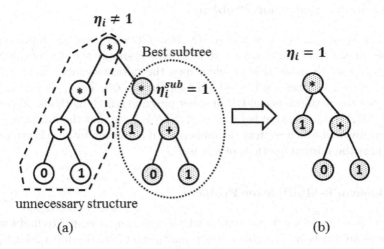

Fig. 3. A node is an instruction of program. (a) An individual with a best subtree. (b) A new individual.

"D", "E", ... with arity 1, 2, 3, 4, 5 respectively. Symbol x is the only terminal symbol. The optimal trees for the Royal Tree problem are given by perfect trees. The fitness of the Royal Tree problem is calculated as

$$f(n_i) = W_{comp} \sum_j (W_{full} \cdot f(child(n_i, j)))$$ (8)

where $W_{comp} = 2$ if the subtree rooted at n_i is a perfect tree, otherwise $W_{comp} = 1$. $W_{full} = 2$ if the label of $child(n_i, j)$ is correct and the subtree rooted at $child(n_i, j)$ is a perfect tree. The optimal tree (perfect tree) of $L = n$ is constructed by an optimal tree of $L = (n - 1)$, where L is the tree depth. The optimal solution of the Royal Tree problem must be a full tree and have no redundant information (introns). There is only one optimal tree. Other problems usually have more than one representation for the perfect tree. Therefore, the Royal Tree problem is the best problem for evaluating if an effective structure is created from a particular combination. In this experiment, the Royal Tree problem of symbol "D" ($L = 4$) and "E" ($L = 5$) were used.

In Royal Tree problems, the fitness value of individual is directly dependent on an effective structure related to its building blocks. Accordingly, an individual without building blocks has a very small fitness value. It is difficult to separate individuals because the fitness of many individuals becomes very small value during an evolutionary computation run. It is reasonable to suppose that the Royal Tree problems are the best benchmark test in flat fitness distributions.

3.2   Symbolic Regression Problem

The classic trigonometric function is used here $COS(2x)$ with 20 input-output test cases drawn from the range $-\pi$ to π. Fitness is simply the sum of the errors in Eq. (6). Success using this problem uses the notion of hits where a hit is achieved when the fitness in Eq. (6) is larger than 0.999. This program is often used to test the performance of GP or other program search methods. According to Koza [19], the function set is $\{+, -, *, \%, SIN\}$. Note that the function "%" is protected division which returns the numerator if a division by zero is attempted and returns the normal quotient otherwise.

3.3   Boolean 6-Multiplexer Problem

This is a classic benchmark problem in which evolution attempts to find the six-input boolean function. The goal of the 6-multiplexer is to decode a 2-bit binary address $(00, 01, 10, 11)$ and return the value of the corresponding data register (d_0, d_1, d_2, d_3). Thus, the 6-multiplexer is a function of 6 activities; (a_0, a_1) determines the address and d_0 to d_3 determine the answer. There are $2^6 = 64$ possible combinations for the 6 arguments, and the entire set of 64 combinations is used for evaluating fitness. For this problem, the goal is to find a boolean expression that fits the given 64 combination data s et. According to Koza [19], the function set $\{AND, OR, NOT, IF\}$ is used.

3.4   Results

The cumulative success probability curves for all selection methods are shown in Figure 4. The graph in Figure 4 (a) illustrates the results of each selection method in the Royal Tree D problem. In this problem, the tournament size of the SFSwT is 50 because individuals do not include intronic structures. This tournament size was determined by experiments in order to have best results. Since individuals do not have structural diversity, it is reasonable to suppose that exploring a wide range in search space can find a good solution efficiently. Figure 4 (a) shows that the SFSwT and the tournament selection outperform the random and the SFS methods in the Royal Tree D problem.

Figure 4 (b) illustrates the results of each selection method in the Symbolic Regression problem. Figure 4 (c) illustrates the results of each selection method in the Boolean 6-Multiplexer problem. In these problems, the tournament size of the SFSwT is 5 because individuals in these problems include intronic structures. Of course, these tournament size were determined by experiments in order to have best results. We see from Figures 4 (b) and (c) that these selection performances are the same. In these experiments, individuals have a variety of fitness values in order to include intronic structures. Therefore, an effective individual is selected easily from the population by any of these selection methods.

The distributions of node connectivities are shown in Figure 5. Only distribution plots for the Royal Tree D problem are included in order to demonstrate

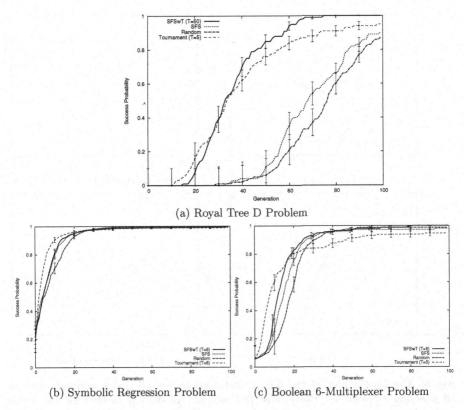

(a) Royal Tree D Problem

(b) Symbolic Regression Problem   (c) Boolean 6-Multiplexer Problem

Fig. 4. Cumulative Success Probability for each selection method on each problem. The error bars on the figures show standard deviation (SD).

the effectiveness of the selection model in a population containing many individuals with similar evaluation values. In spite of similarity between the SFS and the SFSwT, each connectivity distribution is quite different. In Figure 5 (a), an important point is that the SFSwT method generates a scale-free distribution in a small k area in spite of a near uniform fitness distribution. Figure 5 (b) shows that the mating network using the SFS does not have a scale-free distribution contrary to the previous paper; it shows a Poisson distribution related to a random selection. In fitness space, almost all individuals have similar values because many individuals attain higher fitness quickly by generating two offspring by pair selection using crossover operations. On the other hand, in the previous paper, the selected pair generated only one offspring. We would suggest that generating a scale-free distribution is dependent on the shape of fitness space in the SFS model. The distributions of node connectivities in the tournament selection in Figure 5 (c), show that this distribution has a bias different from scale-free. Figure 5 (d) shows the connectivity distribution of obtaining networks by random selection; the mating network appears to follow a Poisson distribution.

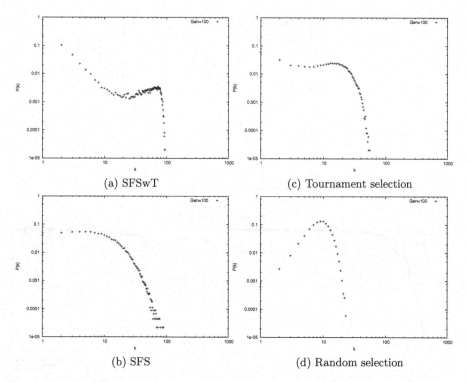

(a) SFSwT   (c) Tournament selection

(b) SFS   (d) Random selection

Fig. 5. The log-log plot of the degree distributions for all nodes' connectivities in all selection methods for Royal Tree D Problem. All data is plotted for the 100th generation.

On each experiment, the connectivity distribution has a different behavior because a successful individual is generated at a different generation. Therefore, an arbitrary experiment is used here in order to investigate a generating process of scale-free distribution. The details of the connectivity distribution of an arbitrary experiment for the Royal Tree D problem appear in Figure 6. Figure 6 (a) shows the distribution of node connectivity at the 10th generation. This distribution is a Poisson distribution because this distribution is influenced by the initial network structure at the 5th generation. Figure 6 (b) shows the mating network structure at the 40th generation. At the 40th generation, a successful individual was discovered, however, the distribution still is a Poisson distribution. Figure 6 (c) shows the scale-free distribution of the mating network structure at the 70th generation. Figure 6 (d) shows the network distribution at the 100th generation is different from the scale-free structure. The network structure changes to the other distribution after the 70th generation. However, in a small and medium area, the distribution has a scale-free distribution with a steeper slope than the 70th generation. This is different behavior compared to theoretical analysis in complex network study [7, 9] because individuals have a dynamic fitness value in the SFSwT. After an individual has achieved a good fitness, other individuals

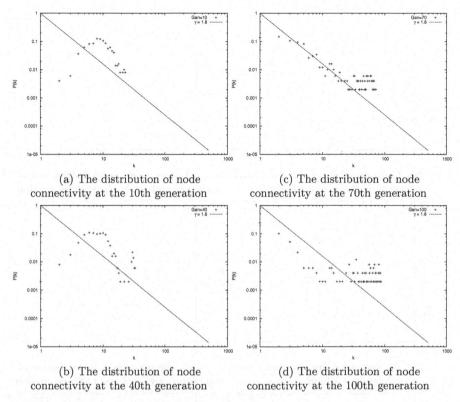

(a) The distribution of node
connectivity at the 10th generation

(c) The distribution of node
connectivity at the 70th generation

(b) The distribution of node
connectivity at the 40th generation

(d) The distribution of node
connectivity at the 100th generation

Fig. 6. Log-log plot of the degree distributions for nodes' connectivities in SFSwT for Royal Tree Problem. (b) At this generation, a successful individual was discovered. The solid line corresponds to $\gamma = 1.8$ in Eq. (1).

link to the fit individual quickly thus many individuals will have good fitness after the 70th generation. Consequently, the scale-free structure changes to the other structure, in order to increase number of individuals with higher fitness. In SFSwT, the point to address is whether the scale-free distribution is related to the performance of evolutionary computation or not by exploring the difference between a successful distribution and unsuccessful distribution in Royal Tree E problem. In this experiment, the parameters in Table 1 are used except for Max Generation (=200).

The detail of connectivity distribution of an arbitrary experiment for the Royal Tree E problem appear in Figure 7. Figure 7 (a)-(c) are successful distributions and Figure 7 (d)-(f) are unsuccessful distributions at the 90th, 130th, and 200th generation respectively. At the 90th generation in Figure 7 (a), a successful individual was discovered, however, the distribution still is a Poisson distribution. Figure 7 (b) shows the scale-free distribution of the mating network structure at the 130th generation. Figure 7 (d) shows the network distribution at the 200th generation is different from the scale-free structure; the network structure

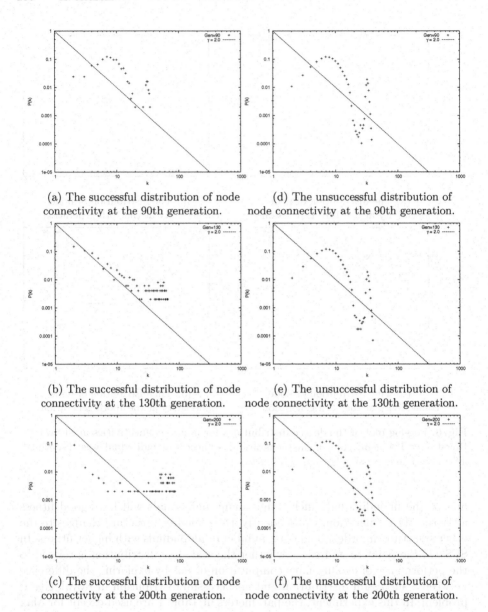

(a) The successful distribution of node connectivity at the 90th generation.

(d) The unsuccessful distribution of node connectivity at the 90th generation.

(b) The successful distribution of node connectivity at the 130th generation.

(e) The unsuccessful distribution of node connectivity at the 130th generation.

(c) The successful distribution of node connectivity at the 200th generation.

(f) The unsuccessful distribution of node connectivity at the 200th generation.

Fig. 7. Log-log plot of the degree distributions for nodes' connectivities in SFSwT for Royal Tree E Problem. (a) At this generation, a successful individual was discovered. The solid line corresponds to $\gamma = 2.0$ in Eq. (1).

changes to the other distribution after the 130th generation. The behavior in Figure 7 (a)-(c) is similar to the experiment of the Royal Tree D problem in Figure 6 (b)-(d). Figure 7 (b)-(d) are similar to the Poisson distribution in a small k area. On the other hand, although Figure 7 (a) and 7 (d) have exactly the same distribution, the other distributions are quite different. The fact that there are different distributions which depend on successful and unsuccessful computation suggests that nodes' connectivity in SFSwT can be used as a performance indicator of evolutionary computation.

4   Discussion

Although each problem is quite different in solution structure, the SFSwT method has better performance than other selection methods in all experiments. In the Royal Tree problem, the fitness value directly relates to the effective structure because the individuals do not have any introns. On the other hand, in the Symbolic Regression problem and the Boolean 6-Multiplexer problem, an individual may have the same evaluation value by an intron effect even if an individual structure is different. It is an important point that this model has a mechanism to handle variable tournament sizes and thus is a more natural mechanism. The results of these experiments clearly show that the new selection strategy is effective in evolution computation according to Eq. (3) and Eq. (5).

Compared to the SFS that was previously proposed, the SFSwT not only improves computational performance, but also significantly reduces the amount of computation. The SFSwT does not need to calculate the fitness value of all individuals unlike the previous SFS model. Hence, this model is a more realistic model because in the real world, individuals do not require the fitness value of all individuals around the world to determine a mating partner.

The fact that the scale-free structure was generated at the 70th generation in Figure 6 (c) and Figure 7 suggests that the scale-free structure is strongly related to the performance of evolutionary computation. Furthermore, that different distributions in Figure 7 depend on the success or failure of computation suggests that SFSwT node connectivity can be used as a performance metric of evolutionary computation. Consequently, it is highly probable that the distribution of node connectivity could be used as an index of search efficiency. The fact that the performance of SFSwT is also very sensitive to the problem suggests that the SFSwT mechanism is a more realistic approach in a mating network. In other words, the individual needs to connect to the various communities in order to solve real world problems.

5   Conclusions

This paper proposes a new selection method named the SFSwT based on complex network study. Different experiments showed that the SFSwT has better performance than other selection methods. The proposed model has improved computational performance as well as reduced computational complexity. Scale-free structures occur in various real world applications, implying close relations to evolutionary computation.

Therefore, the SFSwT may be applied to various problems in GP computation and the optimization problem is relevant to complex network study. Furthermore, the distribution of node connectivity could be used as an index of search efficiency. For future studies, we plan to apply the proposed method to more problems and to investigate more network properties using the proposed method.

References

1. Albert, R., Barabási, A.-L.: Statistical mechanics of complex networks. Reviews of Modern Physics 74 (2002)
2. Barabási, A.-L., Albert, R.: Emergence of scaling in random networks. Science 286, 509–512 (1999)
3. Kivac Oner, M., Garibay, I.I., Wu, A.S.: Mating Networks in Steady State Genetic Algorithms are Scale Free. In: Genetic and Evolutionary Computation Conference 2006 (GECCO 2006), pp. 1423–1424 (2006)
4. Giacobini, M., Tomassini, M., Tettamazi, A.: Takeover Time Curves in Random and Small-World Structured Populations. In: Genetic and Evolutionary Computation Conference (GECCO 2005), pp. 1333–1340 (2005)
5. Lieberman, E., Hauert, C., Nowak, A.: Evolutionary dynamics on graphs. Nature 433(20), 312–316 (2005)
6. Araseki, H.: Effectiveness of Scale-Free Properties in Genetic Programming. In: SCIS-ISIS 2012, pp. 285–289 (2012)
7. Dorogovtsev, S.N., Mendes, J.F.F.: Evolution of Networks: From Biological Nets to the Internet and WWW. Oxford University Press, Oxford (2003)
8. Ohkubo, J., Horiguchi, T.: Complex Networks by Non-growing Model with Preferential Rewiring Process. Journal of the Physical Society of Japan 74(4), 1334–1340 (2005)
9. Ohkubo, J., Tanaka, K., Horiguchi, T.: Generation of complex bipartite graphs by using a preferential rewiring process. Physical Review E (72), 036120 (2005)
10. Baiesi, M., Manna, S.: Scale-free networks from a Hamiltonian dynamics. Phys. Rev. E 68, 047103 (2003)
11. Fang, Y., Li, J.: A Review of Tournament Selection in Genetic Programming. In: Cai, Z., Hu, C., Kang, Z., Liu, Y. (eds.) ISICA 2010. LNCS, vol. 6382, pp. 181–192. Springer, Heidelberg (2010)
12. Blickle, T., Thicle, L.: A Mathematical Analysis of Tournament Selection. In: Proceedings of the 6th International Conference on Genetic Algorithms (ICGA 1995), pp. 9–16 (1995)
13. Bianconi, G., Barabási, A.-L.: Competition and multiscaling in evolving networks. Europhys. Lett. 54(4), 436–442 (2001)
14. Sato, H., Ono, I., Kobayashi, S.: A new generation alternation method of genetic algorithms and its assessment. J. of Japanese Society for Artificial Intelligence 12(5), 734–744 (1997)
15. Erdös, P., Rényi, A.: On the evolution of random graph. Publications of the Mathematical Institute of the Hungarian Academy of Sciences 5, 17 (1960)
16. Dignum, S., Poli, R.: Sub-tree Swapping Crossover and Arity Histogram Distributions. In: Esparcia-Alcázar, A.I., Ekárt, A., Silva, S., Dignum, S., Uyar, A.Ş. (eds.) EuroGP 2010. LNCS, vol. 6021, pp. 38–49. Springer, Heidelberg (2010)

17. Muntean, O., Diosan, L., Oltean, M.: Best SubTree Genetic Programming. In: Genetic and Evolutionary Computation Conference (GECCO 2007), pp. 1667–1673 (2007)
18. Punch, B., Zongker, D., Goodman, E.: The Royal Tree Problem, a Benchmark for Single and Multiple Population Genetic Programming. In: Advances in Genetic Programming, vol. 2, pp. 299–316. MIT Press (1996)
19. Koza, J.: Genetic Programming: On the Programming of Computers by Means of Natural Selection. MIT Press (1992)

Preliminary Study of Bloat in Genetic Programming with Behavior-Based Search

Leonardo Trujillo, Enrique Naredo, and Yuliana Martínez

Doctorado en Ciencias de la Ingeniería, Departamento de Ingeniería Eléctrica y Electrónica,
Instituto Tecnológico de Tijuana, Blvd. Industrial y Av. ITR Tijuana S/N,
Mesa Otay C.P. 22500, Tijuana B.C., México
leonardo.trujillo@tectijuana.edu.mx,
{enriquenaredo,ysaraimr}@gmail.com

Abstract. Bloat is one of the most interesting theoretical problems in genetic programming (GP), and one of the most important pragmatic limitations in the development of real-world GP solutions. Over the years, many theories regarding the causes of bloat have been proposed and a variety of bloat control methods have been developed. It seems that one of the underlying causes of bloat is the search for fitness; as the fitness-causes-bloat theory states, selective bias towards fitness seems to unavoidably lead the search towards programs with a large size. Intuitively, however, abandoning fitness does not appear to be an option. This paper, studies a GP system that does not require an explicit fitness function, instead it relies on behavior-based search, where programs are described by the behavior they exhibit and selective pressure is biased towards unique behaviors using the novelty search algorithm. Initial results are encouraging, the average program size of the evolving population does not increase with novelty search; i.e., bloat is avoided by focusing on novelty instead of quality.

Keywords: Bloat, Genetic Programming, Novelty Search.

1 Introduction

Genetic programming (GP) has shown to be an effective search method for automatic program induction, with noteworthy results in many domains [10]. Nonetheless, most GP practitioners will most likely have to overcome several issues in order to apply the paradigm successfully. For instance, GP has many degrees of freedom that require a proper initialization and parametrization, an issue for all evolutionary algorithms (EAs). However, probably the most studied GP problem, is the way in which solutions grow in size as the search progresses, what is known as bloat. Stated more precisely, bloat is excessive code growth within the individuals of the evolving population without a proportional improvement in fitness. Over the years, many bloat theories have been developed and many bloat control methods have been proposed [31, 32]. One of the most promising attempts at explaining the bloat phenomenon is the crossover bias theory [5, 28], that has lead to a powerful bloat control method called operator equalisation [32]. However, recent experimental results have blurred what is understood regarding the causes of bloat [29], as well as what are the best strategies that can be used to eliminate it from GP runs [7, 29].

M. Emmerich et al. (eds.), *EVOLVE - A Bridge between Probability, Set Oriented Numerics,*
and Evolutionary Computation IV, Advances in Intelligent Systems and Computing 227,
DOI: 10.1007/978-3-319-01128-8_19, © Springer International Publishing Switzerland 2013

This work revisits the fitness-causes-bloat theory of Langdon and Poli [13, 14, 31], that basically states that the search for better fitness will bias the search towards larger trees, simply because there are more large programs than small ones. Silva and Costa [31] state it clearly:

> ... one cannot help but notice the one thing that all the [bloat] theories have in common, the one thing that if removed would cause bloat to disappear, ironically the one thing that cannot be removed without rendering the whole process useless: the search for fitness.

This paper presents a preliminary study that supports the fitness-causes-bloat theory in GP. It is shown that, for the test problems presented here (supervised data classification), bloat can be avoided by abandoning an explicit fitness function. In other words, by not searching for fitness directly the bloating effect is eliminated. Moreover, the proposed approach is not *useless*, in fact it is quite competitive with standard fitness-based search. The proposal is to use a behavior-based search with GP applying the novelty search (NS) algorithm, substituting an explicit fitness function by an optimization criterion that is biased towards solutions that are unique, or novel, with respect to the rest of the population. Indeed, experimental results confirm that by eliminating an explicit fitness bias from the search then GP is not bloated, and the quality of the solutions is not compromised, particularly for hard problems.

The remainder of the paper proceeds as follows. Section 2 provides a brief overview on bloat, discussing recent theories and bloat control methods. The concept of behavior-based search and behavioral space is explored in Section 3, along with the NS algorithm. Section 4 describes a NS-based GP for data classification as a case study for bloat analysis. Afterwards, experimental results are presented and discussed in Section 5. Finally, a summary, conclusions and future work are given in Section 6.

2   Recent Advances in Bloat

Over the last twenty years, many theories have been put forth to explain bloat, attempting to understand why it happens, and to propose strategies to eliminate it. Many bloat control methods have been developed, focusing on modifying different aspects of the evolutionary process, such as the genetic operators, selection and survival strategies and fitness assignment. For a comprehensive review on previous theories and methods, the reader is referred to the work by Silva and Costa [31].

Currently, the most plausible theory for bloat is the crossover bias theory (CBT), proposed by Dignum and Poli [5, 28]. Focusing on standard Koza style GP with a tree representation [11], the CBT states that bloat is produced by the effect that subtree crossover has on the distribution of tree sizes in the population. While the average tree size is not affected by crossover, the distribution of tree sizes is modified. Subtree crossover produces a large number of small trees, and since small trees represent trivial individuals in most problems, their fitness values are usually very bad. Therefore, selection will favor larger trees, causing an increase in the average size of trees within the population, effectively bootstrapping the bloating effect. This theory seems very promising, and experimental data supports it [32]. Moreover, powerful bloat control methods have been developed based on CBT, namely operator equalisation [32]. However, recent studies have made the matter significantly less clear. For instance, Silva [29] suggests that some of the properties of operator equalisation that allow it to provide a bloat-free GP search, might not be consistent with the CBT. Harper [7] also has shown that operator equalisation is not the best bloat control method on some problems, and proposes other, more elaborate, strategies to eliminate bloat from GP runs. Finally, it is important to point out that even though operator equalisation can limit bloat for some problems, it is a computationally expensive algorithm, so developing other methods remains a worthwhile endeavor.

This work revisits another bloat theory, which can be called the fitness-causes-bloat theory (FCBT), developed by Langdon and Poli [13, 14]. The main arguments of FCBT proceed as follows [13, 14, 32]. For a variable length GP representation (such as a tree based GP) many genotypically different programs, of different sizes, can produce the same outputs on a given set of fitness cases. Therefore, all of these programs trees will be assigned the same fitness value. Then, because GP crossover tends to be a predominantly destructive search operator, when improved solutions are difficult to find then selection is biased towards offspring that have the same fitness as their parents. Since there are exponentially more large programs than small ones, an almost unavoidable tendency towards larger, or bloated, programs is present during a GP search.

This work attempts to shed some light on this matter. If bloat is a natural consequence of the search for better fitness, as Langdon and Poli, and Silva and Costa stated; then a natural bloat control strategy would be one where fitness is abandoned. However, the question then becomes: if fitness is not used to determine selection pressure, then what could be used in its stead? This question is not as strange as at first it may appear. Consider that EAs are inspired on biological Darwinian evolution, a natural process where an a priori purpose, or objective, is not present; i.e., there is not an explicit objective function in natural evolution. Natural evolution is an open-ended search process, where the search for fitness is not explicitly carried out, it is a natural consequence of the evolutionary dynamics induced by physical laws and chemical reactions. Inspired by nature, some researchers have also proposed open-ended EAs, dating back to the origins of the field [4], and other more recent examples [6, 9, 27]. In particular, this work studies the effect of bloat on one of the most recently developed open-ended EAs, the novelty search algorithm, where fitness is substituted by a measure of solution novelty. It is hypothesized that, if the FCBT is correct, then a NS-based GP would have to produce a significant reduction in bloat, or possibly a complete elimination of it.

3 Behavior-Based Search

The main goal of any EA, is to search for the solution that achieves the best possible performance; hence, a proper measure of performance needs to be proposed for each problem. In the GP case, performance is normally computed based on a set of fitness cases or training set of data. Traditional fitness-based approaches will provide a single measure that characterizes the performance of an individual; for instance, the mean error with respect to a desired output. This is a coarse view of a program's performance, usually averaging out performance variations that a program might have on different fitness cases. However, it is not the only possible alternative, below two other approaches are discussed: semantics and behaviors.

Semantics in GP describes the performance of a program with the raw output vector computed over all fitness cases [21, 1, 2, 12, 20, 36]. Given a set of n fitness cases, the semantics of a program K is the corresponding output vector it produces $\mathbf{y} \in \mathbb{R}^n$. In GP, many genetically (phenotypically) different programs can share the same semantic output. Therefore, semantics adds another space of analysis in which the search is being conducted, along with genotypic, phenotypic and objective (fitness) space, we can also consider semantic space; where a many-to-one relation will usually exists between genotypic (phenotypic) space and semantic space.

Researchers have used semantics to improve GP in different ways, such as modifying traditional genetic operators to improve the semantic diversity of the evolving population [1, 2, 36], or by explicitly performing evolution within semantic space [12, 21]. In general, all of these works have shown improved results using a canonical GP as a control method, mostly on symbolic regression problems.

However, strictly focusing on program outputs might not be the best approach in some domains. For example, consider the GP classifier based on static range selection (SRS) [38] (it will be further discussed in Section 4 and used in the experimental work), that functions as follows. For a two class problem and real-valued GP outputs, the SRS classifier is straightforward; if the program output for input pattern $\mathbf{x}$ is greater than zero then the pattern is labeled as belonging to class A, otherwise it is labeled as a class B pattern. In this case, while the semantic space description (as defined above) of two programs might be different (maybe substantially), they can still produce the same classification for the input pattern.

Now, consider the case of evolutionary robotics (ER). In ER, evolution is normally used to search for neuro-controllers for autonomous robots [26]. The goal is to find robust solutions with good performance, while introducing as little prior knowledge as possible into the fitness function, such that the search is performed based on a very high-level definition of the task which needs to be solved [25]. Therefore, in ER the correspondence between program inputs, outputs and induced actions is not straightforward. Moreover, in ER fitness evaluation can be performed within real or simulated environments, where noisy sensors and the physical coupling between actuators and the real world, can produce a non-injective or non-surjective relation between program output and robot actions.

Therefore, some researchers have turned towards explicitly considering behavioral space [22, 34]. In robotics, the concept of behaviors dates back to the seminal works of R. Brooks in behavior-based robotics [3]. A behavior is a description β of the way an agent K (program in the GP case) acts in response to a series of stimuli within a particular context $\mathscr{C}$. A context $\mathscr{C}$ includes the description that an agent has about its own internal state and the characteristics of the surrounding environment at a given moment in time. Stated another way, a behavior β is produced by the interactions of agent K, output $\mathbf{y}$ and context $\mathscr{C}$. In behavior-based robotics, behaviors are described at a very high level of abstraction by the system designer. Conversely, in ER some researchers have proposed domain-specific numerical descriptors that describe each behavior β, to explicitly consider behavioral space during evolution as another criterion that helps guide the search. The justification for this is evident, given that the objective function is stated at a high-level of abstraction, then population management should also consider the behavioral features of the evolved solutions, that characterize them based on a high-level description of their performance. Therefore, researchers have proposed diversity preservation techniques [34, 35] and open-ended search algorithms [15, 17]; Mouret and Doncieux [22] provide a comprehensive overview of previous works on behavioral evolution in ER.

In summary, a behavior should be understood as a higher-level description of program performance, compared to the semantics approach that employs a low-level description of performance. An individual's behavior is described in more general terms, accounting not only for program output, but also for the context in which the output was produced. For instance, for the SRS GP classifier described above, context is given by the SRS heuristic rule used to assign class label. Therefore, fitness, program semantics, and behavior can be understood as different levels of abstraction of the performance of a program. At one extreme, fitness provides a coarse look of performance; a single value (for each criteria) that attempts to capture a global evaluation. At the other end, semantics describe the performance of a program in great detail. On the other hand, behavioral descriptors move between fitness and semantics, providing a finer or coarser level of description, depending on how behaviors are meaningfully characterized within a particular domain.

3.1 Novelty Search

Following the behavior-based approach, Lehman and Stanley proposed the NS algorithm that eliminates an explicit objective function [15–17]. The search is not guided by a measure of *quality*, instead the selective pressure is provided by a measure of *uniqueness*. The strategy is to measure the amount of novelty each individual introduces into the search with respect to the progress the search has made at the moment at which the individual is created. Each solution is described by a domain dependent behavioral descriptor, where each individual is mapped to a single point in behavioral space, as described in the previous section.

A known limitation of fitness-based search is the tendency to converge and get trapped on local optima. A common solution to this problem is to incorporate niching or speciation techniques into an EA [19]. However, through the search for novelty diversity preservation introduces the sole selective pressure and can, in principle, avoid search stagnation.

In practice, NS uses a measure of local sparseness around each individual within behavioral space to estimate its novelty, considering the current population and novel solutions from previous generations. Therefore, the novelty measure is dynamic, since it can produce different results for the same individual depending on the population state and search progress. The proposed measure of sparseness ρ around each individual K described by its behavioral descriptor β, is given by

$$\rho(\beta) = \frac{1}{m} \sum_{i=0}^{m} dist(\beta, \alpha_i) \,, \tag{1}$$

where α_i is the ith-nearest neighbor in behavioral space of β with respect to the average distance $dist()$, which is a domain-dependent measure of behavioral difference between two descriptors. The number of neighbors m considered for the sparseness measure is an algorithm parameter. Given this definition, when the average distance is large, then the individual is within a sparse region of behavioral space, and it is in a dense region if the measure is small.

To compute sparseness, the original NS proposal is to consider the current population and an archive of novel individuals. An individual is added to the archive if its sparseness is above a minimal threshold ρ_{min}, the second parameter of the NS algorithm. The archive can help the algorithm avoid backtracking, however it can grow large in size, and make calculating sparseness a computational bottleneck. Finally, at the conclusion of the run after a fixed number of generations, the best solution based on fitness (it is the only moment in which NS uses an explicit fitness function) from both the final population and the novelty archive is returned as the best solution found by the search.

Since its proposal in [15], and later works [16, 17], most applications of NS have focused on robotics, such as mobile robot navigation [15–17], morphology design [18] and gait control [17]. Only until recently has NS been used in general pattern recognition problems, particularly supervised classification [24] and unsupervised clustering [23]. This paper will use the work in [24] as the case study to analyze bloat in a NS-based GP search.

4 Novelty Search for Supervised Classification

This section describes the GP system used in this work to evolve supervised classifiers, and how NS is incorporated into the evolutionary process.

4.1 Static Range Selection GP Classifier

This work uses the Static Range Selection GP Classifier (SRS) described by Zhang and Smart [38]. In a classification problem, a pattern $\mathbf{x} \in \mathbb{R}^p$ has to be classified as belonging to a single class from $\Omega = \{\omega_1, ..., \omega_M\}$, where each ω_i represents a distinct class label. Then, in a supervised learning approach the goal is to build a mapping $g(\mathbf{x}) : \mathbb{R}^p \to \Omega$, that assigns each pattern $\mathbf{x}$ to a corresponding class ω_i, where g is derived based on evidence provided by a training set $\mathcal{T}$ of N p-dimensional patterns with a known classification. In this work, only two-class classification problems are considered. In SRS, $\mathbb{R}$ is divided into M non-overlapping regions, one for each class. Then, GP evolves a mapping $g(\mathbf{x}) : \mathbb{R}^p \to \mathbb{R}$, such that the region in $\mathbb{R}$ where pattern $\mathbf{x}$ is mapped to, determines the class to which it belongs. For a two-class problem, if $g(\mathbf{x}) > 0$ then $\mathbf{x}$ belongs to class ω_1, and belongs to ω_2 otherwise. The fitness function is simple, it consists on minimizing the classification error of g.

4.2 Novelty Search Extension of SRS

As stated above, to apply NS with SRS the fitness function is substituted by the sparseness measure of Equation 1. Therefore, a proper domain specific behavioral descriptor must be proposed [8].

Accuracy Descriptor β^A: The training set $\mathcal{T}$ used by SRS-GPC contains sample patterns from each class. For a two-class problem with $\Omega = \{\omega_1, \omega_2\}$, If $\mathcal{T} = \{y_1, y_2, ..., y_L\}$, then the behavioral descriptor for each GP classifier K_i is a binary vector $\beta^{A_i} = (\beta_1, \beta_2, ..., \beta_L)$ of size L, where each vector element β_j is set to 1 if classifier K_i correctly classifies sample y_j, and is set to 0 otherwise. In [24] an analysis on the β^A descriptor is given, considering the fitness landscape it produces.

Finally, given the proposed binary descriptors, a natural $dist()$ function for Equation 1 is the Hamming distance, that counts the number of bits that differ between two binary vectors. This similarity measure has been used to measure behavioral diversity in ER [22]. It is noteworthy to point out that [24] reports good performance on several synthetic classification problems. In particular, the authors report a clear trend, the performance of NS-based GP improves, relative to a control method, as problem difficulty increases; i.e., NS performs comparatively better on hard problems than on easy ones. The explanation for this observation is that, for easy problems random solutions can perform quite well. Therefore, the selective pressure provided by NS will not necessarily lead the search towards better solution in behavioral space, in fact the opposite might happen. Conversely, for difficult problems random solutions for a two-class problem will roughly produce 50% accuracy. Therefore, the diversity exploration of NS will have to lead the search towards better regions in the search space. In other words, when problem difficulty increases the search for novelty can lead towards quality.

5   Experiments and Results

The goal of the experimental work is to test the NS-GP classifier on two-class classification problems, evaluating its performance based on classification error and the mean size of the evolved population, a good indicator of the effect bloating is having on a GP run [37]. The proposed algorithm will hereafter be denoted by NS-SRS. For comparative proposes, the basic SRS classifier is used as a control method.

Gaussian Mixture Models are used to generate five random synthetic problems, each with different amounts of class overlap and geometry. All problems are set in the $\mathbb{R}^2$ plane with $x, y \in [-10, 10]$, and 200 sample points were randomly generated for each class. The parameters for the GMM of each class were also randomly chosen, following the same strategy reported in [33]. The five problems are of increasing difficulty, denoted as: *Trivial*; *Easy*; *Moderate*; *Hard*; and *Hardest*; these problems are graphically depicted in Figure 1.

The parameters of both GP systems are given in Table 1. Moreover, for NS-SRS the NS parameters are set as follows: (1) the number of neighbors considered for sparseness computation is set to $m = 15$; and (2) the sparseness threshold is set to $\rho_{min} = 40$. Finally, all algorithms were coded using Matlab 2009a and the GPLAB toolbox [30].

Both algorithms are executed 30 times and performance is analyzed based on averages over all runs. For each problem, 200 sample points are created for each class, and 70% of the data is used for training and the rest for testing. In each run data partition is done independently at random. First, Table 2 presents the average classification error on the test data. The results are consistent with those reported in [24], NS-SRS performs well, with small performance differences compared to SRS. However, there is a trend,

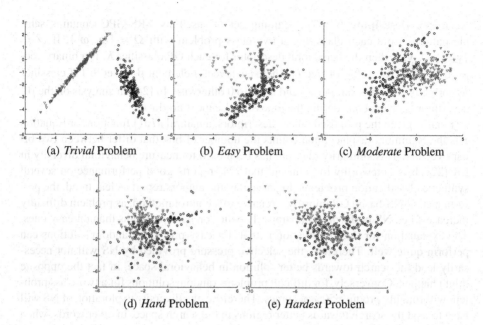

(a) *Trivial* Problem   (b) *Easy* Problem   (c) *Moderate* Problem

(d) *Hard* Problem   (e) *Hardest* Problem

Fig. 1. Five synthetic 2-class problems used to evaluate each algorithm in ascending order of difficulty from left to right

Table 1. Parameters for the GP systems

Parameter	Description
Population size	100 individuals.
Generations	100 generations.
Initialization	Ramped Half-and-Half, with 6 levels of maximum depth.
Operator probabilities	Crossover $p_c = 0.8$, Mutation $p_\mu = 0.2$.
Function set	$\left\{ +, -, \times, \div, \lvert \cdot \rvert, x^2, \sqrt{x}, log, sin, cos, if \right\}$.
Terminal set	$\{x_1, ..., x_i, ..., x_p\}$, where x_i is a dimension of the data patterns $\mathbf{x} \in \mathbb{R}^n$.
Hard maximum depth	20 levels.
Selection	Tournament of size 4.

NS performs well on the more difficult problems, and worse on the easier ones. These results are similar to those reported in [24, 23], with similar conclusions. Basically, the explorative search performed by NS is fully exploited when random initial solution perform badly, under these conditions the search for novelty can lead towards better solutions. Conversely, for easy problems, random solution can perform quite well, thus the search for novelty can lead the search towards solutions with undesirable performance. However, this is not necessarily a limitation for NS-based systems, since for most real-world scenarios difficult problems should be expected.

Therefore, regarding fitness we can say that NS achieves basically equivalent performance compared with fitness-based search on the more interesting problems. Figures 2 and 3, on the other hand, show how the average size of the population evolves

Table 2. Average classification error and standard error of the best solution found by each algorithm

Problem	SRS Ave.	SRS Std.	NS-SRS Ave.	NS-SRS Std.
Trivial	0.004	0.007	0.007	0.008
Easy	0.105	0.040	0.144	0.044
Moderate	0.136	0.033	0.159	0.041
Hard	0.260	0.052	0.266	0.053
Hardest	0.365	0.033	0.370	0.043

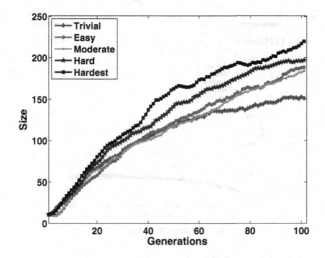

Fig. 2. Evolution of tree size for SRS on each problem; curves represent the average size over thirty runs

across generations. Consider SRS, Figure 2 shows typical GP behavior, with a clear tendency of code growth across generations; i.e., SRS bloats like any fitness-based GP search. Second, in the case of NS-SRS, code growth is analyzed from three different perspectives. First, considering how the average size of the population grows across generations, shown in Figure 3(a). Second, considering how the average size of the individuals in the archive grows, shown in Figure 3(b). Finally, when both the archive and the population are considered concurrently in each generation, this is shown in Figure 3(c). In all cases, it is clear that NS controls code growth quite effectively. A numerical comparison is given in Table 3, with the average population size in the final generation. It is clear that program size is considerably larger with fitness-based search, without the quality of the results being compromised. There are some additional observations that are of interest in the results of Table 3. It seems that standard GP bloats more as problem difficulty increases, which is coherent with the FCBT. On the other hand, NS shows the same average program size for all problems. Moreover, average program size in the archive is slightly smaller than the average program size in the population; i.e., archived individuals are smaller than the population average. This is noteworthy, because it suggests that novel individuals tend to be smaller, an unexpected result.

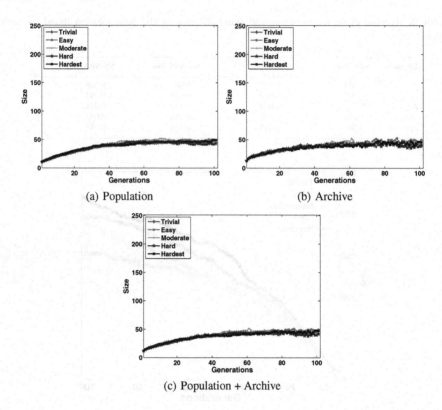

(a) Population

(b) Archive

(c) Population + Archive

Fig. 3. Evolution of tree size for NS-SRS on each problem; curves represent the average size over thirty runs. (a) Considers individuals in the population; (b) considers individuals in the archive; and (c) considers all of the individuals at each generation.

Table 3. Average program size in the final generation for each algorithm. For the NS algorithm, the population (Pop), archive (Arc) and both (Pop+Arc) are considered.

Problem	SRS	NS-SRS Pop	NS-SRS Arc	NS-SRS Pop+Arc
Trivial	151.3	46.17	38.57	42.37
Easy	188.5	49.25	49.89	49.57
Moderate	184	47.04	42.54	44.79
Hard	197	42.93	40.17	41.55
Hardest	220.1	49.05	45.52	47.29

6   Summary and Conclusions

Bloat is one of the main research topics in modern GP literature, and several theories have been developed that attempt to explain its underlying causes. Moreover, given the detrimental effect that bloat has on GP runs, a variety of bloat control methods have been proposed. Nonetheless, a complete explanation and a general effective strategy have not been devised. A general agreement, however, is that the search for fitness is the one core element that leads towards bloated populations in GP; which is explicitly stated by the fitness-causes-bloat theory. If this is so, then bloat would seem to be unavoidable, since fitness based is a requirement of most evolutionary algorithms.

This work, on the other hand, studies bloat in a GP system where fitness is not considered explicitly. The proposal is to evolve GP programs using novelty search, where fitness is substituted my a measure of program uniqueness. Initial results are indeed encouraging, it seems that bloat is eliminated, or at least controlled, by the evolutionary dynamics induced by NS. Experimental tests seem to confirm the FCBT, by showing that when an explicit fitness is abandoned then the evolving population will not bloat. However, these results are preliminary and much work still needs to be done. A complete and comprehensive study on bloat and behavior-based search is still necessary, but this work provides a first approximation at a plausible approach for bloat-free GP.

Acknowledgments. Funding for this work provided by CONACYT (Mexico) Basic Science Research Project No. 178323 and DGEST (Mexico) Research Project No. TIJ-ING-2012-110. Second and third author are supported by CONACYT (Mexico) scholarships, respectively scholarships No. 232288 and No. 226981.

References

1. Beadle, L., Johnson, C.: Semantically driven crossover in genetic programming. In: Proceedings of the Tenth Conference on Congress on Evolutionary Computation (IEEE World Congress on Computational Intelligence) CEC 2008, pp. 111–116. IEEE Press (2008)
2. Beadle, L., John'son, C.G.: Semantically driven mutation in genetic programming. In: Proceedings of the Eleventh Conference on Congress on Evolutionary Computation, CEC 2009, pp. 1336–1342. IEEE Press (2009)
3. Brooks, R.A.: Cambrian intelligence: the early history of the new AI. MIT Press, Cambridge (1999)
4. Dawkins, R.: Climbing Mount Improbable. W.W. Norton & Company (1996)
5. Dignum, S., Poli, R.: Generalisation of the limiting distribution of program sizes in tree-based genetic programming and analysis of its effects on bloat. In: Proceedings of the 9th Annual Conference on Genetic and Evolutionary Computation, GECCO 2007, pp. 1588–1595. ACM, New York (2007)
6. García-Valdez, M., Trujillo, L., de Vega, F.F., Merelo Guervós, J.J., Olague, G.: EvoSpace-interactive: A framework to develop distributed collaborative-interactive evolutionary algorithms for artistic design. In: Machado, P., McDermott, J., Carballal, A. (eds.) EvoMUSART 2013. LNCS, vol. 7834, pp. 121–132. Springer, Heidelberg (2013)
7. Harper, R.: Spatial co-evolution: quicker, fitter and less bloated. In: Proceedings of the Fourteenth International Conference on Genetic and Evolutionary Computation Conference, GECCO 2012, pp. 759–766. ACM, New York (2012)
8. Kistemaker, S., Whiteson, S.: Critical factors in the performance of novelty search. In: Proceedings of the 13th Annual Conference on Genetic and Evolutionary Computation, GECCO 2011, pp. 965–972. ACM (2011)
9. Kowaliw, T., Dorin, A., McCormack, J.: Promoting creative design in interactive evolutionary computation. IEEE Transactions on Evolutionary Computation 16(4), 523–536 (2012)
10. Koza, J.: Human-competitive results produced by genetic programming. Genetic Programming and Evolvable Machines 11(3), 251–284 (2010)

11. Koza, J.R.: Genetic programming: on the programming of computers by means of natural selection. MIT Press, Cambridge (1992)
12. Krawiec, K., Pawlak, T.: Locally geometric semantic crossover: a study on the roles of semantics and homology in recombination operators. Genetic Programming and Evolvable Machines 14(1), 31–63 (2013)
13. Langdon, W.B., Poli, R.: Fitness causes bloat. In: Proceedings of the Second On-line World Conference on Soft Computing in Engineering Design and Manufacturing, pp. 13–22. Springer (1997)
14. Langdon, W.B., Poli, R.: Fitness causes bloat: Mutation. In: Banzhaf, W., Poli, R., Schoenauer, M., Fogarty, T.C. (eds.) EuroGP 1998. LNCS, vol. 1391, pp. 37–48. Springer, Heidelberg (1998)
15. Lehman, J., Stanley, K.O.: Exploiting open-endedness to solve problems through the search for novelty. In: Proceedings of the Eleventh International Conference on Artificial Life, ALIFE XI. MIT Press, Cambridge (2008)
16. Lehman, J., Stanley, K.O.: Efficiently evolving programs through the search for novelty. In: Proceedings of the 12th Annual Conference on Genetic and Evolutionary Computation, GECCO 2010, pp. 837–844. ACM (2010)
17. Lehman, J., Stanley, K.O.: Abandoning objectives: Evolution through the search for novelty alone. Evol. Comput. 19(2), 189–223 (2011)
18. Lehman, J., Stanley, K.O.: Evolving a diversity of virtual creatures through novelty search and local competition. In: Proceedings of the 13th Annual Conference on Genetic and Evolutionary Computation, GECCO 2011, pp. 211–218. ACM (2011)
19. Mahfoud, S.W.: Niching methods for genetic algorithms. PhD thesis, Champaign, IL, USA, UMI Order No. GAX95-43663 (1995)
20. McPhee, N.F., Ohs, B., Hutchison, T.: Semantic building blocks in genetic programming. In: O'Neill, M., Vanneschi, L., Gustafson, S., Esparcia Alcázar, A.I., De Falco, I., Della Cioppa, A., Tarantino, E. (eds.) EuroGP 2008. LNCS, vol. 4971, pp. 134–145. Springer, Heidelberg (2008)
21. Moraglio, A., Krawiec, K., Johnson, C.G.: Geometric semantic genetic programming. In: Coello, C.A.C., Cutello, V., Deb, K., Forrest, S., Nicosia, G., Pavone, M. (eds.) PPSN 2012, Part I. LNCS, vol. 7491, pp. 21–31. Springer, Heidelberg (2012)
22. Mouret, J.B., Doncieux, S.: Encouraging behavioral diversity in evolutionary robotics: An empirical study. Evol. Comput. 20(1), 91–133 (2012)
23. Naredo, E., Trujillo: Searching for novel clustering programs. To appear in Proceeding from the Genetic and Evolutionary Computation Conference, GECCO 2013. ACM (2013)
24. Naredo, E., Trujillo, L., Martínez, Y.: Searching for novel classifiers. In: Krawiec, K., Moraglio, A., Hu, T., Etaner-Uyar, A.Ş., Hu, B. (eds.) EuroGP 2013. LNCS, vol. 7831, pp. 145–156. Springer, Heidelberg (2013)
25. Nelson, A.L., Barlow, G.J., Doitsidis, L.: Fitness functions in evolutionary robotics: A survey and analysis. Robot. Auton. Syst. 57(4), 345–370 (2009)
26. Nolfi, S., Floreano, D.: Evolutionary Robotics: The Biology, Intelligence, and Technology. MIT Press, Cambridge (2000)
27. Ofria, C., Wilke, C.O.: Avida: a software platform for research in computational evolutionary biology. Artif. Life 10(2), 191–229 (2004)
28. Poli, R., Langdon, W.B., Dignum, S.: On the limiting distribution of program sizes in tree-based genetic programming. In: Ebner, M., O'Neill, M., Ekárt, A., Vanneschi, L., Esparcia-Alcázar, A.I. (eds.) EuroGP 2007. LNCS, vol. 4445, pp. 193–204. Springer, Heidelberg (2007)
29. Silva, S.: Reassembling operator equalisation: a secret revealed. In: Proceedings of the 13th Annual Conference on Genetic and Evolutionary Computation, GECCO 2011, pp. 1395–1402. ACM, New York (2011)

30. Silva, S., Almeida, J.: Gplab–a genetic programming toolbox for matlab. In: Gregersen, L. (ed.) Proceedings of the Nordic MATLAB Conference, pp. 273–278 (2003)
31. Silva, S., Costa, E.: Dynamic limits for bloat control in genetic programming and a review of past and current bloat theories. Genetic Programming and Evolvable Machines 10(2), 141–179 (2009)
32. Silva, S., Dignum, S., Vanneschi, L.: Operator equalisation for bloat free genetic programming and a survey of bloat control methods. Genetic Programming and Evolvable Machines 13(2), 197–238 (2012)
33. Trujillo, L., Martínez, Y., Galván-López, E., Legrand, P.: Predicting problem difficulty for genetic programming applied to data classification. In: Proceedings of the 13th Annual Conference on Genetic and Evolutionary Computation, GECCO 2011, pp. 1355–1362. ACM, New York (2011)
34. Trujillo, L., Olague, G., Lutton, E., de Vega, F.F.: Discovering several robot behaviors through speciation. In: Giacobini, M., Brabazon, A., Cagnoni, S., Di Caro, G.A., Drechsler, R., Ekárt, A., Esparcia-Alcázar, A.I., Farooq, M., Fink, A., McCormack, J., O'Neill, M., Romero, J., Rothlauf, F., Squillero, G., Uyar, A.Ş., Yang, S. (eds.) EvoWorkshops 2008. LNCS, vol. 4974, pp. 164–174. Springer, Heidelberg (2008)
35. Trujillo, L., Olague, G., Lutton, E., de Vega, F.F., Dozal, L., Clemente, E.: Speciation in behavioral space for evolutionary robotics. Journal of Intelligent & Robotic Systems 64(3-4), 323–351 (2011)
36. Uy, N.Q., Hoai, N.X., O'Neill, M., Mckay, R.I., Galván-López, E.: Semantically-based crossover in genetic programming: application to real-valued symbolic regression. Genetic Programming and Evolvable Machines 12(2), 91–119 (2011)
37. Vanneschi, L., Castelli, M., Silva, S.: Measuring bloat, overfitting and functional complexity in genetic programming. In: Proceedings of the 12th Annual Conference on Genetic and Evolutionary Computation, GECCO 2010, pp. 877–884. ACM, New York (2010)
38. Zhang, M., Smart, W.: Using gaussian distribution to construct fitness functions in genetic programming for multiclass object classification. Pattern Recogn. Lett. 27(11), 1266–1274 (2006)

Evidence Theory Based Multidisciplinary Robust Optimization for Micro Mars Entry Probe Design

Liqiang Hou[1,2], Yuanli Cai[1], Rongzhi Zhang[2], and Jisheng Li[1]

[1] Xi'an Jiaotong University, Xi'an, ShaanXi 710049, China
houliqiang2008@139.com
[2] State Key Laboratory of Astronautic Dynamics, Xi'an, Shaanxi 710043, China

Abstract. A robust multi-fidelity optimization method for micro Mars entry probe design is presented in this paper. In the robust design, aerodynamic and atmospheric model are assumed to be affected by epistemic uncertainties (partial or complete lack of knowledge). Evidence Theory is employed to quantify the uncertainties, and formulate the robust design into a multi-objective optimization problem. The optimization objectives are set to minimize interior temperature of Thermal Protection Systems (TPS), while maximize its belief value under uncertainty. A population based Multi-objective Estimation of Distribution Algorithm (MOEDA) is designed for searching robust Pareto front. In this algorithm, affinitive propagation clustering method divides adaptively the population into clusters. In each cluster, local Principal Component Analysis (PCA) is adopted for estimation of distribution, and reproducing individuals. Variable-fidelity aerodynamic model management is integrated into the robust optimizations. The fidelity management model uses analytical aerodynamic model first to initialize the optimization searching direction. With the development of the optimization, more data from high-accuracy model (CFD) are put into aerodynamic database. Artificial Neural Network (ANN) based surrogate model is used for reducing the computational cost. Finally, an application of the proposed optimization strategy for a micro probe with diameter no more than 0.8 meter is presented.

Keywords: robust optimization, multidisciplinary design, Mars reentry probe, evidence theory.

1 Introduction

When designing micro Mars entry probe, challenges for the designers are complicated by many factors. Atmospheric pressure on Mars is approximately 1% of that on Earth and varies about ±15% during the year due to condensation and sublimation of its primarily $CO2$ atmosphere. Temperature on the Mars surface might be cold enough that carbon dioxide freezes during the winter and "snows" onto the polar cap. Therefore, uncertain impacts of these factors should not be

M. Emmerich et al. (eds.), *EVOLVE - A Bridge between Probability, Set Oriented Numerics,* 307
and Evolutionary Computation IV, Advances in Intelligent Systems and Computing 227,
DOI: 10.1007/978-3-319-01128-8_20, © Springer International Publishing Switzerland 2013

neglected during the mission design. Particularly for micro Mars entry probe, little change in the parameters may lead some unexpected impacts. Therefore a robust design optimization method taking uncertainties into account is required.

Numerous examples of MDO applications have been found in many areas. [1–3]. Roshanian proposed an integrated approach for multi-disciplinary design of launch vehicle , using response surface method for approximation of the propulsion model[4]. Huang proposed a multi-objective Pareto concurrent subspace optimization method for multidisciplinary design [5]. To reduce computational cost, some researchers introduced variable-fidelity model management into MDO ,particularly for the reentry vehicle and aircraft design [6, 7]. Taking int account uncertainty impacts, Mueller and Joseph B. proposed a robust optimization method for collision avoidance maneuver planning [8]. Lantoine G. proposed a hybrid differential dynamic programming algorithm for robust low-thrust optimization [9]. Vasile M. and Croisard N. proposed a robust mission design method through evidence theory [10, 11].With Evidence Theory (Dempster-Shafer's theory), both aleatory and epistemic uncertainties, coming from a poor or incomplete knowledge of the model parameters, can be correctly modeled [12]. The values of uncertain parameters are expressed by means of intervals with associated probabilities. In particular, the value of belief expresses the lower probability that the selected design point remains optimal (and feasible) even under uncertainties.

In this paper, a population based Robust Optimization (RO) method for micro Mars entry probe design is proposed. The optimization objective is to minimize the interior temperature of TPS, while at same time maximize its belief value. An adaptive clustering MOEDA is developed to implement the multi-objective optimization. In this algorithm, population is clustered using Affinity Propagation method. No predefined number of the clusters is required. In each cluster, Principal Component Analysis (PCA) technique is performed for estimating population distribution, sampling and reproducing individuals. Nondominated individuals are sorted and selected through the NSGA-II-like selection procedure.Multi-fidelity aerodynamic models are integrated into the robust optimization procedure. Variable-fidelity model management is conducted through Artificial Neural Network (ANN) surrogate model. When the optimization goes close to the optima, more data from high accuracy model are put into the aerodynamic database, making the optimization procedure converges to optima quickly while keeping high level accuracy.

The rest of the paper is organized as follows: Section 2 describes dynamic models involved in this paper, including the entry dynamic, aerodynamic and thermal dynamic models. In section 3, uncertainty impacts are analyzed and modeled through Evidence Theory. The multi-objective Robust optimization method is presented in section 4. The Population based MOEDA and multi-fidelity model management strategy are proposed to solve the robust multi-fidelity design problem. Numerical simulation of micro probe (no more than 0.8 m in diameter) design under uncertainty is given in section 5, and conclusions are presented in section 6.

2   Dynamic Models

2.1   Entry Dynamic Equations

In planet-centered frame, the entry dynamic equations with state variable $[r, \lambda, \phi, v, \theta_p, \xi]^T$ are (see [13])

$$
\begin{cases}
\dot{r} = v \sin{(\theta)_p} \\
\dot{\lambda} = v \frac{\cos \theta_p \cos \xi}{r \cos \phi} \\
\dot{\phi} = \frac{v \cos \theta_p sin \xi}{r} \\
\dot{v} = -\frac{D}{m} - g \sin \theta_p \\
\dot{\theta}_p = \frac{L}{mv} \cos \gamma_v - (\frac{g}{v} - \frac{v}{r}) \cos \theta_p \\
\dot{\xi} = \frac{L}{mv \cos \theta_p} \sin \gamma_v - \frac{v}{r} \cos \theta_p \cos \xi \tan \phi
\end{cases}
\tag{1}
$$

where r, λ and ϕ are the distance from the center of the planet to the vehicle, longitude and latitude respectively; θ_p denotes flight path angle , and ξ denotes velocity azimuth angle. Drag D and lift L in the Eq.(1) are given by

$$
\begin{cases}
D = \frac{1}{2} \rho(h) S C_d v^2 \\
L = \frac{1}{2} \rho(h) S C_l v^2
\end{cases}
\tag{2}
$$

Given the initial condition, state variables at each point of the trajectory can be obtained by integrating Eq.1. The aeroshell of the probe is defined by a set of geometric parameters (see Fig.1). The parameters consist of radius of nose R_n, diameter R_b, and semi-apex angle θ. Leeward side of the body is a semisphere.

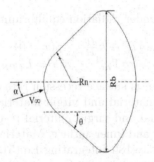

Fig. 1. Geometric definition of the micro Mars probe

With Modified Newtonian Theory, drag coefficient and lift coefficient can be obtained by [14]

$$
\begin{cases}
C_d = C_d(C_{pt2}, \alpha, \gamma, R_n, \theta, R_b) \\
C_l = C_l(C_{pt2}, \alpha, \gamma, R_n, \theta, R_b)
\end{cases}
\tag{3}
$$

where α is angle of attack (AoA), γ is specific heat rate, and pressure coefficient C_{pt2} can be computed with

$$C_{pt2} = \frac{2}{\gamma}(\frac{\gamma+1}{2})\frac{\gamma}{\gamma-1}\left(\frac{\gamma+1}{2\gamma-\frac{\gamma-1}{M_\infty^2}}\right)^{\frac{1}{\gamma-1}} - \frac{2}{\gamma M_\infty^2} \qquad (4)$$

where M_∞ is Mach number.

Suppose that $M_\infty \gg 1.0$, lift coefficients and drag coefficient can be obtained with Eq.(3) and Eq.(4). More accurate data taking into all effects account can be from numerical CFD model. In this paper, the commercial CFD software Numeca® is used to perform high accuracy computation of Reynolds Averaged Navier-Stokes equations. However, the costly computation makes it impractical to rely exclusively on the high-fidelity CFD model for design optimization.

2.2   TPS Model

The thermal protection system (TPS) is composed of SLA-561V, a material widely used in space engineering for thermal protection, and has been used as the primary TPS material on all the sphere-cone Mars entry vehicles sent by NASA. Equation for computing heating flux on the TPS surface is given by[14, 15]

$$\dot{q} = 1.89^{-8}\sqrt{\rho/R_n}v^3 \qquad (5)$$

Suppose that the heat transfer occurs only in one direction and deep into the TPS layer, then the one dimensional heating transfer equation is given by[16, 17]

$$\begin{cases} \frac{\partial}{\partial x}\left(k_c\frac{\partial T}{\partial x}\right) + \dot{m}_p c_p \frac{\partial T}{\partial x} = \rho_c c_p \frac{\partial T}{\partial t} \\ \frac{\partial}{\partial x}\left(k_v\frac{\partial T}{\partial x}\right) = \rho_v c_p \frac{\partial T}{\partial t} \end{cases} \qquad (6)$$

with boundary conditions under radiation equilibrium condition:

$$\begin{cases} \dot{q} = \varepsilon\sigma T_w^4 + k\frac{dT}{dx} & (x=0) \\ k\frac{dT}{dx} = \varepsilon\sigma T_{in}^4 & (x=L_{TPS}) \end{cases} \qquad (7)$$

where c_p is heat capacity, $\dot{m}$ is surface recession rate of TPS materials, ρ_c and ρ_v denote density of char material and virgin material; k_c and k_v are thermal conductivity of char material and virgin material respectively; T_w and T_{in} denote temperatures at outer and inner surface. Material properties of SLA-561V are taken from TPSX website[18]. Integrating Eq.(5), Eq.(6) and Eq.(7), temperature distribution of TPS can be obtained with finite-difference method.

3   Uncertain Impacts and ET Modeling

3.1   Uncertain Impacts

Uncertainties in micro entry vehicle design are from different sources, and can be varied greatly with different formulation modeling. To accurately describe the impacts is difficult, and need large amount of experimental data. Fig. 2 illustrate a case of uncertain impacts of the aerodynamic coefficients. The error model is set to be a Gaussian model, with its magnitude is set to be varied in a linear way. Eq.(8) shows how the uncertainties are modeled

$$\begin{cases} L = L_d + Err(\alpha, M_\infty)C_e(\alpha, M_\infty)L_d \\ D = D_d + Err(\alpha, M_\infty)C_e(\alpha, M_\infty)D_d \end{cases} \tag{8}$$

where $Err(\alpha, M_\infty)$ is a two dimensional Gaussian distribution model, C_e is a linear function related to α and M_∞ whose magnitude is set to be from zero at altitude of 75km to 0.40 at altitude 15km.

Fig.2 shows corresponding probability density distribution of TPS interior temperature and their maximum likelihood estimates (MLEs). The initial entry conditions are same with Mars probe "Spirit", with geometric parameters $R_n = 0.1m, R_b = 0.8m$, $\theta = 35^0$, and $L_{TPS} = 1.4cm$.

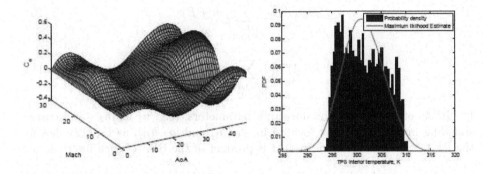

Fig. 2. Uncertain impacts of TPS interior temperature

If uncertainties can be expressed by probability distribution, expected value and variance of the system response can then be obtained with sampling techniques. Multi-disciplinary robust optimization can be implemented in a conventional way to minimize the variance and expected value of TPS temperature. But because of absences of knowledge, such probability distribution can not be given explicitly. Therefore, an uncertainty modeling technique without probability distribution is required.

3.2   Evidence Theory Based Uncertainty Modeling

In this section, an Evidence Theory based uncertainty modeling technique is presented. In this method, as opposed to a single value of probability, bounds for uncertainty quantification are used instead. Propagation of the information is through Basic Probability Assignment (BPA) [10–12]. The total degree of belief in a proposition A is expressed within a bound $[Bel(A), Pl(A)]$ lies in the unit interval [0, 1]. $Bel(A)$ is obtained by the accumulation of BPAs of the propositions that imply proposition A, whereas $Pl(A)$ is the plausibility calculated by adding the BPAs of propositions that imply or could imply the proposition A.

In Evidence Theory, the belief of an uncertain parameter $u \in [a, b]$ is an elementary proposition. The level of confidence an expert has on an elementary proposition is quantified using the Basic Probability Assignment (BPA). The BPA satisfies following rules

$$\begin{cases} m(E) \geq 0, \forall E \in \mathbf{U} \\ m(\emptyset) = 0, \\ \sum_{E \in \mathbf{U}} m(E) = 1 \end{cases} \tag{9}$$

A focal element (FE) is an element of U that has a non-zero BPA. The Belief (Bel) and Plausibility (Pl) functions are [11, 12]:

$$Bel(A) = \sum_{\substack{FE \subset A \\ FE \subset U}} m(FE) \tag{10}$$

$$Pl(A) = \sum_{\substack{FE \cap A \neq \emptyset \\ FE \subset U}} m(FE) \tag{11}$$

For BPAs of more than one uncertain parameters, (e.g. u_1 and u_2, and corresponding intervals they are located in, $[a_1, b_1]$ and $[a_2, b_2]$), as Eq.(12) shown, the BPA of given Cartesian product is product of the BPA of each interval, i.e.

$$m((u_1, u_2) \in [a_1, b_1] \times [a_2, b_2]) = m(u_1 \in [a_1, b_1]) \times m(u_2 \in [a_2, b_2]) \tag{12}$$

Using Evidence Theory for robust engineering design was proposed in 2002 by Oberkampf et al.[19], and was recently applied to robust design of space systems [20, 12] and space trajectory design[11]. Using Evidence Theory, with uncertain parameters $\mathbf{u} = [u_1, u_2, ..., u_m] \in \mathbf{U}$, and design variables $\mathbf{d} = [d_1, d_2, ..., d_n] \in \mathbf{D}$, the robust optimization can be formulated as

$$\begin{cases} \max_{v \in \mathbf{R}, \mathbf{d} \in \mathbf{D}} Bel(f(\mathbf{d}, \mathbf{u}) < v) \\ \min_{v \in \mathbf{R}, \mathbf{d} \in \mathbf{D}} v \end{cases} \tag{13}$$

where v is the threshold to be minimized, and $f(\mathbf{d}, \mathbf{u})$ is the main objective function. Based on this idea, M.Vasile et al. proposed three approaches to solve the OUU (Optimization under Uncertainty) problem: evolutionary multi-objective approach, step technique and clustering approximation method [11]. These approaches are applied to BepiColombo preliminary mission design, and minimize wet mass of the spacecraft taking into account uncertainties. In the following sections, a new multi-objective optimization method is proposed for the robust optimization problem.

4   Multi-fidelity Robust Optimization

4.1   Multi-objective Optimization with Adaptive Clustering Density Estimator

As Eq.(13) shown, robust design optimization can be formulated as a multi-objective optimization problem. In this section, a population based MOEDA is proposed. In this method, individuals are divided into clusters. New individuals in each cluster are reproduced based on the density estimator. In the estimator, local principle component analysis is used for modeling distribution of the individuals. The method is organized as follows:

First, a group of individuals are sampled and preparation for density estimator is implemented.

Step 2, group the individuals by their characteristics. The characteristic is categorized using a method with affinity propagation technique.

Step 3, in each cluster, produce new individuals for selection with density estimator.

Step 4, select individuals with a selection strategy similar to NSGA-II multi-objective optimization method.

1. Clustering Using Affinity Propagation
In the first step, the entire population is divided into clusters. It is easier to identify stochastic models for each cluster separately than for the whole data set. There are several ways to cluster a data set. The popular k-centers clustering technique begins with a set of randomly selected exemplars and iteratively refines this set so as to decreases the sum of squared errors. Zhang proposed using local PCA for partitioning population in EDA [21]. Both the algorithms have the number of clusters predefined before performing the data partitioning.

In this paper, a new clustering technique, affinity propagation technique is used instead [22]. In this algorithm, affinity propagation takes as input a real number $s(k, k)$,"preference", for each data point. Data points with larger values of $s(k, k)$ are more likely to be chosen as exemplar. Real-valued messages are exchanged between data points until a high quality centers and corresponding clusters gradually emerges. The "responsibility" $r(i, k)$, sent from data point i to candidate exemplar point k, reflects the accumulated evidence for how well-suited point k is to serve as exemplar for point i. The "availability"$a(i, k)$ sent from candidate exemplar point k to point i reflects the accumulated evidenced for how appropriate it would be for point i to choose k as its exemplar. Fig.7 shows how the messages of "responsibilities" and "availabilities" work.

314   L. Hou et al.

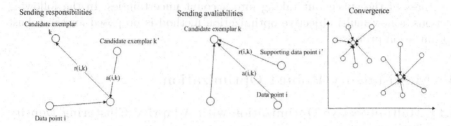

Fig. 3. Messages of "responsibilities" and "availability"

One advantage of affinity propagation clustering method is that the number of clusters need not be specified beforehand. Instead, the appropriate number of clusters will be determined from the message passing and depends on the input exemplar preference. In simulation, the initial preference is set to be the median of the input similarities, which is set to be a negative squared error. For point i and k , the similarity is

$$s(i, k) = -\|x_i - x_k\|^2 \tag{14}$$

The responsibilities are computed according to

$$r(i, k) \leftarrow s(i, k) - \max_{k', s.t. k \neq k'} \{a(i, k') + s(i, k')\} \tag{15}$$

where $s(i, k)$ is the similarity indicating how well the data point with index k is suited to be the exemplar for data point i . Similar is the update rule of availability

$$a(i, k) \leftarrow \min \left\{ 0, r(k, k) + \sum_{k', s.t. k \neq k'} \max\{0, r(i', k)\} \right\} \tag{16}$$

Each iteration of affinity propagation consists of:
   (i) Updating all responsibilities given the availabilities;
   (ii) Updating all availabilities given the responsibilities;
   (iii) Combining availabilities and responsibilities to monitor the exemplar decisions and terminate the iteration when change of decisions falls below a predefined threshold.

2. Modeling Using Principal Component
In each cluster, principle component technique is employed for modeling individuals[21]. For each cluster S^j , boundaries of the subset are given by

$$\begin{cases} a_i^j = \min_{x \in S^j} (x - \bar{x}_j)^T U_i^j \\ b_i^j = \max_{x \in S^j} (x - \bar{x}_j)^T U_i^j \end{cases} \tag{17}$$

where the principal component U^j is computed as a unit eigenvector associated with the sample data covariance matrix C of the points in S^j

$$C = \frac{1}{|S| - 1} \sum_{x \in S} (x - \bar{x})(x - \bar{x})^T \tag{18}$$

where $|S|$ is cardinality of S^j. With the ith largest eigenvalue λ_i^j of C in S^j, we have

$$\sigma_j = \frac{1}{n - m + 1} \sum_{i=m}^{n} \lambda_i^j \tag{19}$$

where m is the number of objectives, n is the dimension of the vector x .

To make the clusters more approximate the Pareto Set (PS), and avoid missing real solutions, extensions on both sides of boundary are made to generate new sampling subspace Ψ^j. The percentage the extension made is 25% of either side:

$$\Psi^j = \left\{ x \in R^n \,\middle|\, x = \bar{x}^j + \sum_{i=1}^{m-1} \alpha_i U_i^j, i = 1, \ldots, m-1 \right\} \tag{20}$$

with

$$a_i^j - 0.25(b_i^j - a_i^j) \le \alpha_i \le b_i^j + 0.25(b_i^j - a_i^j) \tag{21}$$

3. Reproduction by Sampling
Generate τ samples with the probability proportional to its size for subspace Ψ^k:

$$P(\tau = k) = \frac{vol(\Psi^k)}{\sum_{j=1}^{k} vol(\Psi^j)} \tag{22}$$

Uniformly randomly generate a point x' from Ψ^τ. Generate N new individuals $x = x' + \varepsilon$, where ε is a noise vector from $N(0, \sigma_\tau I)$.

4.Selection The selection procedure is based on the non-dominate sorting similar to NSGA-II [23]. Crowding distance of the points in set is defined as the average side length to the largest m-D rectangle in the objective space. Solutions with lager crowing distance will be put into the new population.

Figure 4 shows an analytical robust optimization using the proposed algorithm(see [24] for definition of optimization problem MV1). It can be seen that a stairs-type Pareto front is obtained, and each stair segment represents the optima with the same belief value.

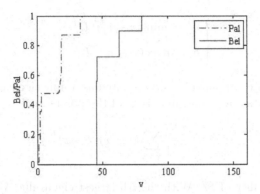

Fig. 4. Bel and Pal value of MV1 problem

4.2   Multi-fidelity Model Management

In Mars entry probe design optimization, direct search methods relying exclusively on high fidelity aerodynamic models is cost-prohibitive. Therefore, researchers proposed different model management strategies for variable-fidelity optimization [25, 26]. In this section, an ANN based multi-fidelity strategy is designed to integrate the models with different accuracy levels into the robust optimization.

The model management works as follows: First,a Latin hype-cube sampling is carried out for different geometric configuration, with which an ANN for response surface fits is initialized by the results from both low and high fidelity models. In the ANN approximator, the input are drag coefficient and lift coefficient from both models, and the corresponding Mach number and AoA, while the output is the difference between the CFD and the modified Newtonian theory. The actual lift and drag data in the optimization procedure are then from the analytical model plus the output of the ANN.

The ANN is updated after every few generations of optimization. Only the individuals of great value are sampled using high fidelity model. The sample sets consist of the centroid of the cluster, individuals located on the lower and higher bounds of the sampling cluster, and other interested individuals , e.g, individuals with highest temperature in the cluster. Putting into the ANN new high-fidelity results and corresponding lower fidelity results and training the ANN, new response surface can then be established.During the optimization, more and more information from high fidelity model is put into the surrogate model, makes the surrogate model closer to the high fidelity model, and finally after several generations, converges to the optimal values with accuracy level of the high fidelity model. Fig. 5 shows flow chart of the robust multi-fidelity optimization.

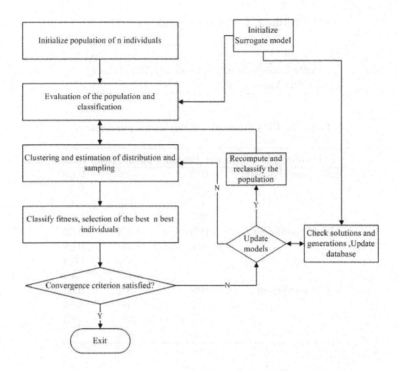

Fig. 5. Evolutionary control and database handler

5   Numerical Results

In this section, robust design optimization of a micro entry probe is presented. The probe cross section diameter is 0.8 meter, with entry mass $m = 12kg$ and TPS thickness $L_{TPS} = 1.4$cm. The design parameters are $\mathbf{d} = [R_n, \theta]^T$. Initial values of the design parameters are given in Tab.1.

Table 1. Initial values of design parameters

Parameters	θ	R_n
Lower bound	35.0⁰	0.04m
Upper bound	75.0⁰	0.15m

Initial entry conditions are set to be same as the Spirit spacecraft, and listed in Tab.2, with uncertain parameters $\mathbf{u} = [\rho, C_d, C_l]^T$. BPA structure of $\mathbf{u}$ is shown in Tab.3.

ANN based surrogate model is trained through Bayesian regularization back propagation. ANN is trained with different geometric configurations and flow conditions. The first round of training is done by putting into ANN different

Table 2. Initial entry conditions [27]

r	v	ξ	θ_p	λ	ϕ
3392.3	5.628	79.025	-11.495	161.776	-17.742
km	km/s	$(^0)$	$(^0)$	$(^0)$	$(^0)$

Table 3. BPA structure of uncertain parameters

Parameters	Lower Bound	Upper Bound	BPA
Atmospheric density	-10%	-5%	0.05
	-5%	0	0.25
	0	5%	0.3
	5%	10%	0.4
Drag coefficient	-10%	-5%	0.05
	-5%	0	0.25
	0	5%	0.3
	5%	10%	0.4
Lift coefficient	-10%	-5%	0.05
	-5%	0	0.25
	0	5%	0.3
	5%	10%	0.4

values of C_d and C_l from the analytical model and CFD software with respect to M_∞ and geometric configuration, 300 samples in total.

Analytical aerodynamic model is used during the first few iterations to learn the cost function. Numeca® discretize computational domain with multi-block structured mesh. The computational meshes consist of 18 blocks with nearly 1.2106 total nodes, is changed by internal scripting based on design parameters. The flow model consist of CO2 and N2, of which 97% by volume is CO2, and 3% by volume is N2. After every 10 generations, the database is updated by inserting new results from the CFD software. Trajectory equation is integrated until altitude reaches to 10 km, or the Mach number lower than 1.3. The objectives of the robust optimization are to minimize the TPS interior temperature and maximize the belief value, with constraint that the maximum overload should be no more than 9g.

The computation is performed on a Linux platform with 32 Core (4x8 3.0GHz AMD 6220), 40GB of memory. Fig. 6 shows optimization results with a population of 60 individuals after 80 generations. Geometric parameters ($[R_n, \theta]$) of each individual on the Pareto front are listed in the figures as well.

Two individuals A and B are selected from Pareto optimal front for illustrating their performance. Individual A has a belief of 0.58 with highest temperature of 295.19k, while probe Bs temperature is 296.459k with a belief of 0.87. To show their performance, individuals A and B are compared with a sub-optimal candidate, the probe C with $\theta = 40^0$ and $R_n = 0.10m$. Corresponding trajectories and TPS temperature are shown in the figure as well.

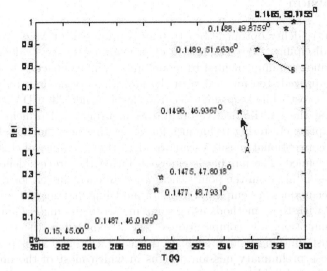

Fig. 6. Pareto front (temperature vs. Bel)

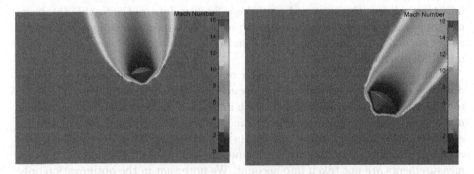

Fig. 7. Optimal solutions: individual A and individual B

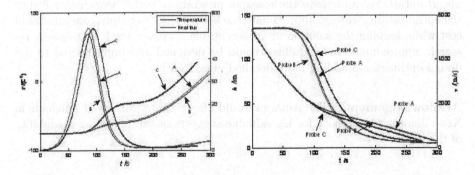

Fig. 8. Trajectory and TPS performance: individual A and individual B

6   Conclusion

A robust multi-fidelity optimization strategy is proposed for Mars entry micro probe multi-disciplinary design. Instead of probability theory, evidence theory based techniques are implemented to model uncertainty impacts. Optimization results are improved step forward with the multi-fidelity models, reducing the computational cost while keeps high accuracy level. A new MOEDA is designed for conducting the multi-objective optimization, using local principle component and adaptive clustering techniques for generating new individuals. Different from the conventional density estimator algorithms, number of clusters does not need predefined. The number is emerged during the process depending on the messages sending from the points each other. In each cluster, local principle component techniques are employed for data modeling and generating.

Compared with those methods using sampling strategies under Gaussian distribution assumption, less computational cost is required in the method. No probability distribution model predefined is needed as well. This is particularly useful for those preliminary mission designs in which most of the uncertainty information is unavailable.

Multi-fidelity model managements are incorporated into the robust multi-disciplinary optimization. ANN surrogate model is trained firstly with low-fidelity results, with more and more high-fidelity data put into the database, the solutions converge to high-fidelity results. The approach proposed provides a way for multi-disciplinary optimization taking into account uncertainties with multi-fidelity models.

In this work, we addressed a design optimization problem involving multi-disciplines, multi-fidelity models and uncertainties. Surrogate model is employed to reduce the computational cost. In each interval, a strategy with SQP searching method is implemented for searching the value in the intervals. For illustrating effectiveness of the algorithm, only parts of the uncertainties are involved. Uncertainties of the entry trajectory parameters such as the position and velocity measurements are not taken into account. We note that in the optimization only a few uncertain parameters could cost high computational resources. With evidence theory, algorithm complexity increases exponentially with the number of uncertainties. We anticipate the research program in future work can advance the optimization strategy, approximate the belief values to reduce computational cost while keeping the solution in an acceptable accuracy level. In this new research, approximate and parallized could be designed and implemented to the design optimization adding more detailed uncertainties.

Acknowledgement. The Authors would like to thank Mr.Javad Zolghadr in Xi'an Jiaotong University for his valuable suggestions on improving readability of the paper.

References

1. Saleh, J.H., Mark, G., Jordan, N.C.: Flexibility: a multi-disciplinary literature review and a research agenda for designing flexible engineering systems. Journal of Engineering Design 20(3), 307–323 (2009)
2. Ravanbakhsh, A., Mortazavi, M., Roshanian, J.: Multidisciplinary design optimization approach to conceptual design of a leo earth observation microsatellite. In: Proceeding of AIAA SpaceOps 2008 Conference (2008)
3. Balesdent, M., Bérend, N., Dépincé, P.: Stagewise multidisciplinary design optimization formulation for optimal design of expendable launch vehicles. Journal of Spacecraft and Rockets 49(4), 720–730 (2012)
4. Roshanian, J., Jodei, J., Mirshams, M., Ebrahimi, R., Mirzaee, M.: Multi-level of fidelity multi-disciplinary design optimization of small, solid-propellant launch vehicles. Transactions of the Japan Society for Aeronautical and Space Sciences 53(180), 73–83 (2010)
5. Huang, C.H., Galuski, J., Bloebaum, C.L.: Multi-objective pareto concurrent subspace optimization for multidisciplinary design. AIAA Journal 45(8), 1894–1906 (2007)
6. Dufresne, S., Johnson, C., Mavris, D.N.: Variable fidelity conceptual design environment for revolutionary unmanned aerial vehicles. Journal of Aircraft 45(4), 1405–1418 (2008)
7. Nguyen, N.V., Choi, S.M., Kim, W.S., Lee, J.W., Kim, S., Neufeld, D., Byun, Y.H.: Multidisciplinary unmanned combat air vehicle system design using multi-fidelity model. Aerospace Science and Technology (2012)
8. Mueller, J.B., Larsson, R.: Collision avoidance maneuver planning with robust optimization. In: International ESA Conference on Guidance, Navigation and Control Systems, Tralee, County Kerry, Ireland (2008)
9. Lantoine, G., Russell, R.P.: A hybrid differential dynamic programming algorithm for robust low-thrust optimization. In: AAS/AIAA Astrodynamics Specialist Conference and Exhibit (2008)
10. Vasile, M.: Robust mission design through evidence theory and multiagent collaborative search. Annals of the New York Academy of Sciences 1065(1), 152–173 (2005)
11. Croisard, N., Vasile, M., Kemble, S., Radice, G.: Preliminary space mission design under uncertainty. Acta Astronautica 66(5), 654–664 (2010)
12. Agarwal, H., Renaud, J.E., Preston, E.L., Padmanabhan, D.: Uncertainty quantification using evidence theory in multidisciplinary design optimization. Reliability Engineering & System Safety 85(1), 281–294 (2004)
13. Roncoli, R.B., Ludwinski, J.M.: Mission design overview for the mars exploration rover mission. In: 2002 Astrodynamics Specialist Conference (2002)
14. Anderson, J.D.: Hypersonic and high temperature gas dynamics. Aiaa (2000)
15. Mitcheltree, R., DiFulvio, M., Horvath, T., Braun, R.: Aerothermal heating predictions for mars microprobe. AIAA Paper (98-0170) (1998)
16. Amar, A.J., Blackwell, B.F., Edwards, J.R.: One-dimensional ablation using a full newton's method and finite control volume procedure. Journal of Thermophysics and Heat Transfer 22(1), 71–82 (2008)
17. Hankey, W.L.: Re-entry aerodynamics. Aiaa (1988)
18. Hartleib, G.: Tpsx materials properties database. NASA,
   http://tpsx.arc.nasa.gov

19. Oberkampf, W., Helton, J.C.: Investigation of evidence theory for engineering applications. In: Non-Deterministic Approaches Forum, 43rd AIAA/ASME/ASCE/AHS/ASC Structures, Structural Dynamics, and Materials Conference (2002)

20. Limbourg, P.: Multi-objective optimization of problems with epistemic uncertainty. In: Coello Coello, C.A., Hernández Aguirre, A., Zitzler, E. (eds.) EMO 2005. LNCS, vol. 3410, pp. 413–427. Springer, Heidelberg (2005)

21. Zhang, Q., Zhou, A., Jin, Y.: Rm-meda: A regularity model-based multiobjective estimation of distribution algorithm. IEEE Transactions on Evolutionary Computation 12(1), 41–63 (2008)

22. Frey, B.J., Dueck, D.: Clustering by passing messages between data points. Science 315(5814), 972–976 (2007)

23. Deb, K., Pratap, A., Agarwal, S., Meyarivan, T.: A fast and elitist multiobjective genetic algorithm: Nsga-ii. IEEE Transactions on Evolutionary Computation 6(2), 182–197 (2002)

24. Vasile, M., Minisci, E., Wijnands, Q.: Approximated computation of belief functions for robust design optimization. arXiv preprint arXiv:1207.3442 (2012)

25. Cheng, Q.S., Bandler, J.W., Koziel, S.: Combining coarse and fine models for optimal design. IEEE Microwave Magazine 9(1), 79–88 (2008)

26. Koziel, S., Ogurtsov, S.: Robust multi-fidelity simulation-driven design optimization of microwave structures. In: 2010 IEEE MTT-S International Microwave Symposium Digest (MTT), pp. 201–204. IEEE (2010)

27. Desai, P.N., Knocke, P.C.: Mars exploration rovers entry, descent, and landing trajectory analysis. In: AIAA/AAS Astrodynamics Specialist Conference and Exhibit, pp. 16–19 (2004)

Author Index